Published for
the Government of Northern Ireland by
HER MAJESTY'S STATIONERY OFFICE

To be purchased from
7 Linenhall Street, Belfast BT2 8AY
49 High Holborn, London WC1
13a Castle Street, Edinburgh EH2 3AR
109 St Mary Street, Cardiff CF1 1JW
Brazennose Street, Manchester M60 8AS
50 Fairfax Street, Bristol BS1 3DE
258 Broad Street, Birmingham 1
or through any bookseller

Price £2 0s 0d net

Printed in Northern Ireland

SBN 337 06009 6

PREFACE TO THE FIRST EDITION

THE GEOLOGY of this Map was investigated by the late Mr. Du Noyer and Messrs. Warren and W. B. Leonard, during portions of the years 1868–69, under the direction of Professor J. B. Jukes, who previous to his decease had also made an inspection of the greater portion of the district.

Under these circumstances, I only deemed it necessary to make a cursory examination of the more important points, especially along the margin of the basaltic table-land; and for this purpose visited Belfast, Lisburn, and Lurgan, in company with Mr. Warren in the Spring of 1869. It will, therefore, be evident that the preparation of this Map and Memoir has been attended with peculiar disadvantages.

The description of the geological details contained in this Memoir has been drawn up by Messrs. Warren and Leonard, with occasional assistance from myself in those portions relating to the Triassic, Rhaetic, and Cretaceous Rocks.

EDWARD HULL
Director of the Geological Survey of Ireland

Geological Survey Office
12*th May* 1871

PREFACE TO THE SECOND EDITION

THE BELFAST (36) Sheet was first surveyed by G. V. Du Noyer, J. L. Warren, and W. B. Leonard in the years 1868-9 and a hand-coloured one-inch geological map was published in 1870. An explanatory memoir by E. Hull, Warren and Leonard followed in 1871. A second edition of the hand-coloured map appeared in 1876 and a third in 1902 (dated 1901) following a revision of the Silurian rocks by F. W. Egan in 1894-6.

A resurvey of the drift deposits of the Belfast district in 1902 by J. R. Kilroe, A. McHenry, H. J. Seymour, W. B. Wright, H. B. Muff, and G. W. Lamplugh was published as a special colour-printed map and accompanying memoir in 1904, which covered the north-east quadrant of this sheet and parts of Sheets 28, 29, and 37.

Resurvey of the area was started by J. A. Robbie and H. E. Wilson in 1953, joined by P. I. Manning in 1954 and was completed in 1956. The one-inch map was published in solid and drift editions in 1966.

The area was covered by a geophysical survey (gravity and aeromagnetic) of the whole of Northern Ireland carried out by the Geophysical Department of the Geological Survey of Great Britain in 1959-61 on behalf of the Government of Northern Ireland.

Palaeontological determinations of the Lower Palaeozoic fossils were by J. D. D. Smith and A. W. A. Rushton, and by Professor O. M. B. Bulman of Cambridge University: the Permian by J. Pattison, Rhaetic and Lias by H. Ivimey-Cook, Cretaceous by R. V. Melville, R. Casey and C. J. Wood, and Recent material by M. J. Hughes, C. J. Wood, and the late A. S. Kennard. The palynology of the Trias was investigated by G. Warrington. The collecting was done by R. Carnaghan and T. P. Fletcher.

Petrographical reports on the rocks were by J. R. Hawkes, P. A. Sabine, and H. E. Wilson. X-ray work was done by B. R. Young.

We acknowledge the assistance rendered in the field by members of the public particularly the owners and workers in the numerous quarries in the district who have allowed unrestricted access to their workings.

This memoir has been written by P. I. Manning, J. A. Robbie, and H. E. Wilson with palaeontological sections by M. J. Hughes, J. Pattison, G. Warrington and C. J. Wood. The chapter on the soils of the area was written by J. S. V. McAllister of the Chemical Research Division of the Ministry of Agriculture, Northern Ireland. The major portion of this memoir has been edited by H. E. Wilson, and a small part in the early stages by J. A. Robbie.

The 'peculiar disadvantages' under which the first edition was prepared have been repeated with this volume as the galley proofs were destroyed by fire during civil disturbances in 1969 and publication has consequently been delayed.

K. C. DUNHAM
Director of Geological Survey in N. Ireland

Institute of Geological Sciences,
Exhibition Road, London, S.W.7.

CONTENTS

ILLUSTRATIONS

TEXT FIGURES

xi

PLATES

LIST OF SIX-INCH MAPS

The following is a list of the six-inch geological maps included, wholly or in part, in the one-inch geological map, Sheet 36, with the initials of the surveyors (J. A. Robbie, H. E. Wilson, and P. I. Manning) and the dates of the resurvey. All are available for public reference in manuscript form in the office of the Geological Survey, 20 College Gardens, Belfast BT9 6BS.

ANTRIM

58 NE	Langford Lodge	H.E.W.	1954
58 SE	Sandy Bay	H.E.W.	1954
59 NW	Crumlin	H.E.W.	1954
59 NE	Dundrod	H.E.W.	1955
59 SW	Glenavy	H.E.W.	1954
59 SE	Leathemstown	H.E.W.	1955
60 NW	Budore	J.A.R.	1954
60 NE	Belfast	J.A.R., P.I.M.	1953–4
60 SW	Collin Glen	J.A.R.	1954
60 SE	Belfast	J.A.R.,P.I.M.	1953, 1955
61 NW	Belfast	P.I.M.	1955
61 SW	Belfast	P.I.M.	1955
62 NW	Selshan	J.A.R.	1952
62 NE	Portmore	H.E.W.	1955
62 SW	Derrymore	H.E.W.	1955
62 SE	Aghalee	H.E.W.	1955
63 NW	Legatirriff	H.E.W.	1955
63 NE	Stonyford	H.E.W.	1955
63 SW	Ballinderry	H.E.W.	1954–5
63 SE	Knocknadona	H.E.W.	1953–6
64 NW	Castlerobin	H.E.W., J.A.R.	1953-4
64 NE	Dunmurry	P.I.M.	1955
64 SW	Lisburn	H.E.W., P.I.M.	1955–6
64 SE	Lambeg	P.I.M.	1955
65 NW	Belfast	P.I.M.	1955
66 NW	Derryhirk	P.I.M.	1954
66 NE	Aghagallon	P.I.M.	1954–5
67 NW	Maghaberry	P.I.M.	1954
67 NE	Halfpenny Gate	P.I.M.	1954–6
67 SW	Inishloughlin	P.I.M.	1954
67 SE	Hasley's Town	P.I.M.	1955–6
68 NW	Lisburn	P.I.M.	1955

ARMAGH

6 NE	Lurgan	P.I.M.	1954
6 SE	Lurgan	P.I.M.	1954

DOWN

4 NE	Belfast	P.I.M.	1955
4 SE	Belfast	P.I.M.	1955–6
8 SE	Sandymount	P.I.M.	1956
9 NW	Milltown	P.I.M.	1956
9 NE	Newtownbreda	J.A.R.	1955
9 SW	Drumbo	P.I.M.	1956
9 SE	Knockbracken	J.A.R.	1955–6
13 NW (& part of 13A)	Legmore	P.I.M.	1954–5
13 NE	Ballycanal	P.I.M.	1954–5
13 SW	Magheralin	P.I.M.	1954
13 SE	Moira	P.I.M.	1954
14 NW	Maze	P.I.M.	1955
14 NE	Sprucefield	J.A.R., P.I.M.	1956
14 SW	Hillsborough	P.I.M.	1955–6
14 SE	Ravernet	P.I.M.	1956
15 NW	Hillhall	P.I.M.	1956
15 NE	Ballynagarrick	H.E.W.	1956
15 SW	Baileysmill	H.E.W.	1956
15 SE	The Temple	H.E.W.	1956
19 NE	Cornreany	P.I.M.	1954
20 NW	Mathersesfort	P.I.M.	1954
20 NE	Kilwarlin Cottage	P.I.M.	1954
21 NW	Titteringtons Bridge	P.I.M.	1956
21 NE	Hillsborough	P.I.M.	1956
22 NW	Ballykeel Lougherne	H.E.W.	1956
22 NE	Lough Henney	H.E.W.	1956

Chapter I

INTRODUCTION

AREA AND LOCATION

THE AREA DESCRIBED in this memoir and represented by Sheet 36 of the one-inch map lies south and west of the city of Belfast, the greater part of which lies within the limits of the sheet. The tidal flats at the head of Belfast Lough, now reclaimed and occupied by harbour and airport, lie at the north-east extremity of the area and some eight square miles (21 sq km) of Lough Neagh are included along the western edge of the map. The land area covered is approximately 208 square miles (539 sq km).

PHYSIOGRAPHY

The district is divided by the low ground of the Lagan Valley which runs diagonally across it from the south-west. To the north the scarp of the Antrim Plateau reaches 1574 ft (479 m) O. D. on the high moorlands of Divis and Black Mountain above Belfast, but falls steadily to the south-west and west giving place to increasingly good arable land below the 700-ft (213 m) contour. The drainage of the western slopes is towards Lough Neagh and converges into the Crumlin and Glenavy rivers. Along the edge of the Lough flat peat-covered bogs are interspersed with so-called islands, once regularly isolated by floods but now well above the lowered level of the lake.

The Lagan Valley, three or four miles in width, includes some of the richest agricultural land in the province of Ulster but is fast being covered by the Belfast-Dunmurry–Lisburn conurbation. The gently undulating and relatively well-wooded land in this valley is still, in its unspoilt areas, of much scenic beauty with parkland landscapes and well-kept farms.

South of the valley the ground rises in an irregular low scarp to the low hills of County Down, rarely exceeding 500 ft (152 m), and falling gradually to the south beyond the gentle and nearly imperceptible watershed. Notable for the abundance of drumlins, the general form of the landscape in this area is often masked by the 'basket-of-eggs' topography. Drainage by a network of small streams, which form small alluvial flats between the drumlins and occasionally open into small lakes, is towards the Ravernet River which flows west to join the Lagan near Lisburn.

HUMAN OCCUPATION AND INDUSTRY

The earliest records of man's arrival in Ireland date from the late Boreal period, about 6000 B.C., and come from the shores of Lough Neagh. Mesolithic flint artefacts are known also from Larne and Strangford Lough though not from the area described in this memoir. There can be no doubt, however, that Mesolithic hunters and fishermen must have roamed the shores of Lough Neagh and Belfast Lough: and by Neolithic times the forests of the Lagan Valley were populated by human communities, with flint 'factories' on Black Mountain and

1

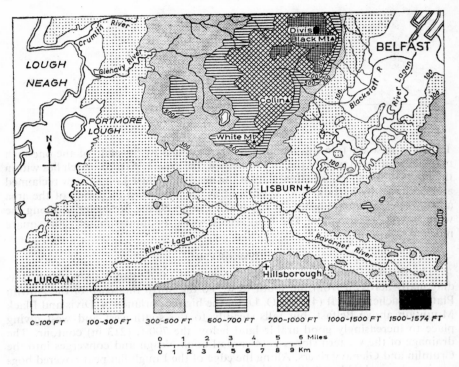

FIG. 1. *Sketch-map showing the physical features of the Belfast area*

Squire's Hill. The Neolithic culture reached its full development some 3500 years ago, when the cultivation of wheat and barley and the use of domesticated animals were established. The relics of the Megalithic civilization, characterized by ritual burial in great stone monuments and the use of pottery, are found all over Ulster. Within the present area the Giant's Ring, a mile south of Shaws Bridge, is the most spectacular—an earth bank 15 ft (4·6 m) high enclosing an area of about seven acres (2·8 ha), near the centre of which is a polygonal chambered grave.

With the early Bronze Age the industrial history of Ulster might be said to have commenced—copper and bronze axes and the moulds for their manufacture have been found abundantly in Antrim and Down and Irish styles of decorated axes travelled widely into Great Britain and the Continent. Bronze Age sites are abundant in the Lagan Valley which, during this period, was increasingly the main way of trade between Ireland and Great Britain and the entry route for successive waves of settlers.

Little is known of the pre-historic Iron Age in Ulster but with the coming of Christianity in the fifth century began the record of religious buildings and carvings. Though none of these are extant in our area, habitation sites of the agricultural communities are abundant as ring forts or 'raths'. These circular earth banks, up to about 100 ft (30 m) across and originally surmounted by palisades, were defended farmsteads, often with several dwellings in the central courtyard or sometimes completely roofed over as at Lissue, near Lisburn. Excavation in them has revealed grain-impressions on pottery, cereal pollens, rotary querns, and iron ploughshares and coulters, indicating a settled agricultural civilization. From this period also date the souterrains, underground stone-built stores or defensive refuges, and the crannogs, or lake dwellings.

From about 800 A.D. these farm dwellers were systematically ravaged by Norse 'Vikings' who sacked the religious communities and pillaged where they could. The Round Towers remain as relics of refuges from these raiders. Viking coins and ornaments have been found in the Lagan Valley while Carlingford and Strangford to the south owe their names to the Norsemen.

For over a century from about 1177 Ulster was subjected to penetration by the Normans who built 'motes'—earthwork mounds with a timber tower on top —later replaced in some places by stone castles. There were several of the former in the Lagan Valley. During the 14th century the Normans were reduced to a foothold on the coast while the Scots were active in Antrim, first as mercenaries and later as settlers.

In the 16th century with the 'plantation' of Ulster much of the Lagan Valley and the areas round it were settled by English colonists who imported to it the characteristic landscape pattern which still survives. Lisburn, Lurgan and Hillsborough were the plantation towns, built by the landowners as centres for their estates. Belfast grew around the Castle, built, perhaps as early as the 13th century, to control the lowest ford across the Lagan.

The clearance of the woods in the area also dates from the 16th century when vast quantities of timber were used to make charcoal for bloomeries or iron works which were numerous in the Lagan Valley. The ore was probably imported from Cumberland though bog-ore may have been used to some extent.

The industrial history of the area becomes important since the early 18th century when the linen industry sprang up under the impetus of Huguenot émigrés and gradually became concentrated in factories which used the water power of the rivers. Cotton spinning in Belfast and the surrounding district, in steam-powered mills, hastened the concentration of textile industry in the region. The ship-building industry in Belfast began in 1853 and, with the manufacture of textile machinery, acted as nucleus for the diverse engineering industries now existing.

At present, manufacturing industry of many and varied types is mainly concentrated in the Belfast–Lisburn conurbation and in Lurgan and apart from a few plants making agricultural by-products there are few rural industries left. The few small towns outside the industrial belt are increasingly becoming dormitory areas, the working population travelling to the larger centres. Agriculture, with emphasis on livestock, remains of major importance over the greater part of the area.

TOWNS AND COMMUNICATIONS

The population in the area is largely concentrated in Belfast which has grown from 13 000 in 1784 to about 450 000 today with a further 150 000 in the outer suburbs. The only other large towns in the district are Lisburn and Lurgan, with smaller communities at Hillsborough, on the road to Dublin across the County Down hills, and in Crumlin and Glenavy at river crossings on the direct road from Dublin to the north of County Antrim. Within the last half-century the Lagan valley between Belfast and Lisburn has become increasingly built up and the village of Dunmurry is now swallowed by the conurbation which includes almost half of the population of Ulster.

With the deforestation of the Lagan Valley in the 17th century and the development of coach roads and turnpikes (1733) across the undulating sandy country from Belfast to Lisburn, the low ground in the centre of the valley became the main axis of communication from the port of Carrickfergus to the hinterland of Ulster. This line has been perpetuated by the Lagan Navigation (Lagan–Lough Neagh canal 1763–1794) now abandoned, the Post Roads built in 1817–19, and by the Ulster Railway from Belfast to Armagh (1839), later extended to Dublin and the west. A new motorway follows the same route, using the abandoned canal in places.

The road to Dublin leaves the valley at Lisburn and runs through Hillsborough across the low hills to the south; a branch railway, now closed, followed the same route. To the north of the valley, roads and railways lead north from Lisburn to Glenavy and Crumlin, while a network of roads leads north and south across the scarps from Belfast.

H.E.W.

GEOLOGICAL SEQUENCE

The formations occuring within the area of Sheet 36 and represented on the map and sections are summarized in the following table:

SUPERFICIAL OR DRIFT DEPOSITS

RECENT
Landslip
Blown Sand
Peat
Lacustrine alluvium
Lacustrine beach deposits
Alluvium of river flood plains
River terrace deposits
Marine and estuarine alluvium

GLACIAL
Glacial sands and gravels and laminated clays
Boulder Clay

SOLID FORMATIONS

			Generalized thickness in feet
TERTIARY	LOUGH NEAGH CLAYS	Clays and sands with some beds of lignite	200+
		Unconformity	
	ANTRIM LAVA SERIES		
	Upper Basalts	Flows of basalt lava . . .	700
	Interbasaltic Bed	Laterite and lithomarge . .	50
	Lower Basalts	Flows of basalt and mugearite lava	1800
		Unconformity	
CRETACEOUS	*Upper Chalk*	Hard white limestone with flint nodules	10–100
	Hibernian Greensand	Glauconitic green and yellow sandstones and grey marls .	0–50

		Unconformity	
JURASSIC	*Lower Lias*	Grey shale and limestone ribs, occasionally shelly . . .	45
		? Non-sequence	
TRIASSIC	*Rhaetic*	Dark grey shales . . .	50
	Keuper Marl	Red, brown, and green marls with gypsum veins . . .	900
	Bunter Sandstone	Red sandstones, marls, and breccia	100
PERMIAN	*Permian Marls*	Red marls with anhydrite, particularly at the base . . .	270
	Magnesian Limestone	Yellowish fossiliferous dolomitic limestone, often oolitic . .	70
	Basal Sandstones and Brockram	Coarse, purple and white sandstones and breccias . . .	145+
		Unconformity	
SILURIAN	*Llandovery*	Greywackes, grits, and shales .	
		? Non-sequence	
ORDOVICIAN	*Ashgill and Caradoc*	Greywackes, grits and shales .	

INTRUSIVE IGNEOUS ROCKS
    Basalt and Dolerite
    Lamprophyre
    Quartz-porphyry

The general geology is shown on the sketch map (Fig. 2).

Lower Palaeozoic rocks of Ordovician and Silurian age crop out in a triangular area in the south-east corner of the map, the older Ordovician beds occurring only as faulted inliers. Both series consist of arenaceous turbidites with subordinate bands of shale.

At the eastern edge of the map the Silurian beds are overlain by Permian sandstones which are succeeded by Magnesian Limestone and Permian Marls. The Permian succession is known only from boreholes. The Permian thins out to the south-west and the Bunter beds of the Triassic, which succeed the Permian without unconformity, overlap on to the Silurian south-west of Lisburn. The later assignment to the Permian of beds classified on the one-inch map as Bunter extends the Permian outcrop. The Bunter beds underlie the whole of the Lagan Valley and are succeeded by Keuper Marls which crop out in a strip along the northern side of the valley, thinning to the south-west and dying out near Magheralin.

Rhaetic shales overlie Keuper in the north-east of the outcrop but are exposed only in the Collin Glen area. In turn they are succeeded by Lias shales which crop out in the same area. Neither formation occurs at outcrop west of Collin Glen but both are known from a borehole at Langford Lodge.

The Upper Cretaceous occurs as a long narrow outcrop on the flanks of the high ground north of the Lagan Valley. The basal Hibernian Greensand is known only from the north-east end of the outcrop, and the overlying Chalk rests directly on the Triassic marls and, finally, sandstone to the south-west.

The north-west quadrant of the map is occupied by the Antrim Lava Series. The Lower Basalts, which include at least one mugearite lava, form the high ground north of Dunmurry and underlie the area almost as far west as Glenavy.

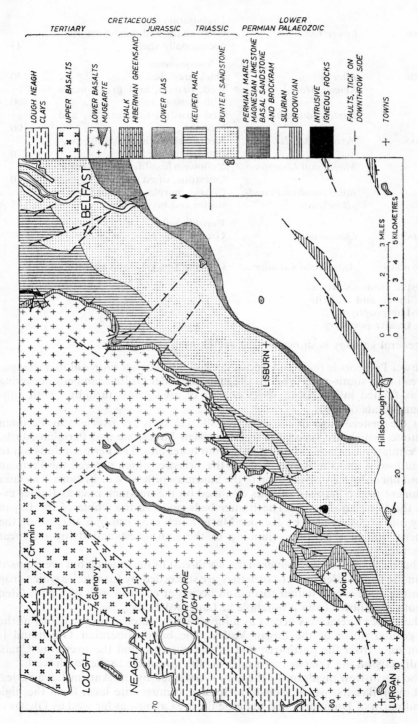

FIG. 2. *Geological sketch-map of the Belfast area*

Down-faulted Upper Basalts underlie the area between Glenavy and Langford Lodge. The Interbasaltic Bed, between the Upper and Lower Basalt series, is known only from a borehole and does not occur at outcrop.

A wide area of low ground bordering Lough Neagh is underlain by Lough Neagh Clays, a series of unconsolidated sands, clays and some lignites, probably of Oligocene age.

The distribution of deposits of Pleistocene and Recent age is shown on the 'Drift' edition of the map. Some of the high ground north of the valley and the inter-drumlin hollows to the south are drift free, but widespread deposits of boulder clay mask the solid rock over most of the district. Glacial sands are confined to the valley though there are scattered eskers to the west. Alluvial deposits are common along the rivers, though rarely extensive, and some river terraces are also seen. Along the eastern shore of Lough Neagh there are small lacustrine beach deposits, sometimes backed by areas of blown sand, and extensive areas of peat and alluvium. Much of central Belfast is underlain by estuarine clay.

PREVIOUS GEOLOGICAL LITERATURE

The district has few spectacular geological features and early references to its geology are few. Probably the earliest is in Barton's treatise on the fossil wood of Lough Neagh (1757)—a subject which recurs frequently in the literature. Berger (1816, pp. 121–222) refers to the area and Buckland (1817, pp, 413–23) described the paramoudras from the Chalk at Moira. Bryce (1852) gave an account of the district to the British Association, but only a synopsis of this was printed. He described the Lagan Valley as separating the Lower Silurian rocks of County Down from the Triassic, Liassic, Cretaceous, and volcanic rocks of Antrim. The estuarine clays of Belfast were described as of Pliocene age. Tate gave the first systematic description of the Lias of Collin Glen (1864, pp. 104–5) and the Cretaceous (1865, pp. 15–44) and effected a wider correlation of the Lias in 1867.

Kelly (1868) recognized that the New Red Sandstone overlies the 'grauwacke' of County Down unconformably, and gave useful descriptions of the sandstone and marl outcrops. He also described chalk quarries at Moira, above Lisburn, at Colinwell and White Rock, with the thickness of the strata worked. He does not mention the basalts in this area, though describing them in detail in other localities, but gives an account of the fossil wood and lignite of Lough Neagh.

The Geological Survey of Ireland map of Sheet 36 appeared in 1870 and the Memoir followed in 1871.

The first useful papers on the Lower Palaeozoic rocks were those by Swanston 1877 (pp. 107–23) and Lapworth (1877, pp. 125–44) which gave details of fossil localities in the area and of the graptolites obtained from them.

Praeger contributed a valuable series of papers on the fauna of the post-glacial deposits between 1887 and 1902, and in 1897 Hume produced a comprehensive paper on the Cretaceous rocks of Antrim.

The first detailed account of the drift geology was given in the special drift map of the Belfast district, and its accompanying memoir, published by the Geological Survey in 1904.

This brief account of the literature has mentioned only the most important of the early works. Details of many other early papers and the numerous later publications are given in the succeeding chapters and the bibliography.

P.I.M., H.E.W.

Chapter 2

LOWER PALAEOZOIC

GENERAL ACCOUNT

LOWER PALAEOZOIC ROCKS occupy most of the ground south and east of the River Lagan and thus cover a triangular area of about 45 square miles (116 sq km) in the south-eastern part of the Sheet. They are part of the extensive Longford–Down massif, a direct continuation of the Southern Uplands of Scotland, which stretches from the Ards Peninsula south-westwards to County Longford and the River Shannon. Within the Belfast area these rocks give rise to a hummocky upland of no great elevation bounded on the north-west by low hills forming the southern side of the Lagan Valley. Mounds of boulder clay, many in the form of drumlins, cover the region which is drained by a few small streams.

LITHOLOGY

Over the whole area the dominant rock-type is sandstone, with subordinate siltstones, mudstones and shales, in an interbedded sequence. The sandstones are sub-greywackes (Pettijohn 1957, p. 316) with over 25 per cent rock fragments and less than 15 per cent detrital matrix, but throughout this account they are described as greywackes as they fall within the class of rocks long given this name —"poorly sorted angular and sub-angular rock and mineral fragments, ranging from sand to conglomerate grade, set in a substantial matrix of finer material" (Helmbold 1952).

The greywackes occur in beds varying from 6 in (150 mm) to several feet thick, averaging about 2 ft (0.6 m), and though usually massive some beds are markedly flaggy. They are commonly grey in colour but purple coloration is seen in the northern part of the area. Usually of sandstone grade, they are locally coarser, containing pebbles of blue quartz and argillite flakes and rafts. A large proportion of the greywackes is calcareous, with a tendency for the coarser beds to be most carbonated. Jointing is common but cleavage is rarely seen.

Sedimentary structures are very poorly developed. Well-displayed grading and cross-bedding are virtually absent; slump structures and sole markings are uncommon. In part, this apparent absence may be due to the paucity of good three-dimensional exposures, most of the drift-free areas being on low ground with only small exposures in streams and ditches.

The argillaceous rocks are fissile, finely micaceous, pale or medium grey mudstones, shales and siltstones usually occurring as thin beds, averaging about 6 in (150 mm) thick, interbedded with the greywackes. Exceptionally, thicker beds of argillite are recorded, up to 67 ft (20 m) in one case; these thick bands commonly have thin arenaceous partings. Dark grey or black graptolitic shales are not common and are only exposed in the southern part of the outcrop. These are thinly bedded, slightly micaceous carbonaceous silty rocks of mudstone grade, fissile along bedding planes and showing no effects of cleavage. Temporary exposures have shown considerable thicknesses of these beds, always much dissected by faulting.

North of a line trending north-eastwards through Mealough House [359 657] and Knockbracken reservoir [364 664] the rocks are mainly purplish in colour with some purplish green mottling in the shales. The purple colour appears to be original as there are a few grey bands in the succession, and it seems probable that the coloration is not the result of staining from a now-removed Carboniferous or Permian unconformity. In the Langford Lodge borehole (p. 199) the greywackes were red-stained for only about 30 ft (9 m) below the unconformity at the base of the Permian, though the mudstone partings showed colouring for about twice that distance.

It is of note that red mudstones have been recorded from the Tarannon Shales of central Wales (Wood 1906, p. 655), and the Upper Browgill Beds of the Lake District (Marr and Nicholson 1888, p. 678), where the colour was regarded as secondary. In the Southern Uplands of Scotland red mudstones in the Hawick Rocks (Wenlock) are ascribed to primary pigmentation (Warren 1963, p. 238; Rust 1965, p. 234) though secondary staining below the Permo-Carboniferous unconformity is also recorded from Wigtownshire (Rust 1965, p. 244). In these areas the primary coloration is confined to the argillaceous beds. Further east the Queensberry Grit Group (Llandovery) of the Eyemouth (Scotland 34) Sheet shows purple staining. Although the origin of this is not yet clear it is probably secondary (D. C. Greig, personal communication).

Structure

The strike of the rocks is in general north-east, the normal Caledonian trend, and angles of dip are usually high. A notable exception is the narrow flat belt (where dips are as low as 25 degrees) trending north-eastwards and occurring about ½ mile (800 m) south-east of Purdysburn House. Poor exposure, the almost complete absence of way-up criteria, and the frequent and apparently random changes in direction of dip make any useful structural analysis impossible. From analogy with south-west Scotland and eastern County Down it is probable that the area has been subjected to more than one fold movement. Lindstrom (1958) and Kelling (1961) have recorded two phases of compressive movement in the Ordovician rocks of the Rhinns of Galloway. Warren (1964) suggested a single fold phase, with cross-folding due to local swings in maximum stress-direction, in the Silurian rocks of the Hawick area, and Rust (1965), working on the Silurian rocks of Wigtownshire, has indicated two phases of Caledonian movement. Both Rust and Williams (1959) have shown evidence of post-Caledonian movements.

In the area north-east of Belfast, A. E. Griffith has shown the polyphase nature of the Caledonian earth-movements [Geology of the Country around Carrickfergus and Bangor. *Mem. Geol. Surv. N.I.:* in preparation]. Three main phases of compression have been recognized. The earliest phase (Phase 1) in which the maximum principal stress was aligned approximately north-west has generated an essentially periclinal swarm of isoclinal folds (F_1) with axes trending north-east, a well developed axial plane cleavage (S_1) and extensive strike faulting. Lamprophyre dykes and wrench faults are associated with later episodic reorientations of intermediate and minimum stress axes.

During the second phase (Phase 2) the stress field was aligned north-east and the principal axis of stress plunged at about 70° S.W. The structures associated with this phase are small folds (F₂) trending north-west and an associated axial plane cleavage (S₂) which dips 30° N.E. This phase is demonstrably later than Phase 1 as it distorts the S₁ cleavage.

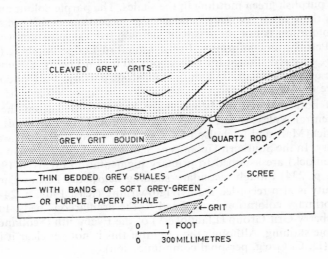

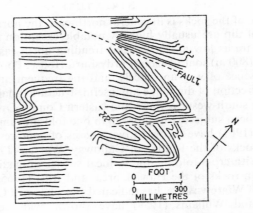

Fig. 3. *Sketch of boudinage structure in Silurian greywackes and shales, Hollands Pool and zig-zag folding, Drumbo Glen*

The third phase of deformation (Phase 3) gave rise to a few small cross folds with axes trending north-north-west and it is noteworthy that they are similar in style and orientation to 'Third-movement phase F₃' folds in the Manx Slate Series of the Isle of Man (Simpson 1963, p. 391).

The styles of folding described from these areas are mainly asymmetric and isoclinal although recent work in the Langholm district has shown that there is no evidence for widespread isoclinal folding, most of the folds being asymmetrical (Lumsden and others 1967, p. 22). It is therefore reasonable to surmise that folds of these types are present in the Belfast area although the flat belt referred to in the previous paragraph suggests the presence of a monocline. The only large fold—amplitude over 50 ft (15 m)—seen within the Sheet is part of a sharp monocline in a quarry at Kiln Quarter [345 637]. Small-scale folding with zig-zag folds of amplitude about 6 in (150 mm) is seen (Fig 3) in dark grey shales in Drumbo Glen [328 651]. In both cases the fold axes follow the north-east Caledonian trend.

Faults, with indications of both vertical and horizontal movement, are common in many of the larger quarry exposures and temporary sections through-out the area, and must be abundant throughout the outcrop. Bands of imbricate crush-rock up to 40 yd (36·5 m) wide have been noted, though crush-zones a few feet wide are normal. The dominant trend appears to be Caledonian, but the paucity of good natural exposures and solid features makes it impossible to trace any faults outside the limits of the man-made exposures. Further, the absence of any palaeontological or lithological evidence by which to establish a detailed stratigraphical succession renders it impossible to calculate the magnitude of such faults as are known.

Petrography

Greywackes. These rocks are regarded by Dr. J. R. Hawkes as sub-greywackes (Pettijohn 1957, p. 316), but they are identical with rocks from Southern Scotland, of similar age, described by several authors as greywackes (Kelling 1961; Warren 1963; Rust 1965, etc.). The limited number of rocks examined consist of sub-angular to sub-rounded quartz grains (0.005 mm to 1.50 mm) with fragments (0.01 mm to 6.0 mm) of devitrified acid (rhyolitic) lava, cherty material, meta-quartzite, perthite, microcline and sodium-feldspar and occasional pieces of intermediate (andesitic) and basic lava, mica-schist and granophyre set in mat-rices of sericite, muscovite, chlorite, clay constitutents, iron oxide (chiefly hematite) and, in some cases, sulphide. Calcite is present in amounts ranging from under 1 per cent to about 5 per cent in the matrices of all the rocks examined except five (NI 2128–9, 2132–3 and 2138). [1]Accessory constitutents sometimes present are schorlite, tourmaline, zircon and sphene. In four specimens (NI 2137-8, 2140 and 2143) piscatite epidote occurs, being relatively abundant in the last two rocks.

One specimen (NI 2137), from an abandoned quarry 140 yd (128 m) N.W. of Clogher School [292 627], contained fragments of quartzite with vermicular chlorite. Similar vermicular chlorite in quartz has been described by Wallis (1927, p. 773) from the Old Red Sandstone of the Bristol district, by Walton (1955, p. 383) from the Silurian greywackes of Peeblesshire, by Warren (1963) p. 228) from the Hawick rocks and by Dearnley (*in* Greig) 1968, pp. 61, 74, 217, from the Longmyndian and Old Red Sandstone of the Church Stretton area-Wallis reported that the material was identified as similar to that found in mica. schists from the Penmynydd zone and quartz augen from the Gwna Green Schists in Anglesey, and to material from the Wrekin in Shropshire. Dr. Warren

---

[1] Numbers preceded by NI refer to rock specimens in the Collection of Geological Survey of Northern Ireland.

(personal communication) has also seen similar material in greywackes from the Hartz Mountains.

Of the derived material in the remaining specimens, rhyolitic fragments are similar to Pre-Cambrian material in the Uriconian Volcanic Series and in the Mona Complex and granophyre is known from Anglesey. Many of the fragments in the sub-greywacke, therefore, may have been derived from a Pre-Cambrian ridge lying to the south-east, but in view of recent studies of the Silurian Palaeogeography of Southern Scotland, this conclusion must be qualified. It seems probable that 'Cockburnland' (Walton 1955; Warren 1963) extended south-westwards across Ulster and that derivation may have been from that direction too.

Argillites. The term argillite is used in the sense proposed by Twenhofel (1937) as a rock derived from siltstone or shale which has undergone a somewhat higher degree of induration than is normal in those rocks.

Sliced specimens are of siltstone, with sub-rounded and subangular quartz fragments from 0.005 mm to 0.5 mm, averaging 0.02 mm, and scattered muscovite flakes averaging 0.03 mm, in a matrix of clay minerals and fine-grained iron ores.

Metamorphism. In both greywackes and argillites the presence of sericite, chlorite and muscovite indicates low-grade regional metamorphism. Quartz grains frequently show signs of re-crystallization, and in some specimens (NI 2050-1, 2053, 2132-4, 2138-9) many of the quartz grains are elongated, giving the rocks a marked schistose texture. A specimen from Purdysburn (NI 2053) is virtually a schistose grit. Schistose texture is also developed locally in specimens from Charity Bridge (NI 2135-6). The sericite appears to have been derived from alteration products of feldspar and the muscovite from amalgamation of sericite flakes.

In one argillite specimen (NI 2052) the quartz in the matrix shows recrystallization and penninitic chlorite is developed after detrital chlorite. The rock is, in effect, a chlorite-phyllite and shows the effects of low-grade metamorphism.

Dr. R. Dearnley has noted (personal communication) similar signs of low-grade regional metamorphism in Lower Palaeozoic greywackes in Scotland and North Wales. In Scotland the metamorphic rocks appear to be confined to relatively narrow belts of country, one of which runs into the sea on the west coast. It is possible that similar localized metamorphic belts may occur in Ulster.

PALAEONTOLOGY

The sparse graptolite fauna from ten localities has been examined by Professor O. M. B. Bulman and Dr. A. W. A. Rushton. In some cases it has only been possible to indicate that the assemblages of poorly preserved forms are of Ordovician or Silurian type.

Ordovician assemblages are recorded as follows:

Bank [272 600] 350 yd (321 m) N.W. of Watson's Bridge. *Climacograptus sp.*, *Orthograptus sp.* or *Diplograptus sp.*

Trench section [241 584] 600 yd (549) m) S.W. of Hillsborough church. ? Amplexograptid, ? Glyptograptid.

Quarry [311 613] 500 yd (457 m) E. of Crossan Post Office. *Climacograptus sp.* (*? C. tubuliferus* Lapworth).

Ditch [366 614] 1320 yd (1207 m) 47° from The Temple. Indet. fragments of Ordovician type.

Trench section [362 590] 200 yd (183 m) N.E. of Killaney House. *Pleurograptus linearis* (Carruthers), *Leptograptus flaccidus* (Hall), *Orthograptus truncatus* (Lapworth), *O. truncatus intermedius* Elles and Wood (Hartley and Harper 1937).

Trench section ([361 587] 130 yd (119 m) S.E. of Killaney House, *Climacograptus* cf. *miserabilis* Elles and Wood, *C. latus* Elles and Wood, *C. supernus ?* Elles and Wood, *Orthograptus* cf. *truncatus abbreviatus* Elles and Wood, *O.* cf. *truncatus truncatus* (Lapworth), *Dicellograptus anceps* (Nicholson), *D.* cf. *complanatus ornatus* Elles and Wood. This assemblage indicates the zone of *D. anceps*, the uppermost Ordovician.

It is of interest to note the zones of *Pleurograptus linearis*, *Dicranograptus clingani* and *Nemagraptus gracilis* have been recognized in an exposure at Yate's Corner some 2 miles (3.2 km) E. of The Temple and a short distance outside the sheet (Pollock and Wilson 1961).

Silurian assemblages are as follows:

Ravernet River [335 586] south of Drumra Hill. *Monograptus gregarius* Lapworth.

Post hole [360 586] Curry's Corner. *Climacograptus sp., Glyptograptus sp.*

Moss View [349 574]. *Climacograptus sp., ? Rhaphidograptus sp.*

Trench [361 587] 100 yd (91 m) S. of Killaney House. *Akidograptus acuminatus* (Nicholson).

Trench [361 586] 160 yd (146 m) S. of Killaney House. *Climacograptus medius* Törnquist, *C. normalis* Lapworth, *C. sp., Glyptograptus sp.*

These last two localities, separated by a faulted block of Ordovician shales, are of the lowest Silurian zones (*Akidograptus acuminatus* and possibly *Cystograptus vesiculosus*), while the *Monograptus gregarius* Zone is also known at Yate's Corner, [Newtownards (37) Sheet] in faulted proximity to the Ordovician shales there.

The coral *Favosites gothlandicus* Lamarck has also been recorded from the same temporary exposure near Killaney House as the graptolites, though its precise locality is uncertain (Griffith 1961, p. 259).

The exposures in the Killaney area indicate the presence in close proximity of the latest zone of the Ordovician and the earliest zone of the Silurian. There is no evidence, palaeontological or stratigraphical, for a break in deposition at this time, and no suggestion that the numerous faults in the trench section are of any great stratigraphical importance (Fig. 4).

The Lower Palaeozoic rocks were considered by early workers to be all of Lower Silurian age and were so designated on the first edition (1870) of the Belfast (36) Map. Later, work by Swanston and Lapworth (1877, pp. 107–48) showed that a subdivision was possible. After a revision by F. W. Egan, a third edition of the Belfast (36) Map, published in 1901, indicated the presence of Lower Silurian (Bala) and Upper Silurian (Llandovery) rocks. The former were delineated as an almost continuous outcrop along the northern margin of the area of Lower Palaeozoic rocks. This division was based on a few fossils from adjoining districts, only one fossil locality being recorded from within the Belfast area. The only other published work on the Lower Palaeozoic rocks of

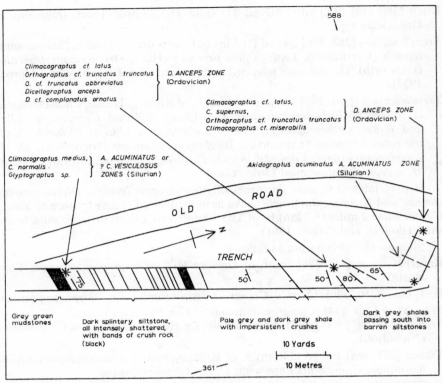

Fig. 4. *Plan of a temporary section at Killaney, County Down, showing the relationship of Ordovician and Silurian beds*

this Sheet records an exposure near The Temple [362 590] of Ordovician shales of Lower Hartfell age (Hartley and Harper 1937, pp. 253–5).

During the resurvey it has been found difficult to make a satisfactory division between the Ordovician and the Silurian, or to subdivide the Silurian, due to the scarcity of fossils, the absence of good continuous sections, and the lithological similarity of the Ordovician and Silurian rocks. In the north it has not been possible to substantiate the outcrop of Ordovician strata delineated by Egan. Fossils are unknown from this stretch of country and the northern outcrop, formerly mapped as Ordovician, is now included in the Silurian. Determinable fossils from the Lower Palaeozoic rocks have been found at only ten localities within the limits of the Belfast Sheet and in a few more exposures a short distance beyond these limits. These indicate the presence of beds of Caradoc and Ashgill (Ordovician) and of Lower Llandovery (Silurian) age.

On the fossil evidence, therefore, it has been necessary to show inliers of Ordovician rocks in the Silurian outcrop and, for want of better evidence, they have been interpreted as faulted inliers. The inliers shown may be only part of the total Ordovician outcrop but none have been inserted unless there is direct evidence of their age. The general picture, therefore, is of folded and faulted Silurian rocks with a few faulted inliers of Ordovician strata.

J.A.R., H.E.W., P.I.M.

ORDOVICIAN

The rocks assigned to the Ordovician consist of alternations of arenaceous and argillaceous beds typically grey or dark grey in colour. The arenaceous beds are fine to medium-grained grey greywackes and are usually subordinate to the argillaceous rock types which consist of mudstones and shales, frequently silty, Though these lithological types can be matched in many Silurian areas it would appear that on the whole the Ordovician rocks are finer grained and do not contain the variety of rock types seen in the Silurian exposures.

There appears to be no difference in the degree of metamorphism between rocks assigned to either Silurian or Ordovician although a difference seems apparent in exposures on the foreshore near Orlock Point in the Carrickfergus and Bangor (29) Sheet.

Samples of shale from the section beside Homra House [275 601] were submitted for bloating tests which proved successful and the shales are potentially useful as a source of lightweight aggregate

DETAILS

Hillsborough Inlier. In the stream-section west of Titterington's Bridge [231 576], 1 mile (1·6 km) S.W. of Hillsborough, unfossiliferous well-bedded greywackes and thin bedded micaceous greywackes, together with grey silty shales, dip to the south at 45°–70°.

From a deep trench for a sewerage scheme on the south side of the Hillsborough-Dromore road [241 584], 600 yd (549 m) S.W. of Hillsborough church, fragmentary ? Amplexograptids and ? Glyptograptids were obtained.

In the stream at the south-west boundary of Hillsborough Castle estate, exposures of grey to lilac coloured shales with subordinate fine-grained greywackes have a southerly dip averaging 40°.

Between Hillsborough and the unnamed tributary of the Ravernet River flowing north beside Homra House [275 601] there are a few exposures of grey shales with some greywacke separated by belts of drift. A small quarry just above the west side of this stream, 350 yd (320 m) N.E. of Watson's Bridge, yielded *Climacograptus sp.*, ? *Orthograptus sp.* or *Diplograptus sp.*

On the horizon of these two fossiliferous localities Professor Bulman comments, "One might hazard the guess that the shales at these two localities are more likely to be Ordovician than Silurian".

Charlesworth (1963, p. 114) gives Ravernet, in conjunction with other localities, as a Lower Valentian (Birkhill) site. This exposure is indicated as the only Lower Palaeozoic fossil locality on the 1901 (revised) edition of the Belfast (36) Map though there is no reference to the locality in the 1904 Belfast Memoir. Clark (1902, pp. 497–500) does not mention the locality and there are no specimens in the Museum Collection of the Geological Survey of Ireland nor in the Ulster Museum.

The section of dark silty shales dipping S.S.E. at 75°–80° along a small path leading from Homra House in to the stream (250 yd (228 m) N.E. of the last locality) yielded no fossils.

North-east of Homra House the Ordovician inlier may extend for another mile (1·6 km) but no further exposures, other than of greyish greywackes some 500 yd (456 m) E. of the house, were seen.

P.I.M.

Ballymacbrennan Inlier. A small shallow quarry [311 613] 500 yd (456m) E. of Crossan Post Office shows almost 100 ft (30 m) of interbedded greywackes and grey shales. The greywacke bands, generally fine grained, range from 1 ft (0·3 m) to 20 ft (6 m) in thickness while the interbedded shales are usually about 1 ft (0·3 m) thick. The shales have yielded a few graptolite fragments including *Climacograptus sp.*, possibly *C.*

tubuliferus. Many of the rocks exposed in this district are probably of the same age, including those in the stream ¼ mile (805 m) along the strike to the south-west. Here a limited exposure, 170 yd (155 m) N. of the road, showed grey micaceous shales, with rare indeterminate graptolite fragments, interbedded with shaly greywackes. Dips in all these exposures are very steep to the south and some bedding planes show strong horizontal or plunging mullions.

South-east of the crossroads near Ballymacbrennan School, small quarries [312 617] show fairly massive, grey, fine to medium-grained greywackes with some shaly layers dipping S.E. at 85°. Nearly ¼ mile (805 m) to the north-east similar greywackes with a dip of 55° S.E. were recorded but the remainder of the inlier to the north-east is drift covered. Up to 32 ft (10 m) of drift are known from well sections.   H.E.W., P.I.M.

The Temple Inlier. An exposure in a deep drain 1320 yd (1207 m) 47° from the Temple crossroads [359 606] and just outside the sheet boundary shows 6 ft (2 m) of fissile dark grey shale with a very steep dip to the south. Fragments of poorly preserved graptolites from this material cannot be definitely named but appear to be of Ordovician type.

The Killaney Inlier. The locality recorded by Hartley and Harper (1937, p. 254) for a temporary section in a pipe trench beside the main road 4·1 mile (6·5 km)N. of Ballyna-hinch appears to be about 200 yd (183 m) N.N.E of Killaney House [361 588] and 600 yd (549 m) S. of St. Andrew's Church [361 595]. The inlier is described as a wedge of black graptolite shales intercalated with flaggy greywackes and the fossils include *Pleuro-graptus linearis, Leptograptus flaccidus, Orthograptus truncatus* and *O. truncatus intermedius.*

A temporary exposure [361 587] in a trench approximately parallel to the road, from 200 yd to 320 yd (183–275 m) north of Curry's Corner was at its northern end cut for 80 yd (73 m) in solid rock. The section was much interrupted by strike faults and yielded graptolites from four bands, two of which gave Ordovician assemblages (Fig. 4).

At the extreme northern end of the trench dark grey shales yielded *Climacograptus* cf. *miserabilis, C. latus, C. supernus ?, Orthograptus* cf. *truncatus abbreviatus.* This assemblage is probably from the zone of *Dicellograptus anceps.*

About 30 yd (27·4 m) S. of the above section, and across a belt of Silurian shales, a second band of dark grey graptolitic shale yielded *Climacograptus* cf. *latus, Climaco-graptus sp., Orthograptus* cf. *truncatus truncatus, O.* cf. *truncatus abbreviatus, Dicello-graptus anceps, D.* cf. *complanatus ornatus.* This assemblage is also from the *D. anceps* zone.   H.E.W.

SILURIAN

The Silurian rocks consist of alternations of arenaceous and argillaceous beds of a predominantly grey colour but with purple, lilac and greenish coloration in places. The arenaceous bands of grit or sandstone texture, usually micaceous, are normally well bedded with beds a foot or so thick and with shaly partings. More massive beds are frequently pebbly and may contain small rafts of mudstone. Thinner flaggy beds of siltier character are not uncommon   P.I.M., J.A.R.

DETAILS

Cregagh Glen. An extensive section in the Silurian rocks is visible in Cregagh Glen [365 709] where almost 800 ft (244 m) of strata are exposed. It is possible that some of the most northerly exposures are of Ordovician age as they appear to have been subjected to a slightly higher degree of metamorphism than the rocks to the south. In one exposure the greywackes are veined with quartz in a manner reminiscent of those on the coast near Orlock Point in the Carrickfergus (29) Sheet.

Entering the glen from Knockbreda Road the first exposure, 190 yd (174 m) S. of the road, is of 18 ft (5·5 m) of purple greywackes, slightly shaly in places and dipping N.N.W. at 80°. South of a gap in the section of about 2 ft 6in (0·75 m), 2 ft (0·6 m) of faulted shales are visible, then 49 ft (15 m) of greywackes, thick-bedded towards the bottom and flaggy and shaly in the top 15 ft (4·6 m). The dip of these rocks varies from N.N.W. at 70° to vertical. Southwards for about 80 yd (73 m), greywackes and shales alternate in almost equal proportions. At 410 yd (375 m) S. of Knockbreda Road, hardened shales are exposed. These rocks are siliceous and slightly gritty in places and at least 62 ft (19 m) thick. Despite prolonged search, no fossils were found, The next important group of beds are the thick-bedded fine-to medium-grained greywackes with shaly partings, totalling 120 ft (36 m) in thickness, which form the waterfall 450 yd (410 m) S. of Knockbreda Road. Another thick group of beds, 67 ft (20 m) of hardened shales, is visible 570 yd (520 m) upstream from Knockbreda Road. These shales are bluish grey and purple with some purple and green mottling; they are twisted in places and are separated by a gap in exposure of 4 ft (1·2 m) from 13 ft (4 m) of closely similar shales. These in turn are separated by 5 ft (1·5 m) of bluish grey and purple greywackes from another 4 ft (1·2 m) of grey and purple shales. For another 200 yd (183 m) southwards there are several exposures of greywackes and shales, mainly purple in colour and dipping 10° to 20° either side of vertical.

Newtownbreda. Exposures are rare in the neighbouthood of Newtownbreda apart from those in the stream to the south which flows westwards to the River Lagan. About 200 yd (183 m) downstream (west) from the road to Purdysburn and Lisburn a few feet of vertical schistose greywackes are visible striking north-east. Upstream from the road purplish grey fine-grained micaceous greywackes are sporadically exposed. The angle of dip varies from 40° to 60° and the direction of dip from S.S.W. to S.E. Purplish, grey and brownish grey shales and purple greywackes dipping S.E. at 80° were at one time visible at the east side of the Newcastle road 20 yd to 90 yd (18–83 m) S. of Common Bridge [355 695]. Continuing upstream where the brook flows in a north-westerly direction there are several exposures of purple greywackes, shaly in places, and associated purple and green mottled shales. The angle of inclination of the beds decreases southwards from 70° to 45°, the direction of the dip being south-easterly.   J.A.R.

Purdysburn district. In Purdys Burn north of the Fever Hospital [338 681] a few feet of purplish grey greywackes are visible for about 100yd (91 m) in the stream bed. About 1500 yd (1370 m) N. of this, in Corbie Wood, a small water-filled quarry shows about 25 ft (7·6 m) of purplish, fine- to medium-grained greywacke with purplish shale partings The greywacke is well bedded in 2 ft to 4 ft (0·6–1·2 m) posts. About 9 in (0·23 m) of purplish fine-grained micaceous greywackes are visible above the water in a drain 150 yd (137 m) N.W. of the quarry.   P.I.M.

In the neighbourhood of the Mental Hospital, Purdys Burn provides an almost continuous section over about 1½ mile (2·4 km). About 450 yd (410 m)N.W. of Purdysburn House [343 680] almost vertical and slightly contorted purplish grey gritty shales and greywackes strike north-north-west, a direction at right angles to the regional strike. About 100 yd (91 m) upstream from these shales and greywackes reddish brown boulder clay rests on purplish grey greywackes with some hardened shales. These beds dip S.S.E. at 45°. Upstream to Purdysburn Bridge [341 677] there are several exposures of greywackes with rare shales. The former are usually purplish in colour but grey in places; they are, on the whole, thickly bedded, even massive at one exposure, and vary from fine to coarse grained. The beds dip S. at 70° and well marked jointing is to be seen 50 yd (45 m) N.W. of Purdysburn Bridge.

Purplish grey, fine- to medium-grained greywackes, generally thick-bedded, are intermittently exposed for 400 yd (365 m) S.E. of Purdysburn Bridge. The direction of dip of the beds is a few degrees east of south and the inclination is 60°. Purplish medium-grained greywackes dipping N. at 78° are exposed 420 yd (383m) S.E. of Purdysburn Bridge.

Continuing upstream (south-easterly) for the next 400 yd (365 m) there are almost continuous exposures. The predominant rock type is purplish greywacke but beds of a grey colour are present. About 6 ft (1·8 m) of bluish shale were noted 20 yd (18 m) upstream from a weir [346 675] and 600 yd (549 m) S.E. of Purdysburn Bridge and red-stained hard shales about 100 yd (91 m) S.E. of the same weir. Beside this weir the rocks are vertical but the dip decreases south-eastwards to a minimum of 25° at an exposure [347 673] 800 yd (732) m S.E. of Purdysburn Bridge. Still proceeding upstream, the direction of dip swings round to south-easterly and the angle of dip increases from 28° to 48°. The rocks exposed are mainly purplish grey, and grey fine-grained, flaggy grey-wackes with some hardened shales, purple and green mottled in places. A specimen of grey micaceous shale [NI 2052] collected 740 yd (676 m) W.N.W. of Charity Bridge [358 671] shows low grade regional metamorphism. A reversal of direction of dip to north-west in gritty shales and shaly greywackes [353 673] 650 yd (607 m)W.N.W. of Charity Bridge suggest the presence of a small syncline trending north-north-east. The axis of another syncline, trending north-east, crosses Purdys Burn [354 673] 550 yd (500 m) W.N.W. of Charity Bridge; the flaggy greywackes and gritty shales on the north-western limb dip S.E. at 64° and similar rocks on the south-east limb dip N.W. at 55°. Between the synclinal axis and Charity Bridge there are several exposures of purple and grey thin-bedded and flaggy greywackes with a few bands of hardened shales. The direction of dip varies between north-west and north- north-west and the angle is usually about 75° although dips of 80° and 40° have been recorded.

In the steeply rising ground to the south of this stretch of Purdys Burn there are many small scars and knolls of bare rock, mainly purplish grey coarse greywackes. This type of rock is exposed in a quarry [357 668] 280 yd (256 m) S. of Charity Bridge.

There are a few small exposures in Purdys Burn 250 yd to 300 yd (228–275 m) S.S.E. of the bridge.

Purple greywackes with some thin bands of grey siltstone and some purple shales are exposed, dipping S.E. at 75°, in the large disused quarry [359 666] on the east side of the road leading south from Charity Bridge.

To the west of Knockbracken Reservoir [364 664] a northwards-flowing tributary of Purdys Burn has cut a deep gulley in greywackes with rare bands of hardened shale. The greywackes are purplish grey and grey, fine to coarse grained and micaceous in places. About 6 ft (1·8 m) of vertical purplish brown siltstone are exposed [361 663] 970 yd (887 m) S.S.E. of Charity Bridge. Along this gully the direction of dip is usually north-north-west at from 70° to 80° but in many of the exposures the beds are vertical.

Moorecrofts Milltown. The stream which flows northwards past the village [347 665] affords an almost continuous section in solid rock for nearly ½ mile (800 m). Much of the rock exposed is basalt, possibly one dyke dislocated by faulting along which the stream has cut a deep gully. However, purple, purplish grey, and grey greywackes are visible at several places. They are usually fine-grained, thin-bedded, locally shaly, and are associated with hardened shales, purple and green mottled in places and usually in thin bands. The direction of the dip varies from north-east to north-west and the angle of inclination from 50° to vertical.

The southern limit of the purple-coloured rocks trends east-north-east through the southern end of Knockbracken Reservoir [364 660]. Thus the greywackes exposed in several knolls [363 657] about ¼ mile (400 m) E. of Mealough House are grey in colour. These fine-grained thin-bedded and flaggy rocks are nearly vertical; at one exposure 550 yd (503 m) E. of Mealough House the angle of dip varies from S. at 80° to vertical.

To the south, exposures are less common but thick-bedded greywackes with rare thin irregular bands of hardened shales are exposed 450 yd (410 m) S.S.E of Mealough House. The beds here dip S. by W. at 60°.   J.A.R.

Ballynagarrick. Much of the townland is drift-free and there are extensive exposures of massive greywackes which have been worked in a number of large quarries. The usual medium-grained greywackes are generally interbedded with coarser beds and there are thin partings of grey mudstones. Only one thick band of mudstones is seen at surface, in a laneway [355 639] 300 yd (275 m) E. of Rockmount House where over 25 ft (7·6 m) of deeply dipping beds are exposed. Horizontal striations (mullions) are seen on some of the bedding planes.

Over the district the beds dip generally to the south-east or south-south-east and the inclination ranges from 30° to nearly vertical, the lowest angle being rare.

Quarry exposures show the beds to be greatly affected by crush zones and faults trending in almost every direction. In Lowe's Quarry [365 647], Carryduff, on the extreme eastern edge of the Sheet, the massive greywackes are so broken by crush belts running west-south-west and north-north-west, and by strong jointing, that it is impossible to determine accurately the strike or succession. Just over ½ mile (800 m) to the south the disused quarry on the west side of the main road in well-bedded greywackes with grey mudstone partings shows a belt of crushed and broken rock running north-west and a small north-north-west crush with quartz veining. In a small quarry [357 641] 750 yd (685 m) E.S.E. of Hill Farm a 2-ft (0·6 m) crush trending north-eastwards cuts a 10-ft (3 m) wide crush belt running approximately east and in another small and abandoned quarry [355 646] small parallel crushes trending north-west are seen in the face. Here, too, the mudstone partings are all crushed and heavily iron-stained. The large quarry [345 638] at Kiln Quarter, in massive slightly calcareous greywackes with soft decalcified nodules up to 1 ft (0·3 m) across, pebbles and streaks of shale, shows in its east face a sharp monoclinal fold the axis of which coincides with a band of broken and irregularly jointed rock. In the south face the intersection of two small northerly crush bands and an easterly fault which parallels the monoclinal axis gives some 40 yd (36 m) of broken rock in the face. Other exposures in the centre of the quarry show small faults trending north-north-east and strong inclined joints with horizontally striated faces trending in the same direction. In the north-east corner of the working a further 10 ft (3 m) wide crush zone trends east-north-east.   H.E.W.

Carr. West and south of Carr School [340 630] the Silurian rocks are practically drift-free. A few drumlinoid mounds with recorded thicknesses of boulder clay of between 31 ft (9·4 m) and 50 ft (15 m) are dispersed over the area, together with irregular patches of peat and alluvium through which bosses of greywacke protrude.

A small quarry [335 634] 650 yd (605 m) N.W. of Carr School showed over 50 ft (15 m) of greywacke in beds up to 12 ft (4 m) thick interbedded with cleaved shales in beds 1 in (25 mm) to 11 in (0·28 m) thick dipping south-east at 85°. A bed of greywacke 16 in (0·4 m) thick, 20 ft (6 m) from the top of the section, showed well developed slump structures.

Smaller sections showing similar lithology and south-easterly or southerly dips of between 40° and 80° were seen throughout this belt.

Drumbo Glen. Near Edenderry the Silurian–Permian boundary has been redrawn east of its former mapped position. At the position of the present sluice gate on the Lagan, a note on the original survey states 'Lower Silurian grits said to have been quarried here in County Antrim—angular fragments of grits seen in bed of canal'. Mr. Lucas, Works Engineer at Edenderry Mill states, however, that at that locality he has seen rockhead which is reddish sandstone.

At New Grove Farm [318 668] no information was available from the owner about a 165-ft (50 m) well, dug for 85 ft (26 m) and bored 5 in (130 mm) diameter for the remainder. Nevertheless, a 15-ft (4·5 m) well 150 yd (138 m) S.E. of this borehole encountered Silurian strata.

One of the most complete and continuous sections of Silurian sediments in the Belfast area is that of Drumbo Glen, particularly the 1000 yd (914 m) stretch of the main glen south-east from the Drumbo–Ballylesson road [320 658]. The beds, purplish grey micaceous greywackes and shales, dip fairly steadily to the south-east at angles varying from 20° to 45° with little apparent structural complication. Boudinage was noted below Holland's Pool where a bed of compact greywacke about 18 in (0·5 m) thick has been drawn out in boudins about 3 ft (1 m) long with quartz rods along the constrictions. At the south-east end of the section, dark shales are folded in tight zig-zag folds (Fig. 4).

East of Belvedere House [315 659], 250 yd (228 m) N.W. of the northern end of the section, a few feet of greywackes are exposed near 'The Spout' and flaggy, grey, shaly greywackes occur 250 yd (228 m) farther north-west.

About 800 yd (732 m) N.E. of Drumbo Glen an unnamed stream running parallel to the Glen from Grey Mare's Tail Waterfall to Black Bridge [324 664] gives an un-interrupted dip section over a distance of more than a mile. North of Black Bridge, beside the drive to New Grove Farm, greyish greywackes with shale bands up to 6 in (0·2 m) thick dip S.E. at 40°. Immediately south of the Bridge and north of Drumbo Rectory [325 662], the dip swings to the south-west and quartz veins up to 4 in (100 mm) thick and parallel to the strike are seen. The beds are predominantly shales with some greywacke; 8 ft (2·4 m) of shale are seen in a bank exposure. 200 yd (183 m) upstream grey micaceous flaggy greywackes dip S. at 34°; about 500 yd (456 m) S.E. of the bridge grey micaceous well-jointed greywackes dipping S. at 22° are faulted against greyish purple greywackes and shales dipping S. at 25°-29°. The fault plane hades S.E. at 44°. About 180 yd (165 m) upstream, 4 ft (1·2 m) of quartz-veined grey shales dipping S.S.E. at 40° are visible under 8 ft (2·4 m) of boulder clay. A farther 100 yd (91 m) upstream contorted shales dip S.E. at 50° and are followed by alternating shales and greywackes to a place where massive greyish purple greywackes form a small gorge. The dip hereabouts is 55° S.E. Above the gorge a group of greenish grey shales is followed by flaggy greywackes and shales, the dip increasing to 75° and finally to vertical. A gap between this exposure and the next, where grey and purple greywackes and shales apparently dip south-west at a shallow angle, is probably occupied by a fault.

Upstream from this, grey and purple greywackes are practically vertical and from here to the Grey Mare's Tail Waterfall the dips are uniformly high and to the south-east. About 450 yd (410 m) N.W. of the waterfall a group of conglomerates containing abundant quartz pebbles dips S.E. at 86°. From here to the waterfall the lithology is that of flaggy greywackes, alternating shales and greywackes and greyish purple mica-ceous greywackes. Some 60 yd (55 m) N. of the waterfall, slickensides on joint and fault planes indicate differential movement in both north–south and east–west directions.

North from the waterfall the coarser beds noted in the stream section crop out 200 yd (183 m) S. of Farrell's Fort [336 661] on the same line of strike (N.E.). Purplish grey-wackes with subordinate shales are exposed sporadically in the stream section running northwards from Farrell's Fort to near Mount Tober Farm and thence north-west to the main Ballylesson–Belfast road.

Hillhall district. Exposures are neither numerous nor extensive in this district. No information was available about Belshaw's deep bore at Ballyskeagh [294 668] though the borehole outside the cottages just west of the Charley Memorial School at Drumbeg [304 669] was said to penetrate [Bunter] Sandstone. Geological Survey boreholes put down south of the road in 1958 opposite the school reached a depth of 41 ft (12·5 m) without reaching rockhead.

At Hillhead Farm Lane Bridge [287 645] the drift was unbottomed at 50 ft (15 m); at Lock View [282 650], dark purplish grey micaceous grits with purplish grey shale layers dipping S.S.E. at 45°–50° were formerly seen. About 1000 yd (914 m) due E. of this locality, boreholes to a depth of 45 ft (14 m) at Sycamore Hill Farm Bridge [290 651] failed to penetrate the drift. No information was available regarding the well at Hilden View which was said to have been dug about 50 ft (15 m) and bored a further 100 ft (30 m).

On the Hillhall Road Bridge site [281 640], 31 ft to 39 ft (9·5–12 m) of drift overlie what was reported as 'hard brown gritstone' and at the Ballynahinch Road Bridge [276 634] 8 ft to 11 ft (2·4–3·4 m) of drift rest on reddish brown grit or greywacke. At Lisburn View Bridge [275 635] the drift cover varied between 13 ft 6 in and 27 ft 6 in (4·1 and 8·4 m); the rock was a fairly coarse, grey and purplish grey micaceous greywacke with some quartz veinlets and a small amount of slickensiding.

About 700 yd (640 m) S. of Drumbeg Orange Hall grey and purplish greywackes dip W.N.W. at 70°–80° in a small group of disused quarries [299 662]. Similar poor exposures occur in a small area 1100 yd (1005 m) to the S.S.E. just west of the Ballyaughlis–Hillhall road.

South of Trench House [311 652] massive greywackes and shales, visible in disconnected exposures, dip N.W. at 78°–80°. About 550 yd (503 m) E. of Trench House a tributary of the Drumbo Glen river exposes greyish silty shales with some flaggy greywackes dipping S.S.E. at 75°–80°. More massive greywackes in exposures close to the road about 500 yd (451 m) W. of Drumbo village are practically vertical and strike north-east. The rock at the village is covered by boulder clay to a depth of 15 ft to 20 ft (4·5–6 m).

A belt of drift-free greywackes and shales extends south-westwards from the upper reaches of Drumbo Glen by Rockmount and Lough View Farm [323 637] through Tullyard townland to Ballymullan Hill rising to over 400 ft (122 m) O.D. at Ballymullan Fort [304 632]. In the eastern half of this belt the rocks dip N.W. at 75°–80°; south of Lough View Farm the dip in vertically cleaved shales and greywackes is N.N.E. at 76° whilst at Cairn Valley Farm [319 632], greywackes dip N.N.W. at 77°. West of Tullyard Hill fairly massive medium- to coarse-grained greywackes with occasional shale layers are exposed in isolated knolls. In a traverse north-west from Cairn View Farm to Manor Farm the dips are at first to the north-east and then in the stream section near Manor Farm, predominantly south-south-east at 50°–60°.

Westwards, in the country north of Ballymullan Fort, the rocks dip S.E. at 75°–80° and south of the Fort in Clogher townland the dips are S. at 70°–75°. Here the rocks are mostly massive grey greywackes. On both sides of the road north-west of Clogher School [292 627] there are small quarries in grey fine-grained greywackes, thin bedded and shaly in places with some thin bands of fine dense shales, which dip S. at 70°.

Exposures of greywackes and greyish purple greywackes and purple shales occur intermittently at the foot of the drift hillock east of Pine Hill Farm [301 625] with north-westerly dips of between 47° and 65°. A similar, practically drift-free, belt flanks a drift mound to the south-east of The Laurels [295 617].

Continuing westwards the next exposures occur in the townland of Taghnabrick where up to 15 ft (4·5 m) of grey silty shales, dipping S.S.W. at 55°–60° and showing small cleaved puckers, are exposed in a small quarry [276 621]. The old quarry 170 yd (155 m) S.S.E. of this, where the rocks were formerly seen dipping S.E. at 60°, is now flooded.

Sprucefield. In the vicinity of Sprucefield [262 623] a number of boreholes put down for the M1, together with a borehole for water (36/350), have pin-pointed the Silurian-Trias boundary in some detail. Beside the Ravernet River 300 yd (275 m) N.E. of Sprucefield, depth to rockhead varied between 3 ft and 27 ft (1–8 m). North of Sprucefield, rockhead was proved between 33 ft and 58 ft (10–18 m) and north-west of Sprucefield another group of borings (36/605) encountered Silurian strata at a depth of 49 ft to 54 ft (15–16 m). About 500 yd (457 m) W. of Sprucefield, on the line of the abandoned canal, borings (36/609) proved 77 ft (23 m) of drift and in a boring for water [205 621] the Silurian floor was met beneath Permian breccias at a depth of 265 ft (81 m).

In the banks of the Ravernet River [272 609], grey, fine-grained greywackes with thin cleaved dark grey shaly mudstone partings dip N.W. at 85°. The cleavage dips in the same direction at 30° and the beds may be inverted.

Hillsborough. North-east of the town, boulder clay proved in wells reaches a thickness of up to 30 ft (9 m); south-west of Ravernet village [268 610], however, there are a few isolated knolls of massive and flaggy grey greywackes.

At Carnbane House [250 608], a well sunk in 1936 was dug for 60 ft (18 m) and bored to 200 ft (61 m). There were 14 ft (4 m) of boulder clay over the rock according to the owner. A manuscript note by the late J. J. Hartley gives the section of this well as:

Well Section at Carnbane House

					ft	m
Boulder Clay	.	.	.	.	14	4·3
? Grit or shale	.	.	.	.	49	14·9
Red Limestone	.	.	.	.	39	11·9
Grit	.	.	.	.	100	30·5
Red mudstone	.	.	.	.	22	6·7
Fine Grit	.	.	.	.	7	2·1
					231	70·4

Core fragments still on the site were all of greyish greywacke.

In a temporary road section south-east of the house, 2 ft (0·6 m) of cleaved greyish mudstone and greywacke were seen under 14 ft to 16 ft (4–5 m) of boulder clay. A small stream section [256 606], showed grey flaggy greywackes and silty shales dipping S.S.E. at 60°.

About 1 mile N. of Hillsborough in a stream section just east of Laurel View [244 606], greyish or purplish massive greywackes with some shale layers dip S.E. at 50°. Between Hillsborough Linen Factory and its mill pond, massive grey greywackes with purple and brown silty shale layers have an easterly dip of between 40° and 48°. Nearer Hillsborough, to the north-west of and in the Park of Hillsborough Castle, grey grey-wackes, fairly massive and with some shale partings, are exposed in the stream that runs north through the Park, in the railway cutting and in knolls to the west of the Hillsborough–Moira road. The beds dip S. to S.E. between 36° and 52° with the exception of part of the stream section adjacent to a dyke where the dip is 35° S.W. About 350 yd (321 m) N.E. of Lisadian House [222 593], grey greywackes dip N.W. at 75° in the bottom of a canal feeder stream.

Between Hillsborough and Titterington's Bridge [231 575], isolated exposures appear as mounds in the interfluve areas as well as in the streams. In the unnamed stream flowing from Titterington's Bridge north-west to Beechfield Farm there is a fairly continuous succession of grey greywackes, massive to thin bedded with dips of 30°–35° S.E. That part of the succession extending 600 yd (549 m) W. of the Bridge is described in the Ordovician section (p. 15).

Just north of Firtree Lodge Farm [203 578], small exposures of greywackes with subordinate shales dipping S.E. at 60°–65° were noted in the stream flowing north to cross the road 830 yd (760 m) E. of Kilwarlin Orange Hall. In a larger stream which joins it near the road, similar exposures with a comparable dip were seen and extend intermittently south-east to 200 yd (183 m) S. of Hollow Bridge.

The westernmost Silurian rocks are overlain by up to 30 ft (9 m) of boulder clay. In Tullyard townland a well, 350 yd (321 m) S. of Tullyard House [172 583], was said to be 12 ft (3·7 m) deep in greywacke. Just south of the well a small exposure of grey, medium to fine greywacke with an apparent dip N.W. at 25° was probably not *in situ*. In the stream flowing north by Tullyard House and 600 yd (549 m) S.E. of the House, grey and purple greywacke with a 2-in band of grey shale dipped S.W. at 75°. In a sub-parallel stream farther east, fine-to medium-grained thin-bedded greywackes with grey shales, red-stained in part, have a northern dip at 65° near Mary Vale Farm [180 578] and 300 yd (275 m) W. of Fort William Farm [183 581] dip varies from vertical to 60°

with a north-west strike. Some 100 yd (91 m) E. of Mary Vale Farm, exposures of massive greywacke up to 5 ft (1·5 m) thick are seen, but with no distinguishable dip. A small exposure of massive medium-grained greyish greywackes with some shale is seen 300 yd (275 m) N.E. of Kilwarlin Cottage [189 576], but no contact between the Silurian sediments and the quartz-porphyry intrusion south of Kilwarlin Cottage was visible. Between the exposures noted above, drift thicknesses of up to 50 ft (15 m) have been recorded in well records. At Fort William Farm two wells to rockhead were 45 ft (13·7 m) and 50 ft (15 m) deep; at Megarrystown Crossroads [187 580] a 35-ft (10·7 m) well penetrated 5 ft (1·5 m) into shale and at Kilwarlin Cottage it was 39 ft (12 m) to rockhead. About 400 yd (365 m) N.80°E. of Kilwarlin Cottage a well was sunk 57 ft (17·4 m) before rockhead was reached.

South-East of Hillsborough Ordovician Inlier. In the grounds of Hillsborough Park (Experimental Farm) massive grey greywackes with shale bands were seen in small quarry exposures; 15 ft (4·5 m) of rock with a dip of 80° S.E. were measured in a quarry [253 575] 350 yd (321 m) E.S.E. of the farm and 400 yd (365 m) E.N.E. of the farm, 12 ft (3·7 m) of interbedded grey greywackes and shales dip S.S.E. at 70° [254 578]. Exposures of similar rock are seen in the woods 300 yd to 400 yd (275–365 m) to the S.E. where dips vary between 72° and 76° and joints trending N.30°E. hade 32° to the W. In Hillsborough Forest, east of the lake, dips are still to the south-east and vary between 50° and 72°. Well bedded greywackes with a southerly dip of 53° contained several 3-in (80 mm) shale bands with a cleavage dipping in the same direction at 73°. Similar greyish greywackes and shales with a southerly dip of 70° occur near Annahilt Presbyterian Church [268 581]. In the lower ground to the north of Annahilt and eastwards to Cabragh Bridge [287 585], grey, massive or flaggy greywackes occur at or near the surface in numerous places with south-east dips of between 65° and near vertical. Some north-east trending quartz veins up to 2 in (50 mm) wide were noted in the greywackes 250 yd (229 m) S.W. of Cabragh Bridge. Greyish, fairly massive greywackes with occasional shale partings are exposed intermittently in stream-side sections along a 1100-yd (1005 m) stretch of the unnamed stream flowing north to join the Ravernet River near Cabragh Cottage [289 602]; 300 yd (275 m) S. of the Cottage the dips range between 55° and 60° and steepen to 75° to 80° farther south.   P.I.M.

Exposures in the Ravernet River near Legacurry Bridge [298 600], and in the drift-free ground extending to the east as far as Eden House [317 600], are all of massive or flaggy greywackes with occasional thin argillite partings. In the stream 200 yd (183 m) S.E. of Grove [315 596], however, there is exposed over 50 ft (15 m) of grey shales with thin greywacke partings dipping nearly vertically to the south.

Westwards of the Ravernet River, exposures are all of massive greywackes with thin highly sheared shale partings, and no exposures of thick shale bands are seen. On the west side of the river extensive exposures through the Deer Park at Larchfield, Low Wood, and McKeown's Plantation are all of very massive greywackes, in places coarse, pebbly and feldspathic, usually calcareous. Farther south, beyond Windmill Hill and at the southern end of the Larchfield demesne, the greywackes contain grey argillite partings. A well about 100 yd (91 m) W. of Jamison's Cross Roads [304 580] is reported to have encountered coal—presumably dark shales, but no shales of this type have been seen at the surface.

In the Baileysmill area [325 586] and south to the edge of the Sheet, the numerous exposures are all greywackes, sometimes coarse-grained. Red jasper grains, common in the area to the west, are seen in places. Exposures in the Ravernet River suggest that shale bands are common—more so than the knoll exposures indicate. From a spoil heap [335 585] beside the river south of Drumra Hill where deepening operations had been taking place in connection with a drainage scheme, a greywacke block yielded a specimen of *Monograptus gregarius*. The grey shale, plentiful in the same locality, proved unfossiliferous.

The country to the west of the main road in the townland of Drumalig is largely drift-free and there are extensive exposures of greywacke with thin bands of grey shale in places. The greywackes are locally calcareous to such a degree that carious weathering is seen on the exposed surfaces. They are generally fine-grained but occasional bands of fine pebbly conglomerate and of greywacke with included shale fragments are seen. In the northern part of the townland, round Colin View [362 626], there is a belt of finer grained rocks with relatively abundant grey micaceous argillite. Spoil from the Mourne Tunnel seen at the air-shaft 200 yd (183 m) S. of Colin View is almost entirely of this argillite and has yielded a few unidentifiable graptolite fragments. Argillites are seen *in situ* west of Barn House and just north of Colin View. With few exceptions all the beds in the area dip south or south-east at angles of more than 45°.

Around The Temple there are many exposures of coarse- and medium-grained grey-wackes, generally calcareous, and all dipping south-south-east at moderate or high angles. Crags of greywacke in a field opposite the old Carricknaveagh School [365 599] display an extensive bedding plane, striking 85°, dip 40° S., on which parallel flute or groove casts give a mullioned appearance. The casts pitch south-south-west, diverging by about 30° from the angle of dip and indicate that the beds are inverted and that the currents producing the bottom structures came from south-west. Spoil heaps from the Mourne Aqueduct, which runs under the area, are to be seen 700 yd (640 m) N.N.W. of The Temple and 440 yd (400 m) N.W. of St. Andrew's Church. In both cases almost all the debris is grey greywacke, with dolerite from an intrusion not exposed at the surface. Grey shale, generally sheared and broken and, in the coarser specimens micaceous, is also present in subordinate amounts. Some of the coarse greywackes contain blue quartz pebbles.

About 200 yd (183 m) N. of Cherryvale [357 589], bands of shale interbedded with calcareous greywackes show a strong slaty cleavage. Cleavage-bedding intersection gives a marked down-dip rodding on bedding planes.

Just north of Curry's Corner [361 585], and 50 yd (45 m) from the main road, the debris dug from a post hole for an electricity transmission line included fissile black shale with graptolites, including *Climacograptus sp.*, *Glyptograptus sp.* and fragments of uniserial stipes. About 150 yd (137 m) N.E. from this locality the temporary trench section mentioned under Ordovician (p. 16) showed over a distance of about 30 yd (27 m) a series of dark silty shales interrupted by numerous strike faults with crush zones up to 9 ft (3 m) wide—about 20 faults in all. A band of relatively unweathered rock near the southern end of this section yielded *Climacograptus medius*, *C. normalis*, *C. sp.*, *Glyptograptus sp.* This assemblage indicates the zones of *Akidograpus acuminatus* or possibly *Cystograptus vesiculosus*, at the base of the Silurian.

Another fossiliferous shale band at the extreme eastern angle of the trench, some 60 yd (55 m) further north, yielded *Akidograptus acuminatus* and Diplograptid fragments, also from the zone of *A. acuminatus*.

Around Rockvale [365 586] there are extensive exposures of massive, strongly jointed grey greywackes, all with prominent red jasper grains, which dip steeply to the south-east, approaching the vertical in places. An anomalous dip of 50° to the N.N.E. is recorded at the southern end of the crags. About 300 yd (275 m) S.W. of Girvan's Lane Ends [365 577] massive greywackes with an uncertain dip which appears to be to the north-east, are exposed in a drainage channel. Vein calcite is prominent here and the greywackes are broken in places. At Moss View, 300 yd (275 m) S. of Robinson's Bridge [350 577] and just off the edge of the Sheet, black graptolitic shales dipping to the S.E. at 60° are seen in the roadway and the foundations of the house. The shales contain abundant *Climacograptus sp.* and specimens suggestive of *Rhaphidograptus sp.*

In the stream west of Robinson's Bridge flaggy greywackes with shale pellets are exposed and south-west along the strike of the beds there are fairly extensive exposures, round a farm, of massive greywackes, usually calcareous and with shale pellets. In a small quarry north-east of the farm 25 ft (8 m) of near vertical flaggy calcareous grey-wackes, with grey shale partings, are seen. [345 577].   H.E.W.

Chapter 3

PERMIAN

ALTHOUGH PERMIAN ROCKS have long been known from Cultra (One-inch Sheet 29), a few miles to the north-east, the first proof of the existence of Permian strata in the Belfast area was a Geological Survey Borehole put down in the grounds of the Avoniel Distillery in East Belfast between November 1959 and February 1960. The Permian now recognized from this key section admits of a threefold division into: (1) Basal Permian Sandstones and Brockram, (2) Magnesian Limestone, (3) Permian Marls. When the one-inch map (1966) was prepared the upper part of the Marls was ascribed to the Bunter but is it now considered that the whole of this group is of Permian age in this area. The revised line of demarcation between Permian and Trias is shown on Fig. 2.

At the time of publication of the original one-inch maps of the Belfast district (Sheet 36, 1870, Sheet 37, 1869) and their accompanying memoirs (1871) the marls beneath the Bunter Sandstone were not known. In 1875 a boring on the right bank of the Connswater proved marl, not sandstone, beneath the drift. This marl was considered to be an outlier of Keuper Marl and was so inserted on later editions of the maps. It was also shown thus on the quarter-inch sheet issued in 1921, with an inexplicable extension to the County Antrim side of the Lagan. Meanwhile, in the 1902-4 revision of the Special Belfast Sheet, Lamplugh had demonstrated that these marls were clearly below the main mass of Bunter Sandstone and suggested a Permian correlation (1904, pp. 22, 145-8) although the drift map (1903) shows no concealed marl outcrop. Hartley (1943, pp. 128-132) termed the marls 'Lower Marls' and gave a sketch map of their outcrop beneath the drift. The Basal Permian Sandstones and Brockram were called 'Lower Sandstone' by Hartley (p. 132) and though he referred them to the Trias and placed the 'Permian' below them in his table of thicknesses, he was aware of the possible correlation with Permian of Cultra. He also observed (p. 130) that "The lower portion of this Lower Marl contains in all cases beds of gypsum and occasional thin bands of limestone have also been reported".

During the resurvey of the Belfast area his borehole records, now in possession of the Geological Survey of Northern Ireland, were re-examined and the probable existence of Magnesian Limestone beneath East Belfast postulated and subsequently confirmed by the Avoniel borehole which penetrated the following succession below the Bunter Sandstone: Permian Marls 271 ft (83 m); Magnesian Limestone 69 ft (21 m); Permian Sandstone and Brockram 145 ft (44 m), unbottomed (Fig. 6).

The exposures of Permian beds on the southern shore of Belfast Lough at Cultra near Holywood, County Down (One-inch Sheet 29), have been described by Bryce (1837), Griffith (1838), King (1853), Binney (1855) and Hull and others (1871). The succession is much attenuated compared with Avoniel—the Basal

Permian Sandstones and Brockram are represented by only 5 ft (1·5 m) of brock-ram and the Magnesian Limestone is only 27 ft 8 in (8·42 m) thick. The Magnesian Limestone of Cultra is of Lower Zechstein Age since it contains *Horridonia horrida* (J. Sowerby) (*see* Stubblefield: discussion *in* Dunham and Rose 1949, p. 39). J. Pattison, after an examination of the faunal collections both from Avoniel and Cultra, favours a correlation with the Lower Magnesian Limestone of N.E. England. This is contrary to the opinion expressed by Fowler (1955, p. 52) for the Permian rocks of County Tyrone.

BASAL SANDSTONES AND BROCKRAM

These beds are known only from bores and have only been seen by the authors in cores from Avoniel, Long Kesh and Langford Lodge. The sandstones are variously pink, white, and orange in colour, usually medium or coarse in grain. The brockrams are usually composed of angular fragments of Lower Palaeozoic greywackes (at Long Kesh the blocks were rounded) embedded in a tough reddish sandstone matrix. Thin beds of gypsum or anhydrite are recorded from several bores, thickening locally to a 12-ft (3·65 m) bed of gypsum at Connswater. Hartley recorded (1943, p. 131) that these beds, though bored through for a maximum thickness of 275 ft (84 m), had never been penetrated. A borehole at Long Kesh in 1955 reached the Lower Palaeozoic floor after proving only 97 ft (30 m) of conglomerate below marls, showing a considerable decrease in the thickness west of Belfast. A conglomerate was noted in a small outcrop near Rosevale at the southern limit of the sheet where it was overlain directly by sandstone of Bunter type. It must be assumed that the conglomerate facies ranges in age from Permian to contemporaneity with part of the Bunter Sand-stone but where the Permian Marls intervene it is possible to demonstrate a Permian Age for the basal sandstones and brockram.

The basal beds are recorded in the following borehole sections: Belfast Ropeworks, Avoniel, Irish Distillery, Castlereagh Laundry, Belfast Co-operative Society, East Bridge Street Power Station, Vulcanite, Linen Research Institute (Lambeg), Long Kesh, Langford Lodge (for locations see Fig. 5 and Appendix 1).

At Belfast Ropeworks 82 ft (25 m) of beds below a depth of 395 ft (120 m) are reddish sandstones or marly sandstones frequently recorded in the driller's log as 'limy'. 'Limestone' 4 ft 3 in (1·3 m) thick at 145 ft (126 m) is probably gypsum or anhydrite. A 1906 water analysis from this borehole gave 166·6 grains per gallon of calcium sulphate. The Ropeworks bore penetrated 23 ft 7 in (7·1 m) of conglomerate before stopping at 500 ft (152 m).

At the Irish Distillery, Connswater, one borehole (not that given in Appendix I) recorded 121 ft (37 m) of light grey sandstone underlain by 90 ft (27 m) of coarse soft red sandstone from a depth of 537 ft (164 m). The grey sandstone was over-lain by a bed of gypsum 12 ft (3·7 m) thick and possibly includes the Magnesian Limestone but the old records are indeterminate. At Avoniel the Magnesian Limestone is underlain by medium- to coarse-grained sandstone, mottled in pinkish shades and passing down into greyish white and orange sandstone. Between 77 ft 6 in (24 m) and 115 ft 6 in (35 m) below the Magnesian Limestone, breccia, composed mainly of subangular fragments of Lower Palaeozoic rocks and chocolate-coloured marl in a sandy matrix, with occasional sandstone bands,

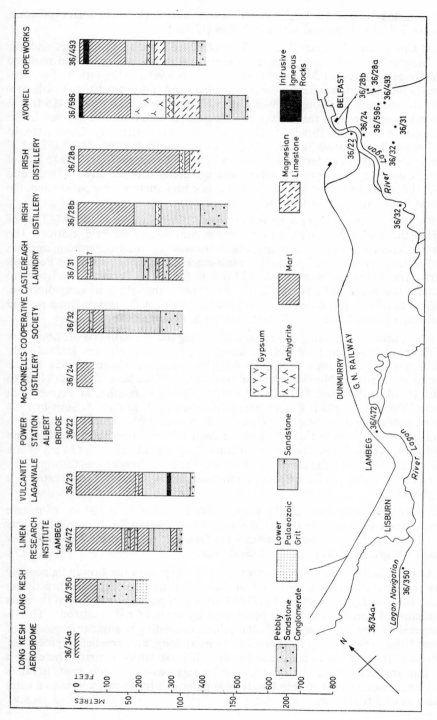

FIG. 5. *Comparative sections of the Permian rocks of the Lagan Valley*

was recorded. It passed down into fairly coarse reddish brown bedded sandstone. The total thickness proved was 155 ft 6 in (129 m).

At Castlereagh Laundry 272 ft (83 m) of Basal Permian Sandstones underlie the Magnesian Limestone. The uppermost 160 ft (49 m) are white and red sandstones with base at 343 ft (104 m). Below this level an alternation of conglomerates and sandstones passes down into marl and 'faky' sandstones with subordinate 'magnesia'. Though a 9-in (0·2 m) limestone is recorded at 415 ft 6 in (126 m), it appears that the thicker development of limestones or gypsum in the Connswater area thins rapidly to the south-west and is virtually absent here. In the Belfast Co-operative Society borehole at Ravenhill Avenue, 236 ft (72 m) of Permian Sandstones underlie the Permian Marls. The uppermost 179 ft (54 m) of the sandstones are red, grey or yellow and are marly between 258 ft (78 m) and 268 ft (82 m). Below these beds 57 ft (17 m) of conglomerate were penetrated, the highest figure recorded in this part of the outcrop.

The 1904 Memoir records 66 ft (20 m) of '? Permian' under 43 ft (13 m) of marl at the Belfast Municipal Electric Power Station in East Bridge Street. The total thickness ascribed to the marls, which are now regarded as Permian, appears low for this part of the outcrop and the sub-marl beds may be part of the Permian Marls. The Vulcanite bore proved 152 ft 8 in (46 m) of Permian Sandstone to underlie the marl at Lagan Vale, the borehole terminating in conglomerate. Shallow boreholes put down for a projected South-East Approach Road proved sandstones beneath the drift 500 yd (476 m) S. of Laganvale.

In the area between Lagan Vale and Drumbeg the Silurian/Triassic boundary on the map has been modified on information from boreholes. Immediately south-west of the Giant's Ring [329 677] boreholes put down in 1951 by the Electricity Board for Northern Ireland proved sandstone beneath 80 ft (24 m) to 95 ft (19 m) of drift and 850 yd (777 m) N. of these boreholes, at Terrace Hill House [326 683], a 300-ft (91 m) borehole drilled by J. Mellon & Sons in 1904 proved sandstone beneath drift. Whilst the Permian Marls outcrop must remain tentative, it is probable that the beds in each of these boreholes are Permian Sandstone. At the Linen Research Institute, Lambeg, about 80 ft (24 m) of the beds penetrated below the marls may be put in the Basal Sandstones Group. They are recorded as red sandstone passing into marl below. The bore terminated in a conglomerate.

Under red sandy marl at the Long Kesh borehole 97 ft (30 m) of coarse conglomerate were penetrated into the underlying Silurian. Most of this hole was drilled with a chisel bit and samples were few, but rounded cobbles of Lower Palaeozoic grit were seen, in a red sandy matrix.   P.I.M.

At Langford Lodge the uppermost 20 ft (6 m) or so of the Lower Palaeozoic grits are somewhat red-stained, and the basal Permian breccia rests directly on these beds at a depth of 4766 ft 3 in (1453 m). The topmost two inches of the grits are disturbed and the joints are penetrated by veins of red sandstone. This is succeeded by a bed of breccia 3 ft (0·9 m) thick, consisting of angular fragments of Lower Palaeozoic grits up to about 3 in (6 mm) long, and smaller fragments of shales and mudstones, in a reddish brown gritty sandstone matrix. There are a few thin irregular veins of gypsum. The breccia passes up into gritty reddish brown sandstone with two thin bands of angular grit fragments and with a total thickness of 5 ft (1·5 m).   H.E.W.

MAGNESIAN LIMESTONE

Though known at outcrop at Cultra and suspected from the journals of some old boreholes in the Connswater area there was no good record of this formation until the Avoniel Distillery borehole (36/596) was drilled 350 yd (320 m) S.W. of the Connswater Bridge in East Belfast in 1959–60.

The borehole proved Magnesian Limestone between 386 ft 0½ in (118 m) and 454 ft 6 in (134 m), a thickness of 68 ft 5½ in (21 m). This thickness is comparable with the range of 63 ft to 70 ft (19–21 m) proved in Tyrone (Fowler 1955, p. 46). For the most part the rock is a yellowish dolomite; occasionally it is greyish pink or creamy grey. Usually massive and cavernous, the dolomite is frequently oolitic. Below 414 ft (126 m) there is a sharp change to a denser dolomite and small crystals or pools of clear anhydrite became common. In the basal 1 ft 6 in (0·4 m) an increasing sand content gives a dolomitic sandstone. Occasional small quartz grains and pebbles were noted at some levels.

The following fossils have been identified from the Avoniel Borehole: Depth 386 ft 6 in (117 m) to 392 ft (119 m): *Omphalotrochus helicinus* (Schlotheim),

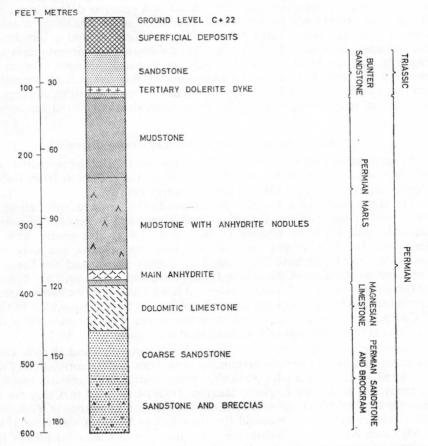

FIG. 6. *Generalized vertical section of the Permo-Triassic beds in the Avoniel Borehole*

Bakevellia binneyi (Brown), *B.* cf. *ceratophaga* (Schlotheim), cf. *Liebea sp.* [juv.] *Schizodus sp.* Depth 408 ft 2 in (124 m) to 413 ft 9 in (126 m): cf. *Bakevellia binneyi*, cf. *B. ceratophaga*, cf. *Liebea squamosa* (J. de C. Sowerby). Depth 436 ft 11 in (133 m) to 451 ft 6 in (137 m): cf. *Cyclobathmus? obtusus* (Brown), *Pleurotomaria sp.*, *Bakevellia binneyi*, *Schizodus sp.*

A bore at the Irish Distillery finished in limestone 27 ft 8 in (8·4 m) thick at a depth of 529 ft (161 m) which was probably the upper part of the Magnesian Limestone, and a bore at the Belfast Ropeworks recorded a variety of limestones and limy sandstones from 332 ft (101 m) to 419 ft (128 m) which probably included both the Main Anhydrite and the Magnesian Limestone (cf. Fig. 6).

At Castlereagh Laundry and at Vulcanite thin beds recorded as 'Magnesia' and as 'Limestone' are thought to be attenuated representatives of the Magnesian Limestone. (See Appendix I).   P.I.M.

Petrography

Sections of the Magnesian Limestone (NI 1503–6) show the upper part to be a dolomitized oolitic limestone. Scattered quartz grains, showing marginal replacement to dolomite, and chert fragments are set in a granular dolomitic matrix. The granules commonly form rounded structures, some containing a single quartz fragment, which are strikingly similar to oolitic structures in limestones. The bottom foot or so of the bed is a sandy dolomite with mineral grains and fragments in a dolomitic matrix. Quartz in rounded and subangular fragments forms 15 per cent of the rock and other detrital material includes hematite, limonite, plagioclase, perthite, sericite, chert, tourmaline and calcite. Aggregates of a clay mineral, probably kaolinite, form green streaks in hand specimens.

J.R.H.

Faunal correlation of the Permian of County Down

In conjunction with the determination of the fossils from the Avoniel Borehole (see above), a study has also been made of the collection from Cultra (Carrickfergus and Bangor (29) Memoir, in preparation).

The fauna from Cultra contains in addition to the forms recorded above, a few species of foraminifera, the brachiopod *Horridonia horrida* (J. Sowerby), ostracods, and a greater number of gastropod and bivalve genera such as *Naticopsis*, *Parallelodon* and *Permophorus*. Thus no bryozoa and only one species of brachiopod have been found in County Down and the fauna is dominated by the bivalves *Bakevellia*, *Permophorus* and *Schizodus*. The Magnesian Limestone of County Tyrone (Fowler and Robbie 1961) has yielded bryozoan remains and no brachiopods but otherwise contains a fauna very similar to that of Cultra. Comparable limited molluscan faunas are also characteristic of the fossiliferous Permian rocks of Cumberland and Lancashire.

These faunas from Northern Ireland and north-west England all have close Zechstein affinities but provide no conclusive evidence for the correlation of the beds in which they occur with particular horizons within the Zechstein rocks of Germany and north-east England. Sherlock (1926) concluded that they are all generally comparable in age with the Lower Magnesian Limestone of Yorkshire and Nottinghamshire; Stubblefield (in discussion *in* Dunham and Rose 1949, p. 39) favoured a Lower Zechstein age for the beds at Cultra because of the presence of *Horridonia horrida*. Several other workers including Hollingworth

(1942), Fowler (1955) and Meyer (1965) have correlated various of these deposits with the Upper Magnesian Limestone.

The molluscan faunas of these beds west of the Pennines are superficially similar to that of the Upper Magnesian Limestone of north-east England but closer examination reveals differences. The fauna of that formation is remarkably uniform throughout the length of its outcrop from north-eastern Durham to south Yorkshire. Three species, *Tubulites permianus* (King), *Liebea squamosa* and *Schizodus schlotheimi* (Geinitz), predominate and when present occur in great numbers. Although all three are recorded west of the Pennines the distinctive assemblage which they form has not been found there, except possibly in the Magnesian Limestone of the Eden Valley.

Tubulites permianus, which was formerly regarded as diagnostic of the Upper Magnesian Limestone, has been found in the Middle Magnesian Limestone of County Durham (Smith and Francis 1967) and can therefore no longer be regarded as proof of Upper Magnesian Limestone age. Moreover, while there is no doubt of its presence in the Middle Magnesian Limestone, most identifications of this species from the Permian west of the Pennines have been doubtful. *Bakevellia binneyi* Brown *sp.* (previously known as *B. antiqua* Muenster *sp.*) is the most common bivalve in the faunas from Northern Ireland and north-west England but is absent from the Upper Magnesian Limestone of north-east England.

The palaeontological evidence thus favours a Lower or Middle Magnesian Limestone age for these Permian rocks in Northern Ireland and north-west England. The faunal assemblage of *Bakevellia* and *Permophorus* associated with small gastropods, common in these beds, is also characteristic of the Lower Magnesian Limestone in parts of Nottinghamshire and south Yorkshire. The geographical position of the beds containing these assemblages suggest that they may have been deposited in similar marginal environments in the Lower Zechstein sea.   J.P.

PERMIAN MARLS

It is now considered that most if not all of the argillaceous rocks below the Bunter Sandstone in the Belfast area are of Permian age. The sub-Bunter marls of the Dungannon area (Sheet 35) were divided into Triassic and Permian on the criterion of the absence or presence of primary gypsum, and when the One-inch Sheet of the Belfast district was prepared these beds were thus divided, on rather limited evidence. Recent developments in the sub-division of the Permo-Trias of northern England suggest, however, that this classification may be incorrect and it is logical to treat this group of beds as a unit and regard them all as Permian.

The Marls are never exposed at outcrop and their description depends on borehole journals. Hartley (1943, p. 130) described them as of red sandy type, but sometimes greyish or greenish in colour and difficult to distinguish from the Keuper Marl. This is true of the material seen from the Avoniel Borehole which was typically a calcareous siltstone or mudstone with much gypsum in veins. Green colour was mainly in spherical 'reduction spots', and sandy bands were common in the uppermost 40 ft (12 m). Current bedding and local auto-brecciation were noted. This group was 272 ft (83 m) thick and included a 16-ft (4.8 m) bed of anhydrite at Avoniel. Maximum thickness recorded is 360 ft (110 m) at Connswater, and like the other Permian rocks, the marls thin out to the west.

The Permian Marls have been recognized in the journals from boreholes at Avoniel, Irish Distillery, Castlereagh Laundry, Belfast Co-operative Society (Castlereagh), Lagan Vale, Dunville's Distillery, Linen Research Institute, Long Kesh, and Langford Lodge.

At Avoniel Distillery an old record of a well drilled in 1897 indicated marl from 119 ft (36 m) to 331 ft (101 m) where 13⅓ ft (4·1 m) 'limestone' was recorded. The Avoniel Borehole (1960) entered marl at 114 ft (35 m) and encountered the Magnesian Limestone at 386 ft (118 m), a thickness of 272 ft (83 m) of marl.

The Main Anhydrite (Fig. 6), from 365 ft to 381 ft (111–116 m), consisted mainly of massive white anhydrite with scattered small ovoid patches of clear anhydrite. In thin section (NI 1509) it is seen to be composed of crystals (0·15 mm) and granules of anhydrite with patches of fibrous gypsum and a few small patches of granular carbonate. The textural relations are not clear but it seems likely that the gypsum is replacing anhydrite.

The anhydrite becomes greyish some 3 ft 4 in (1.0 m) from the base and the lowest part is sandy and calcareous. A thin section of a specimen from 22 in (0·6 m) above the base (NI 1508) showed a fine-grained pink limestone with thin bands of fibrous gypsum along the plane of the bedding. Fibrous gypsum also occurs in irregular pods and veins up to 4 mm wide, which have scattered crystals of anhydrite in them. The limestone contains dispersed hematite granules and subrounded quartz grains.

Between the gypsum bed and the underlying Magnesian Limestone is a 5–ft (1.5 m) bed of calcareous siltstone with patches of gypsum making up 50 per cent of the rock by volume. The gypsum appears to have replaced earlier anhydrite (NI 1507).

At the Belfast Ropeworks marls and sandstones occur immediately below the drift at a depth of 37 ft (11 m). There are a number of sandstones in the upper half with beds up to 29 ft (8·9 m) thick and these beds are possibly better regarded as Bunter Sandstone. Below 146 ft (44 m), however, the beds are mainly marl, usually described as 'limy', and are over 200 ft (61 m) thick. Sandier beds are common below 264 ft (79 m). A 3 ft 1 in bed (0·9 m) of 'limestone' at 331 ft 11 in (101 m) has been interpreted as anhydrite.

Nearby, a boring put down for Sinclair and Boyd on the right bank of the Connswater records 275 ft (84 m) of marl. It was this record which was considered to be evidence of an outlier of the Keuper Marl on the 2nd and 3rd editions of Sheet 36.

One record of a bore at the Irish Distillery gives a thickness of 171 ft (52 m) of marl, while another (Appendix I) quotes 333 ft (102 m). Hartley (1943, p. 130) erroneously records a complete penetration of the series at this locality (placed on the wrong bank of the Connswater in his diagram) with thicknesses of 250 ft (76 m) and 360 ft (110 m). The relative positions of these two sites are not known, but such a rapid variation within a small area may be explained by faulting cutting out part of the succession.

The Castlereagh Laundry and the Belfast Co-operative Society bores both entered the Permian Marls beneath the drift and penetrated the base giving residual thicknesses of marl of 43 ft (13 m) and 74 ft (22 m) respectively. The marl at Castlereagh is recorded as sandy with 'magnesia' and one 6-ft bed (1·8 m)

A.—Knocknadona Quarry   (NI 186)

(For explanation, see p. xiii)   PLATE 1

B.—Kilnquarter Quarry   (NI 239)

of 'magnesia' [? anhydrite] is mentioned. It is probable that the Magnesian Limestone is present in this group of beds. At the Co-operative borehole subordinate sandstones and a 3-ft (0·9 m) conglomerate are recorded. The thick limestones or gypsum of the Connswater area appear to have died out in this district. At Lagan Vale only a 10-in (0·3 m) limestone was noted near the base of 202 ft (62 m) marls and this was regarded as an attenuated representative of the Magnesian Limestone and so recorded on the one-inch map. A number of shallow boreholes for the South–East Approach Road encountered marl beneath the drift 300 yd to 500 yd (274–457 m) S. of Lagan Vale.

Marls are recorded in boreholes at the Ulster Spinning Company [330 734] and Belfast Silk and Rayon Company [322 740].

At the Ulster Spinning Company 149 ft (45 m) of red and grey marls were encountered below 250 ft (76 m) of sandstone and are probably Permian Marls. Five other bores in the immediate vicinity, however, do not enter the marl. 15 ft (4·6 m) of blue marl are recorded below 210 ft (64 m) of sandstone at the Belfast Silk and Rayon Company boreholes. The small scale of Hartley's map (1943, Fig. 1) makes it difficult to identify sites with certainty but he appears to consider that both these marl records are Lower [Permian] Marls.

At the Linen Research Institute at Lambeg at least 156 ft (47 m) of marl were found under 21 ft (6·4 m) of sandstone and this thickness may be 200 ft (61 m) as the record is uncertain, but a thickness of only 64 ft (20 m) is recorded at Long Kesh and the marls probably thin out completely to the south-east beneath the Bunter. A bore at Long Kesh Aerodrome showed 9 ft (2·7 m) of Permian Marls below 61 ft (19 m) of Bunter Sandstone, but the complete thickness was not proved.   P.I.M.

In the Langford Lodge borehole the basal breccia is succeeded by Permian Marls which here are 111 ft 3 in (34 m) thick, and consist of interbedded marls and sandstones with the former predominating. The marls are generally dark reddish brown in colour but small green spots are common. They are rarely fine-grained but earthy and sandy types are common, with in places, quite large quartz grains. Contorted bedding is common and occasionally the marls are broken into discrete fragments embedded in a matrix of similar type. Desiccation cracks are common. The interbedding is often on a small scale, and the sandy beds usually have wisps and thin laminae of marl, sometimes micaceous. Small knots and impregnations of anhydrite were observed at two horizons but they were not plentiful. The sandstones are mostly brownish red in colour but thin white or greenish bands are common. They are generally of medium or fine grain and are occasionally clayey, passing into the marls.

There is not a sharp break between the Permian Marls and the overlying Bunter Sandstones, and the division at 4647 ft (1416 m) in the bore was taken at the point below which the marls were predominant.   H.E.W.

Chapter 4

TRIASSIC

THE TRIASSIC BEDS of the Belfast Sheet may be divided into two major litho-stratigraphical units—the Bunter Sandstone facies and the Keuper Marl facies. On the published one-inch map a group of marls beneath the Bunter was considered to be partly of Triassic age, but it is now regarded as wholly Permian. The revised line of demarcation between the Permian and Trias is shown on Fig. 2.

The Bunter Sandstone forms the greater part of the Trias outcrop and it is known from boreholes, natural stream sections and small quarries. The Bunter may rest conformably on the Permian in the Belfast area; south-west of Lisburn it lies unconformably on Lower Palaeozoic rocks along the north-west flank of the County Down Palaeozoic massif. Several modifications have been made in the re-mapping of the Permo-Triassic/Lower Palaeozoic boundary, notably in the Edenderry area where the junction has been moved to a position over a mile east of that indicated on earlier maps.

Yellowish sandstones occur at the top of the Bunter Sandstone series and were thought at one time to represent the Keuper Sandstone. The uncertainty of the boundary and the occurrence of similar sandstones at apparently low levels in the Bunter Sandstone induced the decision to omit them from the one-inch map. The palynological dating by G. Warrington of the Triassic sequence at Langford Lodge confirms this decision.

The Keuper Marl exposures are limited in extent and nowhere is there visible anything approaching a full sequence. Material from the Langford Lodge borehole, though by no means the thickest sequence proved in Northern Ireland, has yielded abundant spores at 23 levels from the Rhaetic to the upper levels of the Bunter facies. The palynology is discussed by Dr. Warrington (pp. 44–51; Figs. 7 and 8 and Plates 2 and 3). Of prime importance is the first recognition in northern Ireland of beds of Muschelkalk age within the Keuper Marl. The basal portion of the Keuper Marl is also shown to be of Upper Scythian age, the equivalent of Upper Bunter elsewhere. Bunter and Keuper each have maximum thicknesses in the Belfast area of about 1000 ft (305 m).   P.I.M.

BUNTER SANDSTONE FACIES

The Permian Marls pass upwards into an arenaceous series of rocks, the Bunter Sandstone. Typically the Bunter Sandstone is a dull red medium- to coarse-grained sandstone, micaceous in the thinner bedded horizons, and with partings and more rarely thick bands of reddish or chocolate-brown marl, particularly in the upper part. Pellets of shale or marl occur at some horizons. Bands of white or pale grey sandstone are not uncommon in boring records. The sandstones are usually soft when weathered, and some beds are extremely friable, but the bulk of the series is moderately compact when fresh. From an engineering point of view it is rippable and easily worked.

The first edition of the Memoir (1871) indicated that a group of beds 8 ft to 10 ft (2·4–3·0 m) thick at the top of the Bunter were representatives of the Waterstones or Keuper Sandstone. Reynolds (1928, pp. 448–73), however, considered that the sandstone group formed one continuous series in which a Bunter or Keuper horizon was unrecognizable. Nevertheless, a paler yellowish group of sandstones, different from the usual dull red Bunter Sandstone, do appear to characterize the top of the Bunter at several localities.

The full thickness of Bunter Sandstone is estimated at about 1000 feet (305 m) in the Belfast Area and thins to about 550 ft (168 m) south of Moira. At Langford Lodge, in the north-west of the Sheet, it is 887 ft (270 m) thick.

The sandstone has been quarried for building stone at Lurganville, 2 miles (3·2 km) south-east of Moira, and was used principally at Hillsborough. Quarries near Milltown provided sandstone masonry for railway bridges along the Lagan valley. The poor weathering qualities of the Lagan valley rock compared unfavourably with sandstones of the same age from the Scrabo quarries near Newtownards where the sandstones appear to be somewhat indurated by igneous intrusions.

Where the Bunter rests directly on the Lower Palaeozoic rocks west of Lisburn a basal breccia is developed. This crops out also in a small outlier described in the original memoir as Permian. It consists of angular fragments of grit, commonly less than an inch in diameter, set in a sandy matrix usually sufficiently consolidated to make a moderately hard rock, but in places poorly consolidated. It is made up almost entirely of the debris of Lower Palaeozoic rocks.

The Bunter Sandstone is the main aquifer of the district and a large number of boreholes have been drilled, particularly in the Belfast area. Lisburn still obtains a good deal of its water from deep wells (see Groundwater, Chapter 13).

DETAILS

Magheralin Area. On the evidence from the Boxmore Works Borehole (36/39, p. 196) the Keuper Marl–Bunter Sandstone boundary in that district is placed further south than previously mapped. No exposures were seen in the critical area in the vicinity of Newforge and in an effort to confirm the position of the Lower Palaeozoic–Trias boundary an area in the Banbridge (48) Sheet was examined, but there were no exposures and the bulk of the well sections recorded were in drift. Wells in Tullynacross and Drumlis townlands suggest that the boundary passes under Milltown and that the Trias does not extend as far south as Waringstown. The basal breccia formerly recorded in Ballymacbrennan townland was not seen.

In a stream section in Edenmore townland [147 580] a few feet of breccia containing subangular Lower Palaeozoic grit fragments up to 8 in (0·2 m) dip north-north-west at a low angle. At Gartross House [152 580] red sandstone was recorded in an old well and if the regional dip is maintained this sandstone lies beneath the breccia at Edenmore. The breccia appears to lie about 320 ft (98 m) above the local base of the Trias. In the absence of any indication of its lateral extent, the breccia has been mapped as a lens. Hartley (1949, p. 316) records a closely comparable breccia in a boring at Halfpenny Gate. Here the rock was composed of numerous sharp angular fragments of grey and black slate averaging one to two inches in length embedded in a fine-grained marly matrix. The inclusions suggested broken Lower Palaeozoic rock. The breccia was unbottomed after a penetration of 66 ft (20 m). Though both these occurrences are taken to be horizons within the Bunter it is possible that there are concealed structures bringing the Lower Palaeozoic floor near to the surface and that they are actually basal breccia. This is partly substantiated by the record of 'bluish slaty rock' from a well [193 612] south of the Halfpenny Gate inlier.

A borehole (36/56) at the Bovril Factory [136 586] reached a depth of 312 ft (95 m) in Bunter Sandstone. In the Magheralin area the total thickness of Bunter Sandstone is about 550 ft (168 m).

Moira to Lisburn. At Lurganville [175 587] 16 ft (4·9 m) of red and yellow, rather soft, sandstones are poorly exposed beneath 8 ft (2·4 m) of red sandy boulder clay in a small series of disused building-stone quarries. In the sections formerly exposed, purple and white bedded sandstones were seen dipping north-north-west at 5°. Sandstone from this locality can be seen in the wall surrounding Government House at Hillsborough.

Small exposures of fine yellow sandstone dipping north-west at 15° were seen in the stream just south of the quarries. Between these exposures and the breccia exposed further south in the stream, a 70-yd (64 m) belt of Lower Palaeozoic grey grits and shales intervenes. The breccia forms part of a small faulted outlier formerly mapped as Permian. It is now thought to be the local base of the Bunter Sandstone. In the northern-most of the two main exposures up to 8 ft (2·4 m) of breccia are exposed. It contains Lower Palaeozoic micaceous grit fragments in a soft red or yellow sandy matrix and dips south-west at 5°. In the other exposure some 500 yd (457 m) to the south-east, where the stream runs parallel to a minor road, similar red and yellow breccias dip north-north-west at 12°.

At the aqueduct [173 602] where the old Lagan Canal crossed the River Lagan a group of four boreholes (36/599) put down on the site of a motorway bridge penetrated Bunter Sandstone under 34 ft to 45 ft (10–14 m) of drift; 700 yd (640 m) west-south-west of this site a similar group (36/600) proved sandstone under 26 ft to 56 ft (8–17 m) of drift. At Lavery's Bridge [191 597] 1400 yd (1280 m) further east sandstone was met under 10 ft 6 in to 24 ft (3·2–7·3 m) of drift.

Near Halliday's Bridge [207 598], sandstone appears to have been encountered beneath 55 ft to 75 ft (17–23 m) of drift. Wells near the base of the Bunter in this area include one 35 ft (11 m) deep at Corcreeny House said to reach 'freestones' and a well 500 yd (457 m) east of Kilwarlin House with an identical record. The base appears to lie some 300 yd (274 m) east and south-east of that originally mapped.

At Kesh House a well bored in 1949 proved 30 ft (9·1 m) of drift on 52 ft (16 m) of Lower Palaeozoic rock and a group of bores for the motorway at Kesh Bridge [222 606] immediately to the north proved similar rocks beneath 73 ft (22 m) of drift. In addition, in a shallow well at Priesthill 650 yd (594 m) south-east of the latter site, grey shales could be seen *in situ*. These records lie in an area formerly mapped as Trias and the amended boundary runs some 900 yd (823 m) north of the original mapped line.

North of the River Lagan a well at Trummery Cottage [178 615] passed through: red boulder clay, 15 ft (4·6 m); dolerite, 10 ft to 15 ft (3–4·6 m); red and white sandstone, c.60 ft (18 m). This well apparently penetrates an offshoot of the Hollymount intrusion, the main part of which lies 350 yd (320 m) to the east. Three wells just beyond the eastern margin of this intrusion reached Bunter Sandstone beneath the drift. At Hasley's Town [189 611] the old map records that sandstone was met in sinking wells in the area and sandstone is also known from the wells at Upper Broomhedge but the anomalous records of breccia and slate mentioned earlier (p. 35) occur just north-east of this.

In the area between 450 yd (411 m) and 700 yd (640 m) north-east of the Halfpenny Gate boring, a group of boreholes (36/585–8) put down by the Geological Survey in 1958 in search of brick clay encountered only normal Bunter Sandstone.   P.I.M.

A few yards west of Magheragall Mill [213 651] in the overflow channel from the now derelict mill dam, very friable massive red sandstones are exposed, showing good current-bedding which indicates a current direction from the south. The beds dip west-north-west at 10° and over 20 ft (6 m) of sandstone are seen. Just above the weir, and 6 ft (1·8 m) above the sandstone, 6 ft (1·8 m) of dark red marl with thin partings of white sandstone are exposed in the bank of the dam. A ¼ mile (400 m) E. of Magheragall Mill exposures of weathered red sandstone are seen in a stream and a dip of 5° to the north-west was formerly recorded here.

The water-filled quarry [232 654] 600 yd (549 m) E. of Greerstown shows 1 ft (0·3 m) of flaggy red micaceous sandstone over 10 ft (3 m) of massive friable false-bedded yellow sandstone on the west side of the workings. On the east side of the quarry 15 ft (4·6 m) of interbedded yellow sandstones and red shaly sandstones are seen above the water. The reference by Warren in the 1871 Memoir (p. 26) to greenish grey sandstones in this quarry probably referred to material which had been baked by the two dolerite dykes, but no rock of this colour is now exposed. It is interesting to note that Warren referred this exposure to the Lower Keuper Sandstone. 400 yd (366 m) S.W. of the quarry a small stream exposure shows a few inches of yellow sandstone.

The stream north of Fort Cottage [238 661] shows yellow sandstones with marl bands dipping north at 5°.

Two wells near the road running from Kilcorig to Brookmount penetrate white or yellow sandstones. One [216 662] 400 yd (365 m) S.S.E. of Ballyellough School is 30 ft (9·1 m) deep with 6 ft (1·8 m) of boulder clay at the top: the other, 500 yd (457 m) S. of the school, is 40 ft (12 m) deep with the same boulder clay overburden.   H.E.W.

Lisburn to Belfast. There are no surface exposures of sandstone in the area north of Lisburn but thicknesses of up to 400 ft (124 m) were proved in the Lisburn water bores (Appendix I, p. 201–4); 202 ft (62 m) of sandstone beneath 124 ft (38 m) drift were proved in the Clonmore Borehole [274 660]. At Lambeg some 20 ft (6 m) of sandstone were formerly seen in a riverside cliff. Just over ½ mile (800 m) to the north sandstone was seen *in situ* near Wolfenden's Bridge. At the nearby Lambeg Weaving Factory [285 671] the old chimney foundations were in white sandstone beneath alluvium and on the opposite side of the river old quarries formerly exposed the Bunter Sandstone. At few other localities was rockhead seen, or known to exist, at or near the surface along the course of the Lagan.

Just west of Derryaghy House [275 680] red and yellow sandstones, current-bedded, dip to the north-west at a low angle. In the stream section north-west of Milltown, exposures are fairly continuous. In the lower part of the sequence the sandstones are flaggy with intercalated thin mudstones. Higher up the sandstones are red, yellow, or grey, laminated, flaggy and are occasionally false-bedded and échelon-ripple marked. Just below the Keuper Marl red or grey flaggy micaceous sandstones have 1 in to 2 in (25–50 mm) beds of silty micaceous mudstone. The average dip is 15° N.W. Soft red sandstone with sandy shale was seen beneath the Lagmore Reservoir and in the narrow valley below and to the south-east of the reservoir pale yellow and red laminated sandstone with bands up to 4 in (100 mm) of dark red mudstone crop out in intermittent small cliffs for about ⅓ mile (500 m).

In an unnamed stream ¼ mile (400 m) to the north-east a small gorge section is exposed just north of Bogstown [273 692]. Close to the Keuper Marl (exposed higher up the stream) the sandstones are pale grey and yellow, fine-grained and have subordinate thin bands of silty mudstone. Below this a bright red coloration is more typical. A similar sequence is visible in another stream section ¼ mile (400 m) to the north-east.

A small stream bed exposure of yellowish sandstone occurs near Kilwee Cottage [281 702] but there are no exposures in the Collin River between the lowest Keuper Marl seen near Suffolk, and Dunmurry where the lower reaches of the Derryaghy and Collin (Glen) streams exhibit a maximum of 25 ft (7·6 m) of yellow massive false-bedded sandstone with a westerly dip of 8° to 10°. Yellow and red flaggy sandstones appear to underline them. A small section of red sandstone with a 3-in (76-mm) mudstone band dipping 12° N.W. was exposed in the construction of the M1 Motorway in the grounds of Rathmore [298 690]. North-east of the Collin Glen Fault, shallow bores beside the Andersonstown Road (36/1584–5) penetrated red sandstone beneath up to 13 ft 6 in (4·1 m) of the drift close to the remapped boundary [300 713].

Some of the sandstones described above were referred to by the original surveyors (Hull and others) in the 1871 Memoir as Lower Keuper Sandstone or Waterstones but this classification cannot be sustained.

In the Lady Brook stream, flowing eastwards into the Bog Meadows, red and yellow sandstones with purplish bands are noted 300 yd (274 m) W. of Woodlands Bridge. Near the junction of this stream with the Blackstaff 20 ft (6 m) of red and yellow sandstone with a few micaceous mudstone bands were seen in the latter river. Small tributary streams draining into the Bog Meadows area, north and south of Milltown Cemetery, cut into yellowish massive sandstones and red sandstones with west-north-west dips of between 10° and 20°. Intercalated 3-in (76-mm) mudstone bands were well displayed in a small quarry [313 726] 100 yd (91 m) E. of St. Patrick's Industrial School.

Just north of the intersection of Whiterock and Falls Road 6 ft (1·8 m) of red sandstone with silty mudstone bands overlie 3 ft (0·9 m) of yellowish sandstone. Between here and the exposures to the north of the city the Bunter Sandstone has been proved in numerous water and site exploration boreholes (see Appendices 1 and 2). A small selection of these is given in Appendix 1 though few detailed lithological descriptions are available. Thick mudstones within the Bunter sequence were encountered in Telephone House Borehole (36/634) 24 ft (7·3 m) and in boreholes for the Housing Trust development at Cullingtree Road.

To the north of the city the Bunter is exposed at the Grove Playing Fields where it forms a 30-ft cliff (9·1 m) at the margin of the Estuarine Clay. The section [341 767] showed 19 ft (5·8 m) of red and brown sandstones, cross-laminated in the lowest 12 ft (3·7 m).

The junction of the Bunter Sandstone and Keuper Marl in the high level interception sewer tunnel immediately south of Grove School was observed to be faulted.   P.I.M.

Langford Lodge Borehole. Bunter sandstone was proved from 3760 ft (1146 m) to 4647 ft (1416 m). Above the base the sandstone was cored for about 300 ft (91 m) and contained subordinate marl bands up to several feet thick throughout the whole of this thickness. In general the sandstones are fine-grained and reddish brown in colour, but white and greenish grey beds occur sporadically. The marl bands are almost invariably finely interbedded with sandstone layers and the bedding is usually convoluted. Current-bedding is common. Marl pellets and streaks are very common throughout the sandstones, and at four horizons, 4414 ft (1345 m) ,4425 ft (1349 m), 4436 ft (1352 m), 4471 ft (1363 m), the concentration of fragments of marl and small quartz pebbles justifies the description of pseudo-breccia or desiccation conglomerate. These bands are thin, the thickest being only 1 ft 3 in (0·37 m).

The sandstones are much divided by thin marly and micaceous partings, and few posts thicker than a foot or two occur in this part of the succession. The thin marl partings are often broken up into trains of fragments, and deep suncracks filled with coarser material are common. Some of the paler sandstones are slightly calcareous and, on exposure to the weather, cores from the bottom of the succession yielded an efflorescence of salt (sodium chloride).

Because of long-continued drilling difficulties no cores or cuttings were obtained from a depth of 3844 ft (1172 m) to the re-commencement of coring at 4320 ft (1317 m) and nothing is known of the succession over this distance. Above 3844 ft (1172 m) the hole was mainly drilled with a rock bit and only cuttings are available, but an attempt at coring from 3768 ft to 3782 ft (1149–1153 m) yielded two small pieces of compact fine-grained reddish sandstone, current-bedded and micaceous, with brown mudstone partings. This material resembles Bunter Sandstone and though cuttings from depths below 3780 ft (1152 m) contain a high proportion of marl, the Bunter/Keuper boundary has been taken at 3760 ft (1146 m).   H.E.W.

KEUPER MARL FACIES

The Keuper Marl consists predominantly of mudstones and silty mudstones, usually reddish brown in colour but often grey, chocolate, red, or occasionally green. The upper part of the series is locally calcareous, with up to 23 per cent CaO known in calcined samples, and thus some of the beds are true marl; but on the whole the Keuper facies is more accurately described as mudstone.

Subordinate silty or sandy beds (skerries) of a green or greyish colour occur throughout the series, though they are rarely seen at outcrop. These range from a few inches up to 8 ft (2·4 m) thick and have been recognized from cored bore-holes. In the lower part of the succession in the Langford Lodge borehole dolomitic mudstones and sandstones make up a fair proportion of the whole—about 25 per cent in parts—both in beds up to 9 ft (3 m) thick and as wisps, lenses, and thin bands in the marl.

The uppermost 30 ft (9 m) or so of the succession are locally of greyish green mudstones, presumably the equivalent of the Tea Green Marls of the English Midlands. This facies is always present in County Antrim where the overlying Rhaetic occurs.

The thickness of the Keuper is almost 1000 ft (305 m) in the Belfast area, probably thinning markedly towards the south-west.

The middle and lower parts of the succession in the Belfast area contain a moderate amount of fibrous gypsum in the form of irregular streaks, veins and nodules. There is not sufficient to make it worth extracting economically, though it was, at one time, hand-picked from brickyards. Casts of rock salt crystals have been found, but the saliferous beds of the Carrickfergus, Eden and Red Hall areas (Sheet 29) a few miles to the north of the northern margin of the Sheet are absent or are represented by non-saliferous marl.

From an engineering aspect the Keuper is easily ripped yet has high bearing strength and stands up well in excavations. It breaks down on prolonged weathering but is not liable to failure in embankments when adequately graded. Some of the large landslips along the escarpment above Belfast incorporate Keuper Marl and in these circumstances its stability is suspect.

The Keuper Marl was at one time extensively dug for bricks on the north-western outskirts of the city. It was well exposed at the time of the 1902 revision, but many sections have since been built over or used as dumps. Only two pits (Collin Glen and Parkview) were working during the resurvey and only one (Crow Glen) is now (1968) operating. Pit sections and other excavations during the 1902 revision showed the Keuper Marl/Bunter Sandstone boundary to lie east of that originally mapped (Lamplugh 1904, p. 22) and new borehole evidence has now resulted in further modifications to that boundary in the vicinity of Belfast. According to Lamplugh (1904, p. 21) the lowest few feet of the Keuper are of grey micaceous sandstone, and in the Langford Lodge borehole there appeared to be an alternation of mudstones and sandstones at this horizon. The dividing line between Keuper and Bunter is thus not sharply defined and, save in one borehole which proved a sandstone conglomerate, there seems to be a conformable passage from Bunter upwards into Keuper.

DETAILS

Moira to Lisburn. The westernmost record of Keuper Marl in the One-inch Sheet is in a well at Boxmore Works, Dollingstown [111 582] near Moira (36/39) where 116 ft (35 m) of mudstone were penetrated beneath the chalk. The well was deepened from 435 ft (133 m) to the present depth of 530 ft (162 m) in 1958. Near the base of the core some specimens of marl showed autobrecciation. Elsewhere in Northern Ireland autobrecciated horizons are common in the lower part of the Keuper Marl.

East of Dollingstown the Keuper Marl floors the Lagan Valley but the outcrop is completely covered by boulder clay and alluvial deposits, though well records locate the Keuper-Bunter boundary to within a few yards. At Lagan View House [159 594] 30 ft (9·1 m) of 'blue platy clay' were proved beneath 20 ft 6 in (6·2 m) of boulder clay. At Clarehill Farm [158 602] a half mile south-east of Moira, just below the base of the Chalk a well record showed 20 ft (6 m) of boulder clay over 26 ft (7·9 m) of 'freestone'. Only marl fragments were seen in the debris and the rock was thought to be marl.

A well [169 605] in Balloonigan townland proved 26 ft 6 in (8·1 m) of boulder clay on 15 ft 6 in (4·7 m) of blue and bluish red shaly marl. At Boyles Bridge [166 609] a group of four bores (36/598) for a motorway bridge proved white and yellow sandstone below 13 ft to 30 ft (4–9 m) of drift, the sandstone overlying hard blue 'shale' in one borehole. The locality is 900 yd (823 m) N.W. of the Keuper/Bunter boundary and this sandstone later seen in excavations, appears to be about 200 ft (61 m) above the base of the Keuper Marl. Elsewhere in the north-east of Ireland sandstone skerries have been noted at well-defined intervals, but lack of exposure in the Belfast area would render any correlation speculative.

North of the Lagan a well at Thornbrook [173 612] proved 35 ft (10·7 m) of bluish shaly marl under 25 ft (7·6 m) of boulder clay.

At the Ulster Limestone Corporation quarry [135 609] in Legmore townland the basal glauconitic chalk was seen resting on Keuper Marl at one place in the working and rafts of marl (some glauconite-stained) were incorporated in the glauconitic chalk. The section is now under water as the quarry is disused.

At Lisnabilla [166 623] 900 yd (823 m) N.E. of Moira Railway Station, about 40 ft (12 m) of boulder clay overlying 10 ft (3 m) of mottled grey and red shaly marl were recorded in a well, and a well sunk near Primrose Hill Farm [163 632] encountered bluish red marls beneath a thin cover of boulder clay

In the small cut bounding the northern side of Balmer's Glen Quarry [183 631] marls were formerly seen but are no longer exposed. They are, however, now visible beneath the Chalk in Hull's quarries at Mullaghcarton [187 641] at both ends of a drainage tunnel between the quarries and in the 50-ft (15 m) air vent in the southern quarry. The quarries north-east of this were cleaned up by renewed quarrying operations after 1956 and the drainage in the southernmost quarry is said to penetrate to the Keuper Marl, but the marl is not visible. In Laurel Hill Quarry [190 645], however, at a sump hole in the south-eastern corner of the quarry, glauconitic chalk rests on Keuper Marl with a marked angular unconformity. The chalk has a westerly dip of about 10° and the Keuper Marl dips south at about 7°.

Immediately east of Hull's quarries, in Ballynalargy townland, 32 shallow holes put down by a brick company proved about 20 acres (8·08 ha) with 7 ft (2·1 m) or less of drift on marl.

At Beech Hill Farm [202 642] a 42 ft (13 m) shaft was sunk into Keuper Marl but there is no record of what was met with in the old shaft said to have been sunk in search of coal 70 yd (64 m) S. of Mullaghcarton House. A well at Ivy Lodge, 400 yd (366 m) E. of Beech Hill, penetrated blue 'clay' to a depth of 40 ft (12 m) and a well 400 yd (366 m) E. of this entered Keuper Marl within 17 ft (5 m).

In a farm well [205 632] 1100 yd (1006 m) W.N.W. of Peartree Hill the top 49 ft (15 m) were dug in red sandy boulder clay and the remaining 36 ft (11 m) bored in bluish

Keuper Marl. A 150 ft (46 m) borehole at Poplar Vale [222 646] entered Keuper Marl at about 14 ft (4·3 m) and the tongue of Keuper Marl south of the Magheragall Fault was inferred from these two records. More recent boreholes near Red Hill confirm the existence of this outlier. P.I.M.

Lisburn to Belfast. At Old Park Farm, Aghalislone, Lisburn No. 2 Borehole [249 669] encountered 50 ft 6 in (15·4 m) of red, yellow and grey mudstones below 119 ft (36·3 m) of drift. The remainder of the borehole proved Bunter Sandstone.

In the northern part of Aghalislone townland the Keuper Marl may be thinner than further north in the Lagmore Reservoir district. Between these two localities reddish brown and grey silty mudstones are poorly exposed in small unnamed stream sections.

A narrow inlier of Keuper Marl extends from Castle Robin quarries for 1600 yd (1463 m) to the north-east, bounded on the east by the Magheragall Fault. Only small exposures of red mudstone were seen towards the northern end of the inlier. H.E.W.

At the western end of Lagmore Reservoir [265 692] red and pale greenish grey shales with some red sandy shale dipping 30° N.W. were formerly exposed. The exposure lies close to the Keuper/Bunter junction as red sandstone was met in the floor of the reservoir. Between the reservoir and the Chalk outcrop exposures are rare, but red and grey mottled mudstone is seen in the stream bed immediately beneath the Chalk at Mounteagle Glen [256 696] and again 300 yd (274 m) to the north-east. East of Groganstown, red and grey marls are seen at intervals in three unnamed small stream-sections flowing south-east from the main scarp. Hereabouts, the Keuper Marl is about 800 ft (244 m) thick.

In Ballycullo townland the Keuper Marl was dug for brick clay by S. McGladdery and Sons Ltd. Up to 20 ft (6 m) of boulder clay overlie red and green mudstone which is highly calcareous in parts. Westerly and north-westerly dips of up to 15° are recorded. Intermittent sections in Keuper Marl are seen between Collin Glen and Suffolk in the Glen River. The sections are interrupted by numerous basalt dykes and the maximum continuous section is only 20 ft (6 m) in thickness. About 250 ft (76 m) of the uppermost beds in this area are free of gypsum, a feature which has been noted elsewhere in Northern Ireland in boreholes. At the top of the section, and underlying the Rhaetic Beds at Glen Bridge, 10 ft to 14 ft (3–4 m) of Tea Green Marls are exposed. These are blocky-jointed mudstones with a few red silty laminae and their total thickness is probably about 30 ft (9 m). North-east of the Collin (Glen) River for about 2 miles (3·2 km) the Keuper Marl is thickly blanketed with boulder clay. A site investigation for Hannahstown electricity sub-station [288 718] proved Keuper Marl beneath 50 ft (15 m) of boulder clay and at the Brewery [293 721] over 500 ft (152 m) of mudstones were proved beneath 65 ft (20 m) of drift.

A small fault-bounded inlier of Keuper Marl occurs 500 yd (457 m) N.N.E. of Tornaroy Bridge [268 734]. A maximum of 6 ft (1·8 m) of greenish and grey mudstones overlying 8 ft (2·4 m) of red silty mudstones with some coarse quartz grains was seen in the Big River. The beds have a westerly dip and occur at an elevation of nearly 900 ft (274 m) O.D. compared with about 500 ft (153 m) along the main scarp.

At the former Ballymurphy and Clowney brickworks between the Falls and Springfield roads red and grey, highly gypsiferous shaly mudstones with northerly dips of up to 12° were seen, but much of the worked area is now built over. The mudstones dug in this district lie near the base of the Keuper Marl. North of the Springfield Road 35 shallow boreholes (36/910) were put down by Belfast Corporation for the Tracoba housing project and a number showed Keuper Marl at the surface. Subordinate siltstones and dark red fine-grained sandstones were present, but lithologically did not resemble the skerry type of sandstone seen in boreholes elsewhere in Northern Ireland, although they probably represent one of the skerry horizons.

Fairly continuous exposures of marl are seen in an unnamed right bank tributary of the Forth River from Glenview [307 753] to near the base of the Cretaceous. Westerly dips range from 4° to 20°, but may be as high as 75° in the immediate vicinity of numerous dykes. The red and grey mudstone, often silty or shaly, contains a good deal of gypsum in irregular veins up to 3 in (75 mm) thick. Reduction spots are not uncommon. The stratigraphically highest beds seen in the section are red and both the Rhaetic and the Tea Green Marls are absent in this area, probably removed by pre-Cretaceous erosion.

In the upper reaches of the Forth River above Glenbank Bleach Works [306 767] the dips of the marls are northerly and range from 8° to 15°.

At Glenbank Bleach Works a deep borehole for water (36/1) proved 459 ft (140 m) of red and green gypsiferous mudstones beneath 41 ft (12 m) of boulder clay (Appendix 1, p. 198). South of the Bleach Works as far as Woodvale the Forth River has cut down to the Keuper Marl. Greenish blue shales and red mudstones, in places gypsiferous and containing occasional salt pseudomorphs, dip north or north-west at angles between 0° and 70°. There is no continuous section greater than about 30 ft (9 m).

Between the Ballygomartin Road and the Springfield Road a large area along the left bank of the Forth River has been worked over for brick clay. A factory has been built over the Forth River brickworks site, but the Parkview Brickworks are still in operation using glacial clays brought from outside the area. At the time of the resurvey a small unworked area alongside the western margin of Woodvale Park showed a 25 ft (7·6 m) face of comparatively gypsum-free mudstone. Gypsum dumps, however, indicate fairly extensive veining in the worked over area.

Red and green marls were formerly seen in the Oldpark Brickworks, the Belfast Waterworks, Limestone Road Brickworks and Skegoneill Brickworks. The Keuper Marl is at a relatively shallow depth beneath Victoria Barracks housing site. At all these sites the marl was intruded by numerous basalt dykes.   P.I.M.

Langford Lodge. The Keuper facies of the Trias at Langford Lodge is over 1000 ft thick, the top being at a depth of 2735 ft (832 m) and the base at 3760 ft (1146 m). The whole of this thickness was drilled by rock-bit and only cuttings were available for description, but during diversion of the hole because of technical difficulties, cores were recovered from depths of between 3211 ft (979 m) and 3493 ft (1065 m).

The lowest part of the Keuper succession is known only from cuttings which indicate that for 50 ft (15 m) or so above the base the beds include red and white sandstones, though the bulk of the cuttings are of brown and green marl. Because of the time lag involved in drilling at this depth, before the cuttings reach the surface, it is impossible to give any accurate estimate of the proportions of these beds which is sandstone.

Cuttings from the succeeding 200 ft (161 m) are of brown and green marls, sometimes with red marl and red siltstone. The marls are sometimes micaceous. Fragments of anhydrite are common as a small proportion of the whole, and in the upper half of this group there is a little grey or greenish grey dolomitic marlstone among the marl chippings.

The cored succession between 3211 ft (979 m) and 3493 ft (1065 m) is notable for the abundance of dolomitic 'marlstone' which occurs throughout the beds in bands, generally thin, and comprises between 15 and 20 per cent of the total thickness. This compares with similar dolomitic material seen in the Keuper Marl at Mire House, County Tyrone (Fowler and Robbie 1961, pp. 100, 201).

The principal rock type is marl, generally red or green in colour but tending to contain a larger proportion of reddish brown and brown beds in the lower part of the core. Characteristic of the beds above 3400 ft (1037 m) are massive and unbedded red marls which, on weathering, break down into cubical fragments. The texture of the marls varies from fine-grained to coarse, earthy, sandy or silty and fine bands of varying textures are common. Some of the coarser bands have abundant mica flakes. In general, the red-coloured material seems coarser than the green and most of the sandy material is red. In addition there are a few beds of micaceous siltstone or mudstone.

The marls are often interbanded with thin beds of dolomite and of dolomitic mudstone, siltstone and fine sandstone a few inches thick, which appear to be an original deposit as the junctions at the base of these bands are often sharp and transcurrent. These marlstones also occur as lenticles and stringers in the marls and are frequently, but not invariably, surrounded by a narrow aureole of green in the red marls. The marlstones are pale greenish grey in colour and are often intersected by thin partings of marl, generally green, on bedding planes. These partings occasionally show intense contortion.

Petrographic examinations of thin sections of the marlstones by J. R. Hawkes show that they range from porcellaneous dolomite with little detrital material to dolomitic siltstones and sandstone. As well as the dolomite, which is always present, all the clastic specimens contain anhydrite.

The specimen of dolomite which came from a thin band at about 3435 ft (1048 m) in the Langford Lodge bore (NI 852) consists mainly of crystalline carbonate with average grain size 0·01 mm and accessory quantities of sub-rounded grains of quartz, flakes of muscovite or chlorite, and granular iron oxide.

Dolomitic silty mudstone (NI 853) from the same depth contains about 10 per cent of quartz grains in a groundmass of granular carbonate. The average grain size (0·02 mm to 0·03 mm) increases to about 0·05 mm in areas where concentrations of quartz grains occur. There are also some flakes of muscovite and biotite and granules of limonite and hematite, while anhydrite occurs in irregular and lath-shaped crystals.

Most of the marlstones are fine-grained dolomitic sandstones, with poorly sorted subangular or sub-rounded quartz grains ranging from 0·01 mm to 1 mm in diameter cemented by granular carbonate and anhydrite (NI 854-8). In addition to the quartz, other clastic material includes fragments of microcline, sodic plagioclase, microperthite, chert, zircon, muscovite and green biotite, with, occasionally, aggregates and bands of clay materials. One specimen (NI 855) from 3471 ft (1059 m) contains, in addition to the usual clastic debris, fragments of porphyritic rhyolite, trachyte, metaquartzite and tourmaline. This rock also contains rounded 'clay-galls' consisting of iron-impregnated dolomite and a little quartz.

Anhydrite occurs as thin irregular veins and small knots in both marls and marlstone but more frequently in the latter where it may be quite plentiful.

Depositional structures clearly indicate shallow-water conditions of deposition. Mud cracks, filled with coarser material, and the occurrence of galls—fragments of marl and occasionally of mudstone—suggest periodically arid conditions. Desiccation breccias with fragments of marls and marlstone in a matrix which is often rather coarser than the normal marl, and is frequently partly impregnated with dolomite giving a mottled appearance, occur at several horizons and are up to 2 ft (0·61 m) in thickness. The marlstone often shows intense small-scale current-bedding;—slumped and contorted bedding, usually showing very intense convolutions, is common in the marls. Polished and slickensided surfaces are common in some beds, generally on bedding planes, but in places across the bedding.

The upper part of the Keuper from 3211 ft (979 m) to the top at 2740 ft (835 m) is known only from cuttings which show that it consists mainly of red and green marls with appreciable quantities of anhydrite though no indication of the latter occurring in beds was found. In the uppermost 350 ft (107 m) or so the red colour is pronounced but below this level the colour becomes browner and less intense. Dolomitic marlstone was seen in cuttings from as far up as 2810 ft (856 m), only 70 ft (21 m) below the top of the marls, but it is very scarce and marlstone is apparently less abundant in the upper half of the Keuper succession. Cuttings of grey shale found throughout the succession are probably Jurassic or Rhaetic material from higher up the hole, but basalt fragments found at 3160 ft (963 m) may indicate an intrusive sill or dyke.

The cause of the variegated colours in the Keuper marls is not clear. The origin of the material is generally assumed to be lateritic dust from surrounding arid land masses and it is reasonable to conclude that the reddish brown colour is original and that the green is secondary, caused by the reduction of the ferric iron to ferrous iron after deposition. The occurrence of green aureoles around some inclusions, of small green spots—called fish eyes elsewhere—around some small foci, probably mineralogical, and the occasional irregular green patches and mottling, particularly in more arenaceous beds, suggest that the alteration took place some time after deposition. On the other hand the interbedding of red and green bands of marl, often thin but with clear junctions along bedding planes, and the occasional occurence of wisps or lenses of green marl as inclusions in red material, indicates that in at least some cases the reduction of the red ferric compounds must have taken place penecontemporaneously or at latest very soon after deposition.

The term 'marl' has been used in descriptive fashion for all fine-grained rocks in the Keuper whether or not they are calcareous. In fact, a small proportion of these rocks is not noticeably calcareous or dolomitic and of the true marls, over half are dolomitic and do not effervesce with dilute acid.   H.E.W.

PALYNOLOGY OF THE TRIAS OF THE LANGFORD LODGE BOREHOLE

Samples were taken at approximately 50-ft intervals throughout the depth range 2695 ft to 3840 ft (821–1170 m) in the Langford Lodge Borehole; twenty-eight samples were processed and examined for palynological residues. The results indicate the presence of the Rhaetic, Keuper, Muschelkalk and Upper Bunter divisions of the Trias (see Fig. 7). The sample material available, with the exception of a core from the Rhaetic at a depth of 2695 ft (821 m) consisted of rock-bit cuttings; contamination by material from higher in the borehole was noted in all the residues obtained from the latter. Because of this, in the study of assemblages from the cutting samples, the highest occurrence of each species was regarded as being of primary significance in the placing of stratigraphical boundaries.

Rhaetic lithofacies (depth 2678 ft to 2735 ft) (816–833 m):

A core sample from a depth of 2695 ft (821 m) yielded an assemblage including the following species which have previously been recorded only from the Rhaetian:

> *Cyathidites sabuli* Reinhardt 1962
> *Semiretisporis gothae* Reinhardt 1962
> *Rhaetipollis germanicus* Schulz 1967.

The above species were accompanied by the following which have less restricted stratigraphic ranges but support the determination of the age of the sample as Rhaetian (the previously documented stratigraphic ranges are given in square brackets).

> *Apiculatisporis parvispinosus* (Leschik 1955) Potonié and Kremp 1956 [M. Keuper— L. Lias]
> *Anapiculatisporites spiniger* (Leschik 1955) Reinhardt 1962 [M. Keuper—L. Lias]
> *Aratrisporites palettae* (Klaus 1960) Schulz 1967 [L. Carnian—Rhaetic]
> *Ricciisporites tuberculatus* Lundblad 1959 [Norian—L. Lias]
> *Classopollis torosus* (Reissinger) Balme 1957 [Rhaetic—L. Cretaceous]

Keuper Marl lithofacies (depth 2735 ft to 3760 ft) (833–1145 m):

The assemblages obtained from this division indicate that the Keuper Marl facies of the Langford Lodge borehole represents the Upper Scythian, Anisian, Ladinian, Carnian and Norian stages of the Trias. Between 2810 ft and 3210 ft (856–978 m) (inclusive) the microfloras indicate the presence of the Ladinian,

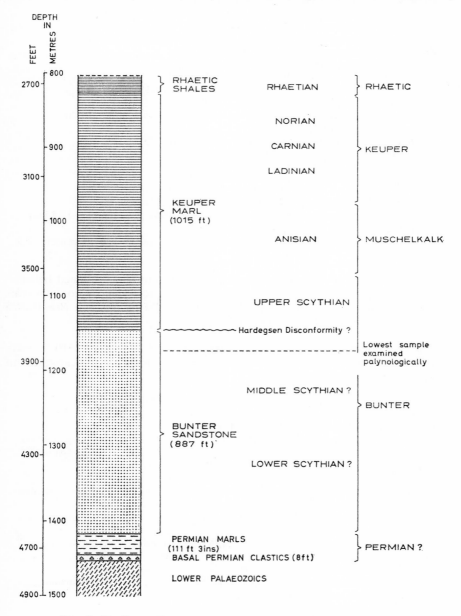

FIG. 7. *The Permo-Triassic sequence in the Langford Lodge Borehole*

Carnian and Norian stages. The assemblages obtained from this depth interval included the following species:

Laevigatisporites globosus Leschik 1955 [Carnian]
Deltoidospora tenuis (Leschik 1955) Mädler 1964 [L. Anisian–Ladinian]
Osmundacidites alpinus Klaus 1960 [Carnian–Rhaetic]
? Camarozonosporites rudis (Leschik 1955) Klaus 1960 [Carnian–Lias]
Aratrisporites palettae
      „   *crassitectatus* Reinhardt 1964 [Ladinian–U. Rhaetic]
Zonalasporites cinctus Leschik 1955 [Carnian]
Succinctisporites grandior Leschik 1955 [Carnian]
Ovalipollis ovalis Krutzsch 1955 [Carnian—L. Lias]
      „   *breviformis* Krutzsch 1955 [Carnian—L. Lias]
Protosacculina macrosacca Schulz 1967 [U. Scythian—Norian]
Vitreisporites subtilis (de Jersey 1959) de Jersey 1962 [Ladinian—Carnian]
      „   *contectus*   „   „   „   „   „   „
? Triadispora crassa Klaus 1964 [U. Scythian—Anisian]
Chordasporites singulichorda Klaus 1960 [U. Scythian—Carnian]
Alisporites parvus de Jersey 1962 [Ladinian—Carnian]
      „   *australis* de Jersey 1962 [Ladinian—Norian]
?   „   *grauvogeli* Klaus 1964 [U. Scythian—? Ladinian]
Microcachryidites sittleri Klaus 1964 [U. Scythian—? Ladinian]
Brachysaccus neomundanus (Leschik 1955) Mädler 1964 [Ladinian—? L. Lias]
Eucommiidites major Schulz 1967 [U. Carnian]
Corollina meyeriana (Klaus 1960) Venkatachala and Góczán 1964 [Carnian—? L. Lias]

The above species were accompanied in all the samples examined between 2810 ft and 3210 ft (856–978 m) by a number of post-Keuper contaminants amongst which *Ricciisporites tuberculatus*, *Classopollis torosus* and *Rhaetipollis germanicus* were particularly common.

The assemblages obtained from the depth interval 3260 ft to 3510 ft (994–1070 m) (inclusive) included the following species which are indicative of an Anisian age:

Punctatisporites triassicus Schulz 1964 [U. Scythian—Anisian]
Cyclogranisporites arenosus Mädler 1964 [Anisian—Ladinian]
Cyclotriletes micrograntifer Mädler 1964 [U. Scythian—Ladinian]
      „   *oligograntifer* Mädler 1964 [„   „   ]
Verrucosisporites jenensis Reinhardt and Schmitz 1965 [U. Scythian—Anisian]
? Spinotriletes echinoides Mädler 1964 [U. Scythian—Anisian]
Apiculatasporites plicatus Visscher 1966 [U. Scythian—Anisian]
Microreticulatisporites opacus (Leschik 1955) Klaus 1960 [U. Scythian—Carnian]
Lundbladispora nejburgii Schulz 1964 [M.—U. Scythian]
Accinctisporites radiatus (Leschik 1955) Schulz 1965 [U. Scythian—Carnian]
Nuskoisporites inopinatus Visscher 1966 [U. Scythian]
Angustisulcites grandis (Freudenthal 1964) Visscher 1966 [U. Scythian—Anisian]
      „   *klausii* Freudenthal 1964 [U. Scythian—Anisian]
      „   *gorpii* Visscher 1966 [U. Scythian—Anisian]
Lunatisporites kraeuseli (Leschik 1955) Mädler 1964 [U. Scythian—Carnian]
Taeniaepollenites jonkeri Visscher 1966 [U. Scythian]
      „   *multiplex* Visscher 1966 [U. Scythian]
Striatoabietites aytugii Visscher 1966 [U. Scythian—Anisian]

Fig. 8. *Distribution of stratigraphically significant spores and pollen in the Trias of the Langford Lodge Borehole*

Depth (in feet)	3840	3810	3760	3710	3660	3610	3560	3510	3460	3410	3360	3310	3260	3210	3160	3110	3060	3010	2960	2910	2860	2810	2695	Species
INST. GEOL. SCIENCES PALYNOLOGY SAMPLE NUMBER (SAL Series)	328	327	326	325	324	323	322	321	320	319	318	317	316	315	314	313	312	311	310	309	308	307		
																							•	*Cyathidites sabuli* Reinhardt 1962
																							•	*Semiretisporis gothae* Reinhardt 1962
																							•	*Classopollis torosus* (Reissinger) Balme 1957
																	•						•	*Aratrisporites palettae* (Klaus 1960) Schulz 1967
																							•	*Rhaetipollis germanicus* Schulz 1967
																						•	•	*Ricciisporites tuberculatus* Lundblad 1959
																•	•	•	•		•	•	•	*Ovalipollis ovalis* Krutzsch 1955
																•	•				•	•	•	*Ovalipollis breviformis* Krutzsch 1955
																	•					•		*Zonalasporites cinctus* Leschik 1955
																		•						*Corollina meyeriana* (Klaus 1960) Ven. and Góczán 1964
																•	•	•	?	•				*Alisporites parvus* de Jersey 1962
		?						?	•							•	•	•					*Chordasporites singulichorda* Klaus 1960
					•	•	•	•	•	•	•	•			•	•							*Microcachryidites sittleri* Klaus 1964
												?	•			•							*Succinctisporites grandior* Leschik 1955
													•	•									*Osmundacidites alpinus* Klaus 1960
														•									*Deltoidospora tenuis* (Leschik 1955) Mädler 1964
														•									*Laevigatisporites globosus* Leschik 1955
	•	•	•	•	•	•	•	•	•	•	•	•	•		?								*Alisporites grauvogeli* Klaus 1964
•	•	•	•	•	•	•	•	•	•	•	•		•		?								*Triadispora crassa* Klaus 1964
														•									*Cyclogranisporites arenosus* Mädler 1964
			•	•			?		•	•	•			•									*Cyclotriletes microgranifer* Mädler 1964
•	•	•	•	•	•	•	•	•	•	•			•										*Alisporites toralis* Clarke 1965
?	•	•	•	•	•	•	•			•	•		?										*Triadispora falcata* Klaus 1964
	•	•	•	•	•	•	•	•		•	•		?										*Triadispora staplini* (Klaus 1964) Visscher 1966
				•	?	•	•	•	•	•	•		?										*Triadispora plicata* Klaus 1964
		•	•	•	•	•	•	•				•	?										*Microcachryidites doubingeri* Klaus 1964
?			•	•	•	•	•	•				•	?										*Angustisulcites grandis* (Freudenthal 1964) Visscher 1966
	?	•					?	•		?	•	?	?										*Tsugaepollenites oriens* Klaus 1964
?	?				•		•		•	•	•												*Cyclotriletes oligogranifer* Mädler 1964
•	•		?	•	•			•	•	•	•												*Apiculatasporites plicatus* Visscher 1966
•	•	•	?	•	•	•		•		?	•												*Voltziaceaesporites heteromorpha* Klaus 1964
•	•	•	•	•	•	•	•	•		?	•												*Colpectopollis ellipsoideus* Visscher 1966
					•			•															*Nuskoisporites inopinatus* Visscher 1966
			?	•	?	•	•	•		•	•												*Lundbladispora nejburgii* Schulz 1964
•				•	•			•				?											*Alisporites microreticulatus* Reinhardt 1964
	•	•			•	•	•	•	•														*Angustisulcites klausii* Freudenthal 1964
		?	?		•	•	•	•	?	•													*Illinites chitonoides* Klaus 1964
				•	•	•	•	•	•														*Illinites kosankei* Klaus 1964
?			•	•	•	•	•	•															*Accinctisporites radiatus* (Leschik 1955) Schulz 1965
				•	•		•	•															*Triadispora epigona* Klaus 1964
			?	•	•		•																*Lunatisporites krauselli* (Leschik 1955) Mädler 1964
•	•	•	•	•	•		•																*Verrucosisporites jenensis* Reinhardt and Schmitz 1965
•							•																*Punctatisporites triassicus* Schulz 1964
			•		•		•																*Sulcatisporites reticulatus* Mädler 1964
•	•	•	•	•	•	?	•	•															*Angustisulcites gorpii* Visscher 1966
•	•					•																	*Taeniaepollenites jonkeri* Visscher 1966
	•					•																	*Taeniaepollenites multiplex* Visscher 1966
?				•	•	•	•																*Brachysaccus ovalis* Mädler 1964
	•			•	•	•	•	•															*Striatoabietites aytugii* Visscher 1966
				•	•																		*Guttatisporites microechinatus* Visscher 1966
		•		•	•	?																	*Lunatisporites puntii* Visscher 1966
	•					•																	*Verrucosisporites pseudomorulae* Visscher 1966
?					•																		*Anaplanisporites protumulosus* (Reinhardt 1964) Schulz 1965
				•																			*Falcisporites snopkovae* Visscher 1966
					•																		*Tubantiapollenites balmei* (Klaus 1964) Visscher 1966
•																							*Falcisporites zapfei* (Potonié and Klaus 1954) Leschik 1956
•																							*Accinctisporites diversus* Leschik 1956
•																							*Klausipollenites schaubergeri* (Pot. and Kl. 1954) Jansonius 1962
•																							*Podocarpeaepollenites thiergartii* Mädler 1964

Depth (in feet)	3840	3810	3760	3710	3660	3610	3560	3510	3460	3410	3360	3310	3260	3210	3160	3110	3060	3010	2960	2910	2860	2810	2695
INST. GEOL. SCIENCES PALYNOLOGY SAMPLE NUMBER (SAL Series)	328	327	326	325	324	323	322	321	320	319	318	317	316	315	314	313	312	311	310	309	308	307	

PALYNOLOGICAL DATING	? MIDDLE SCYTHIAN	SCYTHIAN	UPPER BUNTER (UPPER SCYTHIAN)		MUSCHELKALK (ANISIAN)			KEUPER (LADINIAN, CARNIAN, NORIAN)				RHAETIAN

LITHOFACIES SEQUENCE	BUNTER SANDSTONE		KEUPER MARL									RHAETIC

Triadispora crassa
„ *plicata* Klaus 1964 [U. Scythian—Anisian]
„ *epigona* „ „ [„ „]
„ *falcata* „ „ [„]
„ *staplini* (Klaus 1964) Visscher 1966 [U. Scythian—Anisian]
Illinites chitonoides Klaus 1964 [U. Scythian—Anisian]
„ *kosankei* Klaus 1964 [U. Scythian]
Chordasporites singulichorda
Alisporites grauvogeli
„ *progrediens* Klaus 1964 [Anisian]
„ *toralis* (Leschik 1955) Clarke 1965 [U. Scythian—Anisian]
„ *microreticulatus* Reinhardt 1964 [U. Scythian]
Microcachryidites sittleri
„ *doubingeri* Klaus 1964 [U. Scythian]
* „ *fastidiosus* (Jansonius 1962)Klaus 1964 [U. Scythian—Anisian]
Tsugaepollenites oriens Klaus 1964 [Anisian]
Voltziaceaesporites heteromorpha Klaus 1964 [? Zechstein—Anisian]
Colpectopollis ellipsoideus Visscher 1966 [U. Scythian—Anisian]
Sulcatisporites cf. *reticulatus* Mädler 1964 [U. Scythian—Anisian]
Brachysaccus ovalis Mädler 1964 [U. Scythian]

The above species were accompanied by a number of post-Anisian contaminants amongst which *Ricciisporites tuberculatus, Classopollis torosus, Rhaetipollis
germanicus, Ovalipollis ovalis* and *O. breviformis* were common.

Since a number of species are common in both the Upper Scythian and
Anisian, the base of the latter stage proved difficult to define objectively but is
placed between 3510 ft and 3560 ft (1070–1085 m) where the overall composition
of the assemblage changes to one of Upper Scythian aspect and where species
common in both the Upper Scythian and Anisian but more characteristic of the
former begin to occur in relatively larger numbers than at higher levels in the
borehole. Below 3510 ft (1070 m) the following assemblages, which are indicative
of an Upper Scythian age, occurred:

Toroisporis atavus Reinhardt 1963 [U. Scythian]
Punctatisporites triassicus
Cyclotriletes microgranifer; C. oligogranifer; C. granulatus Mädler 1964 [U. Scythian]
Verrucosisporites pseudomorulae Visscher 1966 [U. Scythian]
„ *reinhardtii* „ „ [„]
„ *applanatus* Mädler 1964 [U. Scythian—Anisian]
„ *thuringiacus* „ „ [? M.—U. Scythian]
„ *jenensis*
Guttatisporites microechinatus Visscher 1966 [U. Scythian]
Spinotriletes senecioides Mädler 1964 [U. Scythian]
Anaplanisporites protumulosus (Reinhardt 1964) Schulz 1965 [U. Scythian—Anisian]
Apiculatasporites plicatus
? *Anguisporites tenuis* Visscher 1966 [U. Scythian]
Lundbladispora nejburgii
Accinctisporites radiatus; A. diversus Leschik 1965 [Zechstein—U. Scythian]
Nuskoisporites inopinatus
Angustisulcites grandis; A. klausii; A. gorpii
Striatoabietites aytugii

**Microcachryidites fastidiosus* (Jansonius 1962) Klaus 1964. Species *M. fastidiosus nom correct.
pro M. fastidioides* (Jansonius 1962) Klaus 1964 *nom imperf.*

Lunatisporites kraeuseli; L. puntii Visscher 1966 [U. Scythian]
Taeniaepollenites jonkeri; T. multiplex
Tubantiapollenites balmei (Klaus 1964) Visscher 1966 [U. Scythian—Anisian]
Triadispora falcata; T. plicata; T. epigona; T. crassa; T, staplini
Illinites chitonoides; I. kosankei
Jugasporites conmilvinus Klaus 1964 [U. Scythian]
Alisporites grauvogeli; A. toralis; A. microreticulatus
Brachysaccus ovalis
Falcisporites zapfei (Potonié and Klaus 1954) Leschik 1956 [Zechstein—U. Scythian]
Falcisporites snopkovae Visscher 1966 [U. Scythian]
Microcachryidites doubingeri; M. fastidiosus; M. sittleri
Sulcatisporites cf. *reticulatus; S. kraeuseli* Mädler 1964 [U. Scythian—Ladinian]
Colpectopollis ellipsoideus
Klausipollenites schaubergeri (Potonié and Klaus 1954) Jansonius 1962 [Zechstein—U. Scythian]
Podocarpeaepollenites thiergartii Mädler 1964 [U. Scythian—Anisian]
Cycadopites trusheimii Visscher 1966 [U. Scythian]
? „ *sufflavus* „ „ [„]

Contamination by younger material was relatively unimportant in the assemblages from below 3510 ft (1070 m) but a few specimens of *Ricciisporites tuberculatus, Classopollis torosus* and *Rhaetipollis germanicus* occurred in most of the preparations.

Bunter Sandstone lithofacies (depth 3760 ft to 4647 ft (1146–1416 m)).

In the lowest samples examined, from 3810 ft and 3840 ft (1161 and 1170 m), the assemblages consisted mainly of species recorded in the lower part of the Keuper Marl facies (Upper Scythian). However, a marked reduction in the abundance and variety of spores and pollen from the Bunter Sandstone horizon suggests that they may consist of material derived by caving from the Keuper Marl facies less than 100 ft (30 m) higher in the borehole. This possibility is supported by the general character of the Bunter Sandstone which is known from numerous cored boreholes in the Coalisland area, Dungannon (Sheet 35). The Bunter Sandstone is lithologically an unlikely source for microfloras since it typically consists of coarse red or reddish brown, sometimes conglomeratic, sandstone with little or no trace of grey or green mudstone intercalations. It is therefore considered unlikely that the assemblages from 3810 ft and 3840 ft (1161 and 1170 m) are indigenous to the Bunter Sandstone.

The lithofacies sequence and palynological subdivision of the Langford Lodge borehole are summarized in Fig. 7. The following is a synopsis of the Triassic sequence in the borehole based on the palynological evidence given above:

Rhaetic . . base at 2735 ft (833 m), 57 ft (17 m) thick
Keuper . . base (of Ladinian stage) between 3210 ft and 3260 ft (978 and 994 m), approximately 500 ft (152 m) thick
Muschelkalk . base (of Anisian stage) between 3510 ft and 3560 ft (1070 and 1085 m), approximately 300 ft (91 m) thick
Upper Bunter . base (of Upper Scythian) possibly between 3760 ft and 3810 ft (1146 and 1161 m). Minimum thickness 200 ft (61 m).

The stage terminology used above (see also Figs. 7 and 8) follows that adopted by Geiger and Hopping (1968) in which the Muschelkalk division of the German

Trias was equated with the Anisian stage of the Alpine sequence. Many European workers, however, adhere to the view that only part of the Muschelkalk is of Anisian age and regard the upper part of the unit (the Trochitenkalk and Tonplatten) as being of lower Ladinian age.

It is possible that the abrupt facies change from sandstone to mudstone at the top of the Bunter Sandstone marks the position of the Hardegsen disconformity (Trusheim 1963) which resulted from a widespread phase of earth movements in northern Europe during the late Middle Scythian. This possibility is supported by the dating of the lower 200 ft (61 m) of the Keuper Marl facies as Upper Scythian. However, until clearly indigenous microfloras have been obtained from the Bunter Sandstone the placing of the Hardegsen disconformity at the top of the Bunter Sandstone remains tentative.

The palynological data from the Langford Lodge borehole demonstrate that the Keuper Marl there is laterally equivalent to the Keuper Sandstone and Keuper Marl facies of the Central Midlands of England (Warrington 1967; in press).

The occurrences of stratigraphically important spore and pollen species recorded during the present study are shown in Fig. 8. With the exception of the

---

EXPLANATION OF PLATE 2

Triassic miospores from the Langford Lodge Borehole

Specimens housed in the Institute of Geological Sciences palynology collection at Leeds; registered numbers are given in square brackets. All photographs at magnification X500.

1. *Toroisporis sp.* [MPK 41]
2. *Verrucosisporites jenensis* Reinhardt & Schmitz [MPK 42]
3. *Guttatisporites microechinatus* Visscher [MPK 43]
4. *Heliosporites altmarkensis* Schulz [MPK 44]
5. *Rhaetipollis germanicus* Schulz [MPK 45]
6. *Classopollis torosus* (Reissinger) Balme [MPK 46]
7. *Triadispora crassa* Klaus [MPK 47]
8. *Rhaetipollis germanicus* Schulz [MPK 48]
9. *Ricciisporites tuberculatus* Lundblad [MPK 49]
10. *Triadispora plicata* Klaus [MPK 50]
11. *Microcachryidites doubingeri* Klaus [MPK 51]
12. *Microcachryidites fastidiosus* (Jansonius) Klaus [MPK 52]
13. *Alisporites grauvogeli* Klaus [MPK 53]
14. *Ovalipollis ovalis* Krutzsch [MPK 54]

EXPLANATION OF PLATE 3

Triassic miospores from the Langford Lodge Borehole

Specimens housed in the Institute of Geological Sciences palynology collection at Leeds; registered numbers are given in square brackets. All photographs at magnification X500.

1. *Nuskoisporites inopinatus* Visscher [MPK 55]
2. *Taeniaepollenites jonkeri* Visscher [MPK 56]
3. *Striatoabietites aytugii* Visscher [MPK 57]
4. *Alisporites microreticulatus* Reinhardt [MPK 58]
5. *Brachysaccus neomundanus* (Leschik) Mädler [MPK 59]
6. *Triadispora falcata* Klaus [MPK 60]
7. *Tsugaepollenites oriens* Klaus [MPK 61]
8. *Colpectopollis ellipsoideus* Visscher [MPK 62]

PLATE 2

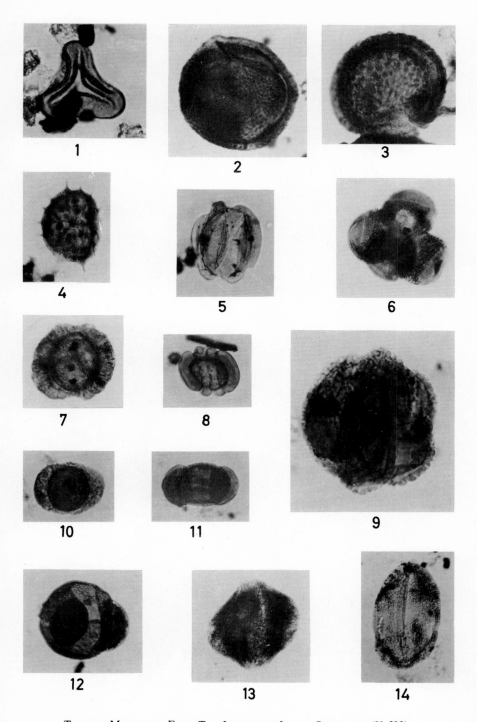

TRIASSIC MIOSPORES FROM THE LANGFORD LODGE BOREHOLE (X 500)

PLATE 3

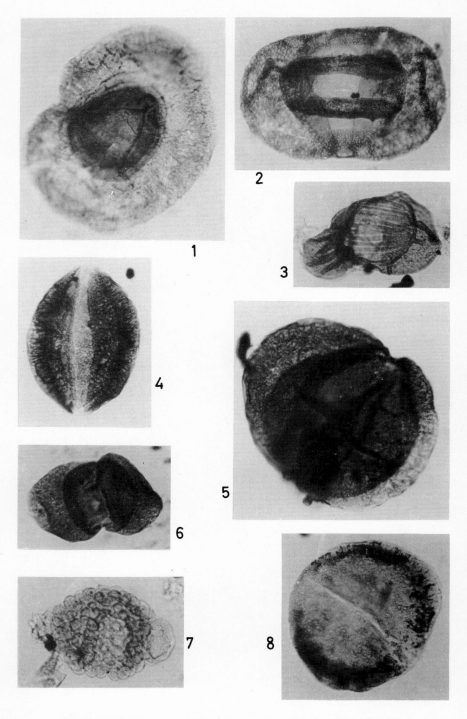

TRIASSIC MIOSPORES FROM THE LANGFORD LODGE BOREHOLE (X 500)

assemblages from 3810 ft and 3840 ft (1161 and 1170 m) only those occurrences which are considered to be indigenous are shown in the table.

The assemblages examined from the Langford Lodge borehole are stored in the Palynology Collection of the Institute of Geological Sciences at Leeds, numbers SAL. 307-328 and SAL. 339. Typical miospores are illustrated in Plates 2 and 3.

G.W.

Information on the ranges of the species quoted is given in the following publications:

BALME, B. E. 1957. Spores and pollen grains from the Mesozoic of Western Australia. *C.S.I.R.O. Phys. Chem. Surv. Nat. Coal. Res., Tech.– Com.* **25.**

CLARKE, R. F. A. 1965. Keuper miospores from Worcestershire, England. *Palaeontology,* **8,** 294–321.

COUPER, R. A. 1958. British Mesozoic microspores and pollen grains. *Palaeontographica,* B. **103,** 75–179.

FREUDENTHAL, T. 1964. Palaeobotany of the Mesophytic, I : Palynology of the Lower Triassic rock salt, Hengelo, the Netherlands. *Acta Bot. Neerl.,* **13,** 209–235.

GEIGER, M. E. and HOPPING, C. A. 1968. Triassic stratigraphy of the southern North Sea Basin. *Phil. Trans. Roy. Soc.,* B. **154,** 1–36.

JANSONIUS, J. 1962. Palynology of the Permian and Triassic sediments, Peace River Area, Western Canada. *Palaeontographica,* B. **110,** 35–98.

JERSEY, N. J. De 1959. Jurassic spores and pollen grains from the Rosewood Coalfield. *Qld. Govt. Min. Journ,.* **60,** 344–366.

———————— 1962. Triassic spores and pollen grains from the Ipswich Coalfield. *Geol. Surv. Qld. Publ.,* **294.**

———————— 1964. Triassic spores and pollen grains from the Bundamba Group. *Geol. Surv. Qld. Publ.,* **321.**

KLAUS, W. 1960. Sporen der Karnischen Stufe der ostalpinen Trias. *Jb. geol. Bundesanst., Wein, Sonderband,* **5,** 107–184.

———————— 1964. Zur sporenstratigraphischen Einstufung von gipsführenden Schichten in Bohrungen. *Erdöl-Zeitschrift,* **80,** 119–132.

KRUTZSCH, W. 1955. Über einige liassische und angiospermide Sporomorphen. *Geologie,* **4,** 65–76.

LESCHIK, G. 1955. Die Keuperflora von Neuewelt bei Basel; II, Die Iso- und Mikrosporen. *Schweiz. paläont. Abh.,* **72,** 1–70.

————————. 1956. Sporen aus dem Salzton des Zechsteins von Neuhof (bei Fulda). *Palaeontographica,* B. **100,** 122–142.

LUNDBLAD, B. 1959. On *Ricciisporites tuberculatus* and its occurrence in certain strata of the "Hollviken II" boring in S. W. Scania. *Grana Palynologica,* **2,** 1–10.

MÄDLER, K. 1964a. Die geologische Verbreitung von Sporen und Pollen in der Deutschen Trias. *Geol. Jb. Beih.,* **65,** 147 p.

————————. 1964b. Bemerkenswerte Sporenformen aus dem Keuper und unteren Lias. *Fortschr. Geol. Rheinld. Westf.,* **12,** 169–200.

POTONIÉ, R. and KLAUS, W. 1954. Einige Sporengattungen des alpinen Salzgebirges. *Geol. Jb.,* **68,** 517–546.

————————. and KREMP, G. 1956. Die Sporae dispersae des Ruhrkarbons, Teil II. *Palaeontographica,* B. **99,** 85–191.

REINHARDT, P. 1962. Sporae dispersae aus dem Rhät Thüringens. *Mber. deutsch. Akad. Wiss., Berlin,* 3, (1961), 11/12, 704–711.

————————. 1964. Über die Sporae dispersae der Thüringer Trias. *Mber. deutsch. Akad. Wiss., Berlin,* **6,** 1, 46–56.

———————— and SCHMITZ, W. 1965. Zur Kenntnis der Sporae dispersae des mitteldeutschen Oberen Buntsandsteins. *Freiberger Forsch.,* C. **182,** *Paläontologie,* 19–28.

SCHULZ, E. 1962. Sporenpaläontologische Untersuchungen zur Rhät–Lias–Grenze in Thüringen und der Altmark. *Geologie*, **11**, 308–319.

————— 1964. Sporen und Pollen aus dem Mittleren Buntsandstein des germanischen Beckens. *Mber. deutsch. Akad. Wiss.*, Berlin, **6**, 597–606.

————— 1965. Sporae dispersae aus der Trias von Thüringen *Mitt. zent. geol. Inst.*, **I**, 257–287.

————— 1967. Sporenpaläontologische Untersuchungen rätoliassischer Schichten im Zentralteil des Germanischen Beckens. *Paläont. Abh.*, **B. II**, 427–633.

VENKATACHALA, B. S. and GÓCZÁN, F. 1964. Spore-pollen flora of the Hungarian "Kossen Facies". *Acta. Geologica*, **8**, 203–228.

VISSCHER, H. 1966. Palaeobotany of the Mesophytic, III. Plant microfossils from the Upper Bunter of Hengelo, the Netherlands. *Acta. Bot. Neerl.*, **15**, 316–375.

————— and COMMISSARIS, A.L.T.M. 1968. Middle Triassic pollen and spores from the Lower Muschelkalk of Winterswijk, (the Netherlands). *Pollen et Spores*, **X**, 161-176.

WARRINGTON, G. 1967. Correlation of the 'Keuper' Series of the Triassic by Miospores. *Nature*, **214**, 1323–1324.

————— Stratigraphy and Palaeontology of the 'Keuper' Series of the Central Midlands, England. *Quart. J. geol. Soc. (In press)*.

RHAETIC

In Northern Ireland remnants of a once-widespread thin and fairly uniform succession of Rhaetic strata owe their preservation to folding and faulting which took place prior to the deposition of the Cretaceous rocks. Within the Belfast (36) Sheet deposits of Rhaetic age occur at outcrop only in the neighbourhood of Collin Glen. Pre-Cretaceous movement along the Collin Glen Fault has cut out the Rhaetic rocks to the west, where Cretaceous strata rest on red Keuper Marl. Farther west, however, deposits of Rhaetic age have been proved in two boreholes, one at Langford Lodge which lies within the Belfast Sheet, and the other at Mire House, to the west of Lough Neagh, in the Dungannon (35) Sheet (Fowler and Robbie 1961, p. 104). To the north of Collin Glen the Rhaetic beds are overstepped by Lias near Hannahstown where Lias rests on red Keuper Marl, which implies the removal of the Tea Green Marls as well as the Rhaetic. Just north of the One-inch Sheet boundary at Crow Glen (Robbie 1964, p. 225) Cretaceous strata rest directly on Keuper Marl. Nevertheless, there are further occurrences of Rhaetic deposits in the Cavehill area (Antrim [28] Sheet) and at intervals along the scarp towards Larne. Their presence or absence is due to structural controls which were active in the post-Lias/pre-Cretaceous interval and which later had a marked effect on Cretaceous sedimentation.

The Tea Green Marls at the top of the Keuper are succeeded by the well-known black *Rhaetavicula contorta* Shales with thin limestones and sandstones. The line of demarcation is sharply defined and where the junction has been seen in the north of Ireland it is erosional though the non-sequence is not of great magnitude. In the lower part of the *contorta* Shales vertebrate remains occur, principally in thin bone beds, while bivalves are more common at higher levels. The succession in these beds is remarkably similar to that of the Westbury Beds of England with which they may be correlated.

Above the dark shales and limestones some grey arenaceous shales are also considered to be of Lower Rhaetic age. The Upper Rhaetic is represented only by a few feet of micaceous shales, poorly exposed.

The Rhaetic has a maximum thickness of about 46 feet (14 m) in Collin Glen whilst 57 feet (17 m) of the shales proved in the Langford Lodge Borehole have been assigned to this formation.

Portlock (1843, pp. 49, 107) was the first to note the presence of transition beds between the Keuper Marl and the Lias in Collin Glen but Tate (1864, pp. 103-11), working on the Liassic strata near Belfast, was the first to recognize that shales containing *Rhaetavicula [Avicula] contorta* were really of Rhaetic age. This paper by Tate is still the only major work on the Rhaetic of Northern Ireland.

<h3 style="text-align:center">DETAILS</h3>

Collin Glen. Sections in the Rhaetic are now confined to a few small exposures in and near Collin Glen. The section recorded in detail by Tate (1864, p. 105) cannot now be identified with any degree of certainty and appears to have been largely obscured by the time of the original survey in 1868. It may have occurred immediately downstream from Glen Bridge [270 719] or, more probably, at the bend in the river about 100 yards (91 m) above the bridge.

At one time a fairly extensive section was visible beside Glen Bridge but this is now almost completely covered by the concrete apron and a retaining wall and only a foot or so of dark shales is exposed, 50 yards (46 m) below the bridge (Fig. 10), dipping north-west at 25°.

About 20 yards (18 m) upstream from Glen Bridge there is a small exposure of vertical shales with thin limestone ribs. Silty micaceous shales dipping south-south-west at 32°, visible in the boulder-strewn bed of the river about 50 yards (46 m) upstream from the bridge, also appear to be *in situ*, and a few yards further upstream, in the east bank, pale greenish grey marl is seen.

The best exposure of the Lower Rhaetic in Collin Glen, presumably the one recorded by Tate (1864, p. 105), occurs at the bend in the river 100 yards (91 m) above Glen Bridge. About 6 yards (5·5 m) north-west of the low cliff of Tea Green Marls there is a discontinuous section of about 22 feet (6·7 m) of strata. Recently (1964) this section [270 720] was cleared of debris and measured in detail by T. P. Fletcher who found numerous fossils including two thin bone beds.

<h3 style="text-align:center">Collin Glen, 100 yd N.W. of Glen Bridge</h3>

	ft	in	m
Marl, light buff/grey with thin darker grey laminae near base: *Rhaetavicula contorta* (Portlock) and fish fragments; seen	4	0	(1·02)
Mudstone, black, laminated, very fossiliferous: *Eotrapezium ewaldi* (Bornemann), *Protocardia rhaetica* (Merian) and *Rhaetavicula contorta*	0	7	(0·17)
Mudstone, black, laminated, with fine-grained, bluish grey limestone lenses up to 3 in thick and some thin paler silty laminae: *Chlamys valoniensis* (Defrance), *Eotrapezium sp.* and *Rhaetavicula contorta*	0	6	(0·15)
Mudstone, black, slightly micaceous, laminated with very thin silty bands and occasional current-bedded limestone lenses. Pyrite present as fossil coatings. *Eotrapezium ewaldi, E. sp., Protocardia rhaetica, Rhaetavicula contorta* high in the bed and *Cercomya sp., Eotrapezium concentricum* ? (Moore), *E. ewaldi* and *Protocardia rhaetica* in the lower part	1	6	(0·45)
Marl, dark grey, extremely silty, slightly micaceous with much granular calcite giving a pronounced gritty texture, some fish fragments, slickensided throughout	1	0	(0·30)

Fault, throw small   *ft   in   m*

	ft	in	m
Marl, dark grey as above with nodular calcite more prominent: *Gyrolepis alberti* J. L. R. Agassiz	1	6	(0·45)
Limestone, grey, fine grained, interbedded with thin wavy lens of 'gritty' marl	0	1¾	(0·04)
Limestone, bluish grey, fine grained with thin wavy lens of 'gritty' marl in middle: *Chlamys valoniensis, Eotrapezium ?* and *Modiolus sp.*	0	3½	(0·09)
Limestone, grey, fine grained, interbedded with thin wavy lens of 'gritty' marl	0	3	(0·08)
Marl, dark grey as above	0	4½	(0·11)
Limestone, grey to buff, medium grained, lenticular, with fibrous calcite and thin wavy dark grey marl partings	0	6½	(0·16)
Mudstone, dark bluish grey: *Eotrapezium ewaldi* and *Rhaetavicula contorta*	0	3	(0·08)
Fibrous calcite vein	0	0¾	(0·02)
Mudstone, shaly, bluish grey, micaceous with very thin silty laminae: *Eotrapezium ewaldi, E. sp.* and *Rhaetavicula contorta*	0	2½	(0·06)
Marl, greyish black, very silty, often with 'gritty' texture, many thin calcite stringers throughout, highly slickensided	1	5	(0·43)
Sandstone, pale grey, fine grained, essentially angular quartz with some calcite and biotite. Pyrite disseminated throughout, often in massive irregular patches. Bedding well developed and lenticular	0	2	(0·05)
Mudstone, grey to black, very silty at top, laminated, with a fish bed in the top part containing *Birgeria acuminata* (J.L.R. Agassiz), *Gyrolepis alberti, Hybodus ?* and fish fragments. In the lower part the following fossils were found: *Eotrapezium ewaldi, E. sp., Lyriomyophoria postera* (Quenstedt), *L. sp.* and *Rhaetavicula contorta*	0	6	(0·15)
Mudstone, shaly, greyish black, with a prominent 4-in very fine grey limestone (flat and nodular, with calcite-lined cavities along well developed rectangular joints) at top. '*Cylindrites*' ?, '*Natica*' *oppelii* Moore, *Eotrapezium ewaldi, Protocardia ?, Rhaetavicula contorta* ,and *Gyrolepis alberti* in the top part, *E. ewaldi, E. sp., Lyriomyophoria postera* and *Rhaetavicula contorta* in the lower part	1	4	(0·41)

Fault, throw small

	ft	in	m
Mudstone, shaly, slightly silty, grey, micaceous, iron stained. In the top part *Eotrapezium concentricum, E. ewaldi, E. sp., Lyriomyophoria sp., Pteromya ?* and *Rhaetavicula contorta;* in the middle part *E.concentricum, E. ewaldi, Lyriomyophoria postera, Pteromya?, Rhaetavicula contorta, Acrodus minimus* J. L. R. Agassiz, *Gyrolepis alberti* and *Hybodus sp.;* in the lower part *Eotrapezium ewaldi, E. sp., Protocardia rhaetica, Rhaetavicula contorta, Dapedium ?* and *Gyrolepis alberti*	1	8	(0·51)
Mudstone, interbedded with paler grey calcareous silty material, shaly in places, micaceous: *Eotrapezium concentricum ?, E. ewaldi, E. sp., Modiolus* cf. *minimus* J. Sowerby, *Protocardia rhaetica, Rhaetavicula contorta, Gyrolepis alberti* and *Hybodus sp.*	0	6	(0·15)

	ft	in	m
Sandstone, very fine grained, silty, pale grey, slightly micaceous, loose, irregularly bedded, iron stained with some pyrite: '*Gervillia*' *praecursor* Quenstedt, *Rhaetavicula contorta*, *Gyrolepis alberti* and fish fragments	0	8	(0·2)
Sandstone as above becoming more compact and platy: *Eotrapezium ewaldi*, *E. sp.*, *Gyrolepis alberti* and fish fragments	0	2	(0·05)
Mudstone, shaly, very silty, dark bluish grey, slightly micaceous	0	4	(0·1)
Sandstone, very fine grained, silty, pale grey, slightly micaceous, iron stained, loose: *Protocardia rhaetica*, *Rhaetavicula contorta*, *Gyrolepis alberti* and fish fragments	0	4	(0·1)
Limestone, bluish grey, fine grained, lenticular with indeterminate bivalves and fish fragments	0	2	(0·05)
Shale, very silty, grey, micaceous	0	3	(0·08)
Mudstone, greyish black, *Rhaetavicula contorta* common	0	0¼	(0·01)
Sandstone, very silty, grey, compact, with disseminated pyrite and prominent fish bed at base: *Eotrapezium sp.*, *Rhaetavicula contorta*, *Acrodus minimus*, *Gyrolepis alberti*, *Hybodus sp.*, and *Tetragonolepis ?*	0	2	(0·05)
Mudstone, shaly, dark grey, slightly micaceous, very fossiliferous: *Eotrapezium ewaldi*, *E. sp.*, *Rhaetavicula contorta* and fish scales and fragments	0	5¼	(0·13)
Fibrous calcite in two prominent thin bands	0	1½	(0·04)
Mudstone, shaly, dark grey, often black, alternating 1 in to 2 in layers of firm and soft, very fossiliferous: '*Natica*' *oppelii*, *Eotrapezium concentricum*, *E. ewaldi*, *E. sp.*, *Protocardia rhaetica*, *Rhaetavicula contorta*, fish fragments and coprolites	2	0	(0·61)
Clay, black, soft (9 in seen) estimated	1	0	(0·3)
	22	5½	(6·8)

The uppermost bed seen in this section, pale marl with *R. contorta*, is probably Tate's Bed 9 and is also seen downstream. Though included in the 'White Lias' by Tate this bed is of Lower Rhaetic age.

The Upper Rhaetic, as now seen, consists of micaceous marls, with a prominent 5-inch (0·13 m) band of compact marl, which appear to represent Tate's Beds 5–7 and are seen 30 yd (27 m) upstream from the top of the measured section. The lower part of this group contains *Eotrapezium ewaldi* and *Modiolus ?*—and the upper part *Protocardia rhaetica* (Merian).

It is unlikely that a complete section of the Rhaetic will be visible in future, and for purposes of comparison the section as recorded by Tate (1864, pp. 105, 108) is also given here (see also Fig. 9).

Section in Collin Glen recorded by R. Tate (1864)

	ft	in
Lower Lias resting on White Lias		
5. Arenaceous Marls with *Cardium Rhaeticum*	1	0
6. White Limestone	0	4
7. Grey Marls	6	0
8. Red Marls	9	0
9. Grey Arenaceous Shales, passing down into Bed 10, with *Axinus cloacinus*, *Cardium Rhaeticum*, etc.	10	0

Avicula-contorta Zone	*ft*	*in*
10. Black Shales with *Axinus cloacinus, Avicula, Cardium Rhaeticum, Placunopsis*	0	11
11. Argillaceous Limestone	0	5
12. Black Shales with *Pecten Valoniensis, Modiola, Avicula, Placunopsis alpina*	1	7
13. Marly Shales with *Axinus cloacinus, Avicula contorta* . . .	1	3
14. Blue Argillaceous Limestone	0	7
15. Marly Shales	0	5
16. Brown Argillaceous Limestone	0	6
17. Black Shales	1	2½
18. Micaceous Sandstone	0	1
19. Soft Shales	0	9
20. Argillaceous Limestone	0	0½
21. Stiff Shales with *Axinus cloacinus, Cardium Rhaeticum, Avicula contorta*	1	9
22. Argillaceous Limestone	0	2
23. Shales	0	6
24. Soft Micaceous Sandstone	0	1
25. Stiff Shales with *Axinus cloacinus*	0	8
26. Compact Calcareous Sandstone	0	1
27. Soft Shales	0	6
28. Stiff Shales with *Pecten Valoniensis, Cardium Rhaeticum, Axinus cloacinus, Avicula contorta*	0	5
29. Compact Sandstone	0	2
30. Black Shales	0	7
31–4. Shales and Micaceous Sandstones	0	4
35. Arenaceous Shales: Fish-bed	0	2
36. Soft Shales	0	5
37. Thinly laminated stiff Shales with scattered fish remains, *Natica oppelii, Trochus Waltoni, Avicula contorta*	0	5
38. Arenaceous Shales	4	6
39. Stiff Black Clay	1	3
Blue Marls of the Keuper below [Tea Green Marl ?]	—	
	46	1

Tate remarked (p. 107) that the red and grey marls forming much of his White Lias "have a close resemblance to the variegated marls of the Keuper; and, but for the well-marked succession and the presence of *Cardium Rhaeticum,* one would be very much disposed to refer them to that period". In a boring made at Larne in 1962, 14 ft 3 in (4·3 m) of drab brown mudstone of typical Keuper Marl lithology were found in shales considered to be of Rhaetic age.

The section measured by Mr Fletcher recognizes two faults not previously recorded but they would appear to be of a minor nature. The thickness of the arenaceous beds near the base of the Rhaetic is greater than that recorded by Tate but the amount of shale below them is less, and there are some other minor differences. The main diagnostic feature of this section, however, is the limestone forming a reef partly across the river.   J.A.R.

H. C. Ivimey-Cook reports that the fauna, containing *Rhaetavicula contorta, Eotrapezium* [*Axinus*] *concentricum, E. ewaldi* and *Protocardia rhaetica,* is typical of the Westbury Beds of the Lower Rhaetic and of the "Zone" of *Rhaetavicula contorta.* Fish remains occur throughout the shales, mudstones and argillaceous mudstones. *Chlamys valoniensis* occurs especially in the more calcareous shales and the limestones.

The Upper Rhaetic of Collin Glen was identified as 'White Lias' by Moore and accepted as such by Tate (1864, p. 108) for his Beds 5–9. As the lowest bed (9) is now found to contain *Rhaetavicula contorta* this part of the identification cannot be upheld as this fossil is characteristic of the Lower Rhaetic.

East of Collin Glen. The only other exposure of Rhaetic within the Belfast Sheet is seen on the east bank of a small stream [273 720], 430 yd (390 m) east-north-east of Glen Bridge. About 6 ft (1·8 m) of black, brown, and pale grey shales are cut by a decomposed basalt dyke at least 2 ft (0·6 m) wide. The shales dip west-north-west at 15° and are invaded by decomposed, finely amygdaloidal basalt in several irregular detached sill-like masses lying approximately parallel to the bedding. These masses are usually about 9 in (0·23 m) thick but occur up to 2 ft (0·6 m) thick.

Though not seen at surface, 'coaly shale' said to have been got in the bottom of an 18-ft (5·5 m) pit, 350 yd (320 m) south-south-east of Hannahstown School, may belong to the Rhaetic, but the outcrop was continued on the One-inch Map only a short distance eastwards beyond the site of this pit, as a well [285 723], 1025 yd (937 m) east, 2° north of Hannahstown School appears, from the debris on the dump, to have passed from Lias shales into red Keuper Marl.

Langford Lodge. The Langford Lodge Borehole was drilled with a rock-bit through most of the Mesozoic rocks and only cuttings were obtained, but a core taken from 2694 ft to 2701 ft (821–823 m) with a recovery of only 3 ft 4 in (1 m) was of dark grey shale of Lower Rhaetic age. The positions of the top and base of the Rhaetic are suggested by the electrical logs of the borehole, but these figures are by no means precise. It is probable that the base of the Lias is at about 2678 ft (816 m), though the gamma-ray log would suggest a point about 15 ft (4·5 m) lower. The base of the Rhaetic is indicated at about 2745 ft (836 m) from the cutting sample, or 2735 ft (833 m) from the gamma-ray log which shows a sharp decrease in radioactivity at that level: the latter figure has been accepted as correct. The total thickness is probably about 57 ft (17 m).

The 3 ft 4-in core recovered was as follows:

Langford Lodge Borehole; core from 2694 ft to 2701 ft

	ft	in	m
Shale, greyish brown, finely laminated with little mica			
Shale or shaly mudstone, grey, silty, very micaceous with *Eotrapezium ewaldi* and *Modiolus langportensis* (Richardson & Tutcher)	0	6	(0·15)
Shale, black with *Eotrapezium concentricum ?, E. sp., Protocardia rhaetica* and *Rhaetavicula contorta*			
Limestone, grey, shaly	0	3	(0·08)
Shales, black, interbanded, fine grained and coarsely micaceous with some calcareous threads. *Chlamys ?, Eotrapezium concentricum, E. ewaldi, E. sp., Protocardia rhaetica, Rhaetavicula contorta* and an indeterminate fish scale	2	7	(0·78)
	3	4	(1·01)

The fossils were examined by Dr. Ivimey-Cook and the lithology of the grey and greyish brown shale suggests a position very high in the Lower Rhaetic or the base of the Upper Rhaetic, the black shales being Lower Rhaetic.   J.A.R., H.E.W.

Chapter 5

JURASSIC: LOWER LIAS

WITHIN THE BELFAST (36) Sheet Liassic rocks crop out only in the neighbourhood of Collin Glen. On the One-inch Map published in 1870 an outcrop of combined Rhaetic and Lower Lias was shown as being more or less confined to Collin Glen and truncated to the south-west and north-east by faults. On the original field maps the outcrop was continued from Collin Glen to beyond the northern boundary of the Sheet although there were at that time no exposures of Lias between Collin Glen and the Sheet boundary. When the One-inch Map of the Belfast District (Drift Series) was published in 1904, an outcrop of combined Rhaetic and Lower Lias was inserted, extending from Collin Glen to the northern limit of the Sheet.

During the resurvey it was confirmed that the Lias does not outcrop west of Collin Glen. The outcrop is bounded by the Collin Glen Fault, which, though it affects the Tertiary lavas, was also active in pre-Tertiary times. The outcrop was traced from Collin Glen eastwards to the flanks of Black Hill where the Lias appears to overlap the Rhaetic on to red Keuper Marl, but beyond that, to the northern limit of the Sheet, there was no evidence of the presence of any Jurassic rocks. In fact, in Crow Glen, about 110 yd (100 m) N. of the Sheet boundary, Triassic and Cretaceous rocks are exposed in such close proximity as to preclude the presence of any Liassic strata (Robbie 1964, pp. 224–9). Far to the west the formation has been proved at depth in the Langford Lodge Borehole [090 748] on the eastern shore of Lough Neagh.

During the resurvey it was not possible to estimate the total thickness of the Lower Lias in Collin Glen due to the lack of continuous exposures. Tate (1864, p. 109) apparently saw a complete section which he recorded as being 38 ft 3 in (12 m) thick. In Langford Lodge Borehole 30 ft (9 m) of grey shales [2648 ft to 2678 ft (807–816 m) depth] have been assigned to the Lower Lias.

Although Liassic strata were recorded from north-east Ireland by Sampson as long ago as 1802 it was not until 1816 that the rocks were described and the extent of the outcrop indicated by Berger and Conybeare (1816, pp. 131, 164–6). The sections in the Belfast region were described in detail by Tate (1864, pp. 103–11).

Recent collections have been examined by H. C. Ivimey-Cook and Professor D. T. Donovan. The zonal succession now proved or, in the case of the lowest zone, inferred, may be summarized as follows (see also Fig. 9):

Stage	Zone	Subzone
Hettangian	*Schlotheima angulata*	
	Alsatites liasicus	
		P. (Caloceras) johnstoni
	[*Psiloceras planorbis* inferred]	*P. planorbis,* including pre-*planorbis* Beds

58

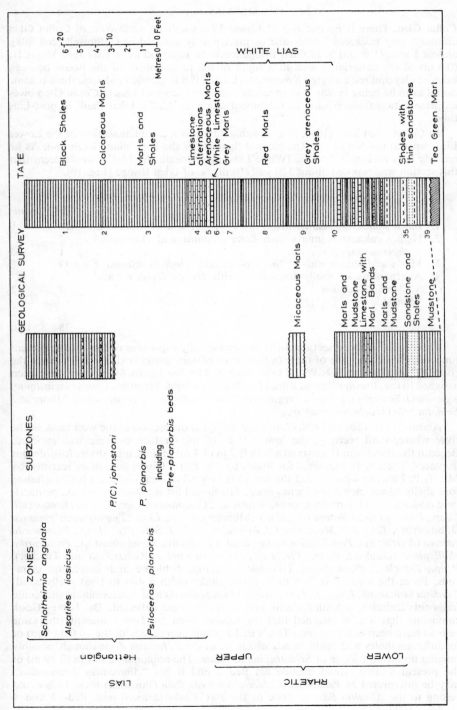

Fig. 9. Zonal sequence of the Rhaetic and Liassic rocks of the Belfast area

DETAILS

Collin Glen. There is no outcrop of Lower Lias south or south-west of Collin Glen although grey micaceous mudstone or marl poorly exposed at Collinwell [268 708], almost 1 mile (1·6 km) S. of Glen Bridge might be regarded as of Lias age. About 1ft (0·3 m) of the material is seen in a small ravine to the north of the house between boulder clay and red and grey Keuper Marl. Even if it is Liassic mudstone there is some doubt as to its being *in situ*. It is apparent that the outcrop of Lias in Collin Glen owes its preservation from erosion to movement along the Collin Glen Fault in post-Lias times.

In Collin Glen Tate (1864, p. 109) apparently saw a complete section of the Lower Lias between the top of the Rhaetic and the base of the Hibernian Greensand. As he records beds of Upper Rhaetic [White Lias] age beneath the Lias it would seem that this section was exposed about 120 yd (110 m) N. of Glen Bridge (Fig. 10).

Section in Collin Glen recorded by Tate (abbreviated)

	ft	*m*
1. Black shales, unfossiliferous	16	4·9
2. Highly calcareous marls with *Lima gigantea* and *Ammonites intermedius*	1	0·3
3. Marls and shales with *Echinus*-spines and *Modiola minima* ?	17	5·2
4. Alternations of shelly limestones with *Ostrea liassica* ? and *Modiola minima*	4	1·2
5. White Lias	—	—
	38	11·6

Three small isolated sections of shale are the only exposures of Lias at the present time in Collin Glen. One of these (a foot or so of dark shale) is visible just west of the river 200 yd (182 m) N.N.W. of Glen Bridge. The few fossils found here have been assigned to the *Alsatites liasicus* Zone: *Cardinia hennocquii* Terquem, *Lima* (*Plagiostoma*) *gigantea* (J. Sowerby), *Liostrea irregularis* (Münster), *Pteromya crowcombeia* Moore and *Schlotheimia* (*Waehneroceras*) *sp.*

About 240 yd (220 m) N.N.W. of Glen Bridge is the section in the west bank of the river where, until recently, the lowest beds of the Cretaceous were well exposed. Beneath the Hibernian Greensand are 18 ft 7 in (5·7 m) of dark grey shales fossiliferous in places. These were examined for fossils by Mr. F. G. Dimes and, more recently, by Mr. T. P. Fletcher who divided the top 16 ft (4·9 m) of strata into ten beds including four shelly bands each a few inches thick. The faunal list is: *Diademopsis sp.*, pentacrinoid ossicles, echinoderm fragments, *Astarte sp.*, *Camptonectes sp.*, *Cardinia hennocquii*, *C. sp.*, *Chlamys sp.*, *Liostrea irregularis* (Münster), *L. sp.*, *Lima* (*Plagiostoma*) *gigantea* (J. Sowerby), *Lucina* ?, *Mactromya* ?, *Modiolus laevis* J. Sowerby, *M. sp.*, *Palaeoneilo galatea* (d'Orbigny), *Parallelodon hettangiensis* (Terquem), *Pseudolimea sp.*, *Protocardia phillippiana* (Dunker), *P. sp.*, *Pteria sp.*, *Pteromya tatei* ? (Richardson and Tutcher), *P. sp.*, *Thracia* ?, ?*Schlotheimia* (*Waehneroceras*) *sp.*, Schlotheimiids indet. and ostracods. From the lower 2 ft 7 in (0·79 m) of shales (with a 4-in (0·1 m) shelly band) *Cardinia hennocquii*, *Lima sp.*, *Liostrea sp.*, *Modiolus laevis* and indeterminate ammonite fragments including a Schlotheimiid and ostracods were collected. Dr. Ivimey-Cook comments that if it is assumed that the section given by Tate represents the same beds as have been collected then Tate's Bed 1 would correlate with the 16 ft (4·9 m) of fossiliferous shales with shelly bands which are of the *A. liasicus* Zone though possibly ranging up into the Zone of *Schlotheimia angulata*. The remaining 2 ft 7 in (0·79 m) of the present section would include his Bed 2 and if his "*Ammonites intermedius*" may be interpreted as *Psiloceras* (*Caloceras*) *intermedium* (Portlock) these beds could belong to the *Alsatites liasicus* Zone or the *P.*(*C.*) *johnstoni* subzone. Beds 3 and 4 of Tate might then be included in the *Psiloceras planorbis* Zone. This correlation can only be very tentative as the horizon of the Cretaceous unconformity may differ between the various exposures and a greater lithological similarity between the sections would be expected. This difficulty of reconciling the sequence as seen today with that recorded by Tate is a very real one and was recognized when the Rhaetic was being examined (pp. 55–57).

Recent collecting has only indicated beds of the Hettangian Stage. The most definite zonal evidence is for the *Alsatites liasicus* Zone. There is no positive evidence for the *Schlotheimia angulata* Zone but the fragmentary ammonites recovered from these beds are such that Professor Donovan would not exclude the possibility that the higher zone may also be present.

The Upper Rhaetic/Lower Lias boundary has not been rediscovered and the recent collections appear to be from the higher parts of this Lias sequence; thus no new evidence has been found of the presence of *Psiloceras planorbis* Zone in Collin Glen.

Tate classified his Beds 3 and 4 as *P. planorbis* Zone in 1867 (he considered *Ammonites johnstoni* and *A. intermedius* were in the *A. angulata* Zone though we would now include them in the *Psiloceras (Caloceras) johnstoni* Subzone of the *P. planorbis* Zone) but no ammonite evidence was given to justify this so that the existence of the *P. planorbis* Subzone rests on the record (Charlesworth and Preston 1960) that *P. planorbis* occurs in the Lias underneath the concrete apron of Glen Bridge. Charlesworth (1960, pp. 438–9) also identifies the pre-*planorbis* beds in Collin Glen but this appears to be inferred from Tate's Beds 3 and 4 as he cites *O. liassica* [=*Liostrea hisingeri* (Nilsson)] and *Modiolus minimus*. Both these fossils also occur in higher and lower beds.

No evidence has been found for the report (Charlesworth and others 1960, p. 438) of the presence of the *bucklandi* to *obtusum* zones in Collin Glen.

On the east bank of the river, 345 yd (315 m) N.N.W. of Glen Bridge, a few feet of shales and shaly mudstones are seen to be cut by a 5-in (0·13 m) basalt dyke.

East of Collin Glen. Dark shale, believed to be of Liassic age, was dug from a pit [277 719], 350 yd (320 m) S.E. of Hannahstown School and Lias shale was recognized in the material dug from an 85-ft (26-m) well [286 723] 900 yd (823 m) W.N.W. of Fairview. As none of the shale seen on this dump appeared to belong to the Rhaetic it is believed that this well passed directly from the Lias into red Keuper Marl.   J.A.R.

Langford Lodge. In the Langford Lodge Borehole, grey shale was detected in the cuttings when at a depth of 2660 ft (811 m) but in view of the results from electrical logging of the borehole it is now presumed that the top of the Lias is at 2648 ft (807 m).

The electrical logs, particularly the section gauge which indicates the extent of caving in the hole, suggest that the base of the Lias is at about 2678 ft (816 m) and the thickness of the formation here is about 30 ft (9 m). Only cuttings were recovered from the bore at this level and no macrofossil evidence was obtained.   H.E.W.

Chapter 6

UPPER CRETACEOUS

INTRODUCTION

THE CRETACEOUS ROCKS of Northern Ireland are all of Upper Cretaceous age, ranging from Cenomanian to Maestrichtian, but the succession is not complete and there are non-sequences and unconformities at several levels. The Cenomanian, Turonian (?), Coniacian and lowest Santonian beds where present are mainly arenaceous, notably rich in glauconite, and constitute the Hibernian Greensand which is up to 73 ft (22 m) thick within the Belfast (36) Sheet. In this area the Turonian has not been certainly proved. They are overlain by the Chalk or White Limestone, a hard, compact, fine-grained white limestone of Santonian to Maestrichtian age. Apart from regular bands of flint nodules it is remarkably pure, and only in the lowest beds and at various non-sequences are there appreciable quantities of quartz and glauconite grains and clay minerals. In Sheet 36 the White Limestone is from 15 ft (4·6 m) to 100 ft (30 m) thick and the Maestrichtian is absent. West of the Groganstown fault the Hibernian Greensand, like the underlying Lias and Rhaetic, is absent and the White Limestone rests directly on the Keuper Marl. The lowest bed of the limestone here is sandy, with glauconite and quartz grains, pebbles of quartz and lenses of Keuper Marl. Within the limits of the Sheet the base of the limestone appears to be at a higher horizon in the west than in the east, although the precise details remain to be worked out.

Upper Cretaceous rocks underlie the north-western half of the area and are overlain by Tertiary formations. The outcrop, extending from near Magheralin to the northern margin of the Sheet north-west of Belfast, is narrow for most of its length and only in the Moira district is an extensive area of Cretaceous rock exposed.

The earliest references to Cretaceous rocks in Northern Ireland are the brief comments of Whitehurst and Hamilton (1786), descriptions of a few sections by Berger (1816) and observations by Conybeare and Buckland (1816) on some coastal exposures. In the early 1850's McAdam (Tate 1865, p. 15) made extensive collections from the Cretaceous rocks in Northern Ireland and Tate (1865), who continued his work, reviewed the previous literature and proposed the first lithological subdivisions, based on a previous attempt by Bryce (1853). In the Geological Survey of Ireland description of the Cretaceous Series in the Sheet 36 memoir (1871) most of the information related to the Chalk. Further progress in the subdivision of the Cretaceous rocks was made by Barrois (1876) and Gault (1877).

A comprehensive study of the Cretaceous of County Antrim by Hume (1897) was mainly concerned with the lower glauconitic sequence. Hume's descriptions were followed in the memoir on the Belfast District (1904).

After the last war attention was again turned to the Cretaceous rocks in Northern Ireland and McGugan (1957, 1959, 1964) and Reid (1958) described respectively the use of foraminifera and sponge assemblages as potential strati-graphical indices. Hancock (1961) reviewed the sequence in the light of present-day knowledge and although the glauconitic sediments received most attention he recognized many of the zones in the White Limestone. Further observations have been made by Reid (1959, 1962, 1963, 1964) and by T. P. Fletcher, who is engaged in a regional study of the Cretaceous rocks of East Antrim.

One result of all the research work has been the bewildering assortment of names for the stratigraphical units and the marked differences in the zonal classification.

P.I.M., J.A.R.

HIBERNIAN GREENSAND

When the Cretaceous sea invaded Northern Ireland in Cenomanian times conditions of sedimentation were closely similar to those which had marked the close of the Lower Cretaceous in the south of England. Thus the arenaceous and glauconitic deposits of the Belfast area are similar lithologically to the Upper Greensand (Albian) of the south-west of England and Berger (1816, p. 169) did in fact correlate the two groups. Tate (1865, p. 20) realized, however, that the arenaceous beds should be equated with the 'Etage Cénomanien' of d'Orbigny. He therefore proposed the name Hibernian Greensand for this group of strata. The Hibernian Greensand as now defined includes all the Upper Cretaceous rocks from the base of the Cenomanian to the top of the Zone of *Micraster coranguinum* (Plate 4). It thus includes all the arenaceous beds which occur in Island Magee but not the highest beds below the Chalk in the Belfast area. The latter comprise the upper part of Hancock's Upper Glauconitic Beds (1961). The major palaeontological gap in the succession, which was recognized by Tate (1865, p. 35), has its maximum development in the Belfast area, where it involves most, if not all of the Turonian, and all the Coniacian and Santonian stages. Hancock (1961, pp. 12, 17) notes that there is seldom a marked lithological break here.

The outcrop of Hibernian Greensand shown on the One-inch Map south-west of Groganstown is now thought to be terminated by the Groganstown Fault. Hancock (1961, pp. 30-1) considered the absence as due to erosion rather than to non-deposition and as supporting evidence quoted the Glauconitic Sands and the Yellow Sandstones and Grey Marls as having been deposited some distance from the shore. The Groganstown Fault may, however, have been active at several periods and could have formed a barrier, so the attenuation of the White Limestone in the neighbourhood of this fault may not be mere coincidence.

Tate (1865, pp. 20-4) proposed names for the three subdivisions of the Hibernian Greensand which had been recognized by Bryce (1853). In ascending order these were 'Glauconitic Sands' ,'Grey Marls and Yellow Sandstones with chert' and 'Chloritic Sands and Sandstones'. Despite modifications suggested by recent authors (McGugan 1957, pp. 30-1; Reid 1958, p. 237; Hancock 1961, pp. 11-18), Tate's nomenclature has been found to be aptly and comprehensively descriptive and provides the most workable grouping for mapping these rocks within the Belfast area. In adopting his terms the word 'glauconitic' has been substituted for 'chloritic' and, as yellow sandstones predominate over grey marls, the term for the middle group has been modified to Yellow Sandstones and Grey Marls.

To the three groups just mentioned must be added a fourth, namely Quartzose Sand, to cater for the pure quartz sand which is exposed on the west flank of Collin Glen. The lithological groups, with maximum thicknesses in the Belfast area, are:

	ft	m
Quartzose Sand	17½	(5·3)
Glauconitic Sands and Sandstones	12	(3·7)
Yellow Sandstones and Grey Marls	35	(10·7)
Glauconitic Sands	8½	(2·6)

Glauconitic Sands. These form a basal group of the Hibernian Greensand, and consist of partially indurated dark green, almost black, medium-to coarse-grained sand. The grains are almost entirely of glauconite with subordinate quartz (see analyses in Hume 1897) set, in places, in a grey clayey matrix. Small pebbles occur up to 12 in (0·3 m) from the base but are mainly concentrated in the lowest 2 in to 3 in (0·07 m) where there is a considerable amount of clay matrix. These pebbles are mostly of quartz, but many are dark brown or black and highly polished and closely comparable with the lydites found at the base of the Gault in Dorset. Small patches of grey clay, free of glauconite grains, occur, these being interpreted by Hancock (1961, p. 12) as probable burrow-fills. Throughout the deposit bedding is apparent only in the orientation and layering of the pebbles and shell bands. In the top 9 in (0·2 m) there is less glauconite and there is an upward transition to a darkish grey and grey fine-grained sandstone.

The Glauconitic Sands contain a rich fauna of bivalves, commonly concentrated in discrete shell-bands; a full list for this and the succeeding subdivisions is given in Hume (1897). The commonest fossil is *Exogyra obliquata* (Pulteney) [=*E. conica* J. Sowerby sp.]. The fauna also includes serpulids, the pectinids *Chlamys aspera* (Lamarck), *C. fissicosta* (Etheridge) and *Entolium orbiculare* (J. Sowerby), *Gryphaeostrea canaliculata* (J. Sowerby), *Inoceramus sp.*, and the following ammonites: *Acanthoceras sp.*, *Mantelliceras sp.*, *Schloenbachia* cf. *subvarians* Spath and *S.* cf. *varians* (J. Sowerby). Some of these species are illustrated on Plate 6. Hancock (1961, p. 14) considered that the dominance of species of *Schloenbachia* indicated a Lower Cenomanian age, but that the presence of *Acanthoceras* suggested that the Middle Cenomanian may also be represented. A Cenomanian age is also indicated by several species of foraminifera recorded by McGugan (1957, p. 331). Reid (1958, pp. 237–8) has commented that many of the fossils of the Glauconitic Sands are facies forms and typical also of the Upper Greensand in England.

Yellow Sandstones and Grey Marls. Within the Belfast (36) Sheet this group of beds consists almost entirely of fine-grained (commonly silty and clayey) grey and brownish grey sandstones which weather to a characteristic yellow colour. Elsewhere, as for example at Magheramorne, about 16 miles east-north-east of Belfast, beds of sandstone alternate with irregular beds of dark grey silty marl.

In Collin Glen, the type section for the area, the deposit is unevenly bedded with irregular doggers of hard sandstone. The rocks are calcareous and the doggers may contain 60 to 80 per cent of limestone and not more than 15 per cent of clay and silt (Hancock 1961, p. 15). On the north side of Windy Gap, north-east of Collin Glen, the rocks are friable sandstone, much less silty, of a pale

		ft	m
Yellow Sandstones and Grey Marls	Sandstone, pale brown weathered with some grey-hearted doggers, fairly fine-grained, bedding not well-defined. Harder greyish brown sandstone occurs in the bottom half of the section where bedding is more apparent.	15	(4·6)
		21–27	(6·4–8·3)

Fossils include *Orbitolina concava* (Lamarck), serpulids, incomplete terebratulids, *Exogyra columba* (Lamarck), *Limatula fittoni* (d'Orbigny), *Neithea* (*Neitheops*) *aequicostata* (Lamarck), *N.* (*Neitheops*) cf. *sexcostata* (S. Woodward) and pectiniform bivalves.

Grey-hearted soft silty sandstone (Yellow Sandstones and Grey Marls) dipping S. of W. at 10° is visible in a small stream 420 yd (384 m) W.N.W. of Glen Bridge. About 60 yd (55 m) upstream is a small exposure of hard glauconitic sandstone (part of the Glauconitic Sands and Sandstones).

The most extensive section of the Hibernian Greensand and what is generally regarded as the type section for the area, was until recently exposed on the west bank of the Collin River, 450 yd (412 m) upstream from Glen Bridge.

Collin Glen, 450 yd (412 m) above Glen Bridge (B, Fig. 10)

		ft	m
	Basal bed of the Chalk	—	—
Glauconitic Sands and Sandstones	Glauconitic sandstone, pale grey, medium grained, with a little mica and much glauconite . . . . about	3	(0·9)
Yellow Sandstones and Grey Marls	Sandstone, pale yellowish grey and grey, fine to slightly medium grained with a few grains of glauconite in the top 1 ft. Serpulids and oyster fragments . . .	7	(2·1)
	Sandstone, dark grey and brownish grey, fine grained, with a little mica, fossiliferous. Some irregular nodules of hard, slightly calcareous, fine-grained sandstone . .	12	(3·7)
Glauconitic Sands	Glauconitic sand or sand-rock, greenish black, composed almost entirely of glauconite. The bottom 12 in contain small ($\frac{1}{4}$–$\frac{1}{2}$ in diameter) pebbles, many dark or black and highly polished (cf. the lydites at the base of the Gault in Dorset); these pebbles occur mainly in the lowest 2–3 in where the matrix is clayey. The top 9 in have less glauconite grains where the rock passes up into dark grey and grey fine-grained sandstone. *Exogyra obliquata* fairly common 6 ft up from base . .	8½	(2·6)
		30¼	9·3

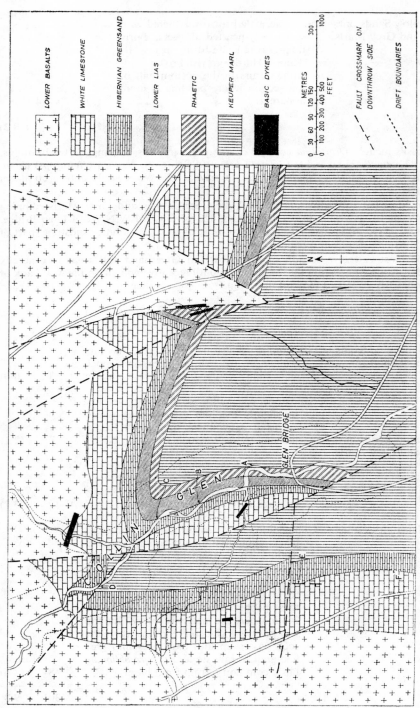

Fig. 10. *Sketch-map of the geology of Collin Glen*

The Glauconitic Sands yielded *Acila* sp., indet. cardiids, *Chlamys aspera* (Lamarck), *Entolium orbiculare* (J. Sowerby), *Exogyra obliquata, Gryphaeostrea canaliculata* (J. Sowerby), *Inoceramus sp., Thetironia laevigata* (J. Sowerby), *Schloenbachia subvarians* Spath, *S.* cf. *varians* (J. Sowerby), fish debris and vertebrae.

The following fossils were obtained from the lower group of the Yellow Sandstones and Grey Marls: *Flabellina sp., Rotularia concava* (J. Sowerby), *Holaster sp.,* cf. *Toxaster murchisonianus* (Mantell), indeterminate rhynchonellid, *Arctostrea colubrina* (Lamarck) juv., *Gryphaeostrea canaliculata* (J. Sowerby), *Neithea quinquecostata* (J. Sowerby), *Panopea ?* and *Pycnodonte vesicularis* (Lamarck), whilst *Rotularia concava, Arctostrea colubrina, Anomia sp., Chlamys aspera, Exogyra columba, E. obliquata* and *Neithea aequicostata* were collected from the higher group of beds. Four of these species, the *Rotularia, Arctostrea, Exogyra columba* and the *Neithea* were found in the same beds 25 yd (23 m) upstream.

The Yellow Sandstones and Grey Marls form a cliff on the west side of the river for about 50 yd (46 m) and then on the east side for a further 100 yd (91 m). The section visible on the east side of the river is as follows:

East Bank Collin River, 380 yd (347 m) N.N.W. of Glen Bridge (C, Fig. 10)

	ft	m
Base of Chalk	—	—
Sandstone, greenish, with glauconite (Glauconitic Sands and Sandstones)	5	(1·5)
Sandstone, darkish grey and brownish grey, fine-grained, with some doggers (Yellow Sandstones and Grey Marls)	20	(6·1)
Sandstone, bluish black, glauconitic, with *Exogyra obliquata* (Glauconitic Sands)	4	(1·2)
	29	(8·8)

In this neighbourhood the beds dip W. at 30°–40° but at the confluence of the two streams (Fig. 12) there is a sharp north-westerly trending anticline causing the Hibernian Greensand outcrop to swing eastwards out of Collin Glen.

Continuing up the main river the next important exposures occur at the waterfall depicted in the special Belfast Memoir (1904, plate IV). At the time of the resurvey an attempt to photograph this section was unsuccessful, as the waterfall was almost completely obscured by fallen tree trunks.

The cliff causing the waterfall (D, Fig. 10) is about 31 ft (9·5 m) high and formed of the Yellow Sandstones and Grey Marls which have here a total thickness of 35 ft (11 m). The beds dip N.W. at 12°. On the east bank of the river a few yards downstream from the waterfall these rocks yielded only serpulids and a rhynchonellid. Beneath the sandstones at the water's edge is a small exposure of the Glauconitic Sands; these beds are also exposed on the west bank of the river, on the south-west (upthrow) side of the fault (Fig. 12). Here about 7 ft (2·1 m) of greenish black soft glauconitic sandstone with *Exogyra obliquata* rest on red Keuper Marl. Also visible, resting on the Glauconitic Sands, are about 10 ft (3 m) of dark grey and greyish brown earthy sandstone with doggers in the bottom 1 ft (0·3 m).

Upstream from the waterfall the higher beds of the Hibernian Greensand lie in an inaccessible gorge but at the junction with the overlying Chalk 70 yd (64 m) from the waterfall they become accessible. The Glauconitic Sand and Sandstones are represented by two separate beds. At the bottom are 2 ft 6 in (0·8 m) of firm to hard, green, glauconitic sandstone in which were found *Chlamys robinaldina* (d'Orbigny) and an elasmobranch tooth; this bed is overlain by 2 ft (0·6 m) of green glauconitic sandstone.

Eastwards along the outcrop there is a small exposure of green glauconitic sand just beyond the large Hannahstown chalk pit and 580 yd (530 m) S.E. of the school [275 722].

A few yards north-eastwards the Hibernian Greensand/Chalk junction is exposed in the side of an old track. The section is:

Section 620 yd (568 m) E.S.E. of Hannahstown School

		ft	m
	Calcareous sandstone or sandy chalk, full of glauconite and pebbles of quartz; (basal bed of the Chalk)	2¼	(0·7)
Glauconite Sands and Sandstones	Sandstone, glauconitic, crowded with small pebbles of quartz with many *Exogyra columba* and fragments of other fossils seen	3	(0·9)
	Small gap		
Yellow Sandstones and Grey Marls	Sandstone, grey, fine-grained, with a little glauconite . . . . . . seen	1	(0·3)
		6¼	(1·9)

From this neighbourhood as far as the north margin of the Sheet the outcrop trends N.N.E. and the next exposures occur about 1 mile (1·6 km) distant. The first of these is visible on the north side of Allen's or The Doon Ravine [290 734]. Here 7 ft (2·1 m) of yellowish brown sandstone with doggers (Yellow Sandstones and Grey Marls) are overlain by 7 ft (2·1 m) of glauconitic pebbly chalk. It seems that the latter represents the basal beds of the Chalk or White Limestone and that the Glauconitic Sands and Sandstones are absent.

About 500 yd (457 m) northwards along the outcrop is an exposure of grey sandstone and a further 25 yd (23 m) to the N. is a small exposure of green glauconitic sand, representative of the lowest beds. Some 540 yd (494 m) W. by S. of the Luther Church [299 738], Whiterock Road and 35 yd (32 m) N. of a small exposure of sandstone is a 30-ft (9·1 m) section in pale grey and yellow sandstones. These beds, which apparently represent the Yellow Sandstones and Grey Marls, are paler in colour and more friable and open in texture than normal and in some ways they are comparable with the off-white sand in John Crow's Pit (p. 66). This variety does not extend very far, however, as at the next exposure, 735 yd (672 m) N.W. of the Luther Church, 15 ft (4·6 m) of the usual pale greyish brown and yellowish sandstone are visible.

Apart from some sandstone rubble just north of the townland boundary of Ballymurphy and Ballymagarry and a small exposure of green sand near the prehistoric flint factory, 870 yd (796 m) N.N.W. of the Luther Church, there is only one other exposure of the Hibernian Greensand within the Sheet. About 100 yd (91 m) S. of the northern margin 8 ft (2·4 m) of brown sand and sandstones are visible.

Although not in the area being described an important section on the south bank of the Forth River should be mentioned (Robbie 1964, pp. 224–9). This section lies 100 yd (91 m) N. of the northern margin of the Sheet and is as follows:

South Bank, Forth River [Crow Glen]

		ft	m
	Chalk, glauconitic, with small pebbles of grit and quartzite, hard and massive (basal bed of the Chalk)	1¾	(0·5)
Glauconitic Sands and Sandstones	Sandstone, chalky and glauconitic, rubbly .	2¼	(0·7)
Yellow Sandstones and Grey Marls	Sandstone, pale greyish brown, fine- to medium-grained, fairly soft, nodular in places with *Serpula* and pectinids seen .	3	(0·9)
		7	(2·1)

About 10 ft (3 m) of the brown sandstone are visible in the river-bed whilst about 30 yd (27 m) downstream (to the east) there is a small exposure of green sand (the Glauconitic Sands).   J.A.R.

CHALK (WHITE LIMESTONE) AND RELATED DEPOSITS

The narrow outcrop of the White Limestone, offset by numerous small faults, extends from near Magheralin in the south-west of the area to the northern margin of the Sheet north-west of Belfast. Natural exposures are uncommon but there are many good sections in the numerous active and disused quarries along the outcrop. The thickness of the formation is variable due, usually, to erosion prior to emission of the Antrim Lava Series (Plate 4). At Dollingstown it is known to be about 15 ft (4·6 m) thick increasing to about 100 ft (30 m) around Magheralin and Moira. North-eastwards along the outcrop from Trummery the average thickness is about 80 ft (24 m) and at Scott's Quarry 1¼ miles (1·8 km) north-east of Maghaberry, where the formation is wholly exposed, it is 58 ft (18 m). For about 5 miles (8 km) north-north-eastwards along the outcrop there is a marked decrease in thickness e.g., at Rockville Quarry, ¾ mile (1·2 km) south-west of White Mountain the White Limestone is 31 ft 10 in (10 m) thick and just south of Groganstown it is about 15 ft (4·6 m) thick. Thickening again rapidly to the north it is over 100 ft (30 m) at Collin Glen. In the Langford Lodge Borehole it was estimated to be 60 ft (18 m) thick.

The rapid thinning south-west of Magheralin appears to coincide with the transgression of the base of the White Limestone on to the Bunter Sandstone which may have formed a topographic high on the floor of the Cretaceous sea. Similarly the thin outcrop of White Limestone south of Groganstown appears to coincide with the south-westerly absence of the Hibernian Greensand suggesting the possible presence of a ridge limiting the Hibernian Greensand sea (p. 63). There appears to be no significant variation in thickness of the individual lithostratigraphical units with the exception of the thickness of sediment between the uppermost and lowest 'green bed' horizons (excluding minor discontinuities at Knocknadona and Legmore). At Moira the thickness of this relatively flint-free chalk sediment averages 3 ft 6 in (1·1 m) at Knocknadona 5 ft 8 in (1·7 m) whilst at Collin Glen about 6 ft 8 in (2·1 m) of chalk with nodular and tabular flints are present between two green-stained surfaces. North of the One-inch Sheet only one green-stained surface persists; it is thought to be the lower one.

The dip of the Chalk varies in direction and amount but is generally towards the west-north-west at from 5° to 10°.

The basal bed of the White Limestone is in general littoral in character and shows some local variation in lithology and thickness. Where the White Limestone overlies the Hibernian Greensand there is a distinct basal bed (the Conglomerate Bed). The lowest 2 ft to 5ft (0·6–1·5 m) of the chalk (the Glauconitic Chalk) contain a variable amount of glauconite in grains accompanied, in places, by grains of quartz. The amount of glauconite decreases upwards. West of Groganstown where the Hibernian Greensand is absent, the basal bed is 6 in (0·15 m) to 1 ft (0·3 m) thick and consists of highly glaucontic chalk with a few phosphatized quartz pebbles and glauconite-stained or unaltered fragments of Keuper Marl up to 9 in (0·2 m) across. The bed is generally poorly fossiliferous though irregular ramose markings, probably of organic origin, are common. Where clearly seen, as at Laurel Hill Quarry or at Hull's Quarry, the junction with

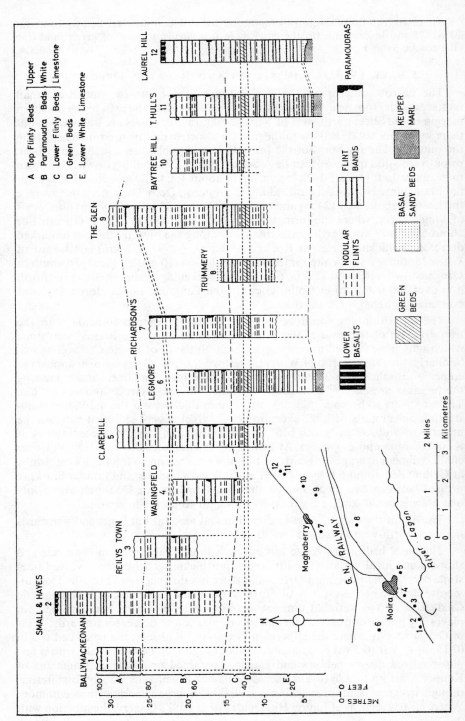

Fig. 11. *Comparative sections of the Upper Chalk rocks of the Moira area*

the underlying Keuper Marl is irregular and the latter may contain lenses of calcareous material several inches below this surface. The bed is rather more sandy than usual at Hull's Quarry. The Chalk may be discoloured up to 1 ft 6 in (0·5 m) above the basal pebble bed. In this area, south-west of Groganstown, faunas earlier than Senonian have not been found, indicating that the westward transgression of the Cretaceous sea was initiated in early Senonian times.

The main mass of the Chalk is a pure hard white limestone. All analyses made by Hume (1897, p. 578) give well over 99 per cent of material soluble in dilute hydrochloric acid, and most of this would be calcium carbonate. The White Limestone becomes pure carbonate rock at a comparatively short distance above the base. Slightly glauconitic chalk returns, however, nearly 30 ft (9 m) above the base.

The distinctive hardness of the Chalk in Northern Ireland has attracted many observers. Stylolites are widespread and jointing is marked, in common with most limestones of Carboniferous age and with the harder chalk of England. Well preserved fossils with few crushed specimens are fairly abundant.

In the past silicification has been held responsible, perhaps because of Jones's comparison (1878, pp. 739–43) of the chalk of Larne with the silicified chalk of Beinn Iadain in Morvern, and from the comparison by Peach (1901, pp. 238–9) between the Antrim chalk and the partially silicified chalk of Arran. That the hot Tertiary basalts might be the causative agent, an early hypothesis, is still brought forward from time to time (see discussion in Black 1953, p. lxxxiii) though the possibility has been disposed of by several writers. Jukes (1868, pp. 345–7) pointed out that the hypothesis would involve alteration of the clay and lignite in the clay-with-flints which overlies the chalk; Hardman (1873, pp. 434–8) said that the chalk would be altered in composition whilst Black (1953, p. lxxxv) and Charlesworth (1953, p. 128) showed that the hypothesis failed to explain the hardness of the chalk elsewhere, where there is no associated vulcanism, as in Yorkshire. Similarly, the superincumbent load of basalts cannot have compressed the chalk since some of the Cenomanian deposits are relatively unindurated.

The density of the White Limestone is much higher than the normal English Chalk. Parasnis (in Cook and Murphy 1952, pp. 8–9) gives the relative densities of the two as 2·60 to 2·64 compared with 1·95. Expressed in terms of a void ratio (volume of pores: volume of minerals) the difference is much clearer (5 per cent compared with 58 to 72 per cent). By some mechanism the pore space of the White Limestone appears to have been filled with calcite. Black (1953, p. lxxx) has suggested "the regional increase in hardness thus seemed to be connected with approach to a shoreline, and probably resulted from the incorporation of a little aragonite in the original sediment". Jukes-Browne and Hill (1903, pp. 291–2) have suggested that "The process by which many Chalk Marls and Chalks become hardened seems to be due to the gradual growth of these minute crystals [calcite]. In many samples the arrangement of these minute crystals suggests that they have continued to increase until they have enclosed the coarser ingredients of the mass, and interlocking with each other have converted the whole into a compact semicrystalline rock".

The hardening, at any rate, appears to be due to penecontemporaneous recrystallization. It has been suggested that part of the recrystallization has been from aragonite to calcite. Aragonitic organisms such as gastropods are present at some levels, indicating that the White Limestone was deposited within a depth

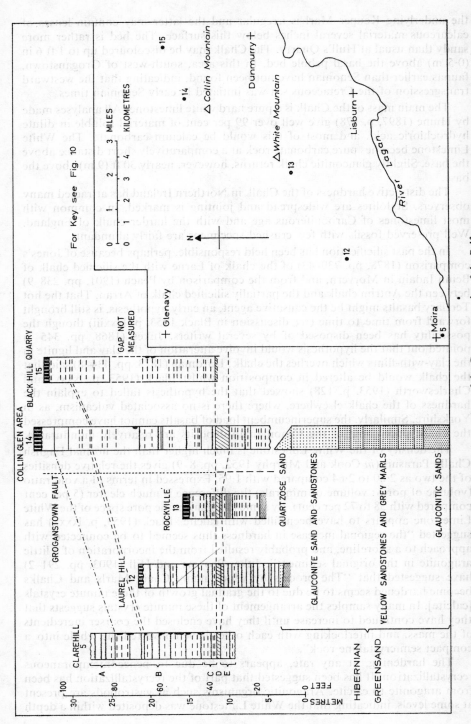

FIG. 12. *Comparative sections of the Cretaceous rocks of the Belfast area*

at which aragonitic organisms could be buried before dissolution and dispersal and redeposition of the material. In thin section, however, the White Limestone appears to be indistinguishable from English Chalk though some patches and infillings of clear recrystallized calcite in foraminiferal chambers are present. Calcispheres are occasionally abundant.

Hancock (1963, pp. 157–64) has recently summarized previous work and hypotheses on the hardness of the Irish Chalk and concluded that it was due to the pore space being occupied by calcite.

Dr. M. J. Wolfe kindly made available the work (now published, 1968) in which he describes in detail the petrology and diagenesis of the Chalk of the Belfast area. He states that the Chalk is a foraminiferal coccolith allo-micrite and describes a number of bioturbation features, both at or near the green-stained intraformational surfaces and at other levels. At Legmore quarry he notes micro-breccia horizons of subaqueous origin. Wolfe now suggests that the Irish Chalk represents a normal carbonate mud–limestone transition and that the English Chalk is abnormal.

From the spherical quartz grains in some of Hume's samples, Bailey (1924, pp. 102–116) adduced that the shores of the Cretaceous seas were desert, but recent work (Humphries 1961, p. 69 *et seq.*) has shown that this sphericity is of no great significance as an indicator of aeolian desert conditions.

Flints occur throughout the White Limestone sequence but are rarer near the base or at the Green Beds horizon. They are present as rows of nodules of varying size and shape or as tabular sheets. All the flint beds are conformably inter-stratified with the chalk, the thickness of the beds swelling or contracting within short distances. Isolated nodules occur between the regular rows. Occasionally lenticular patches of limestone are enclosed by bifurcating leaves of flint, especially in the case of the 'book' flint (p. 78), a characteristic flint of the Lower Flinty Beds in the area west of Collin Glen. There is often a gradation between chains of nodules, semiconfluent nodules and thick nodular bands. Though individually each flint band may not retain its morphology over a wide area, occasional bands such as the twin nodular band (within the Paramoudra Beds) can be readily recognized over some distance.

In all cases the flint, grey, bluish grey or black in colour, with a thin white cortex, is sharply defined from the enclosing chalk from which it is readily separated with the hammer. Colour variations appear to be of little significance. Dr. Wolfe estimates that flint constitutes 12 to 21 per cent of the total thickness of the White Limestone. Names have been given, usually by the quarry workmen, to some of the flints with pronounced shapes.

In the upper part of the White Limestone large barrel-shaped flints with a central core, termed paramoudras, are a characteristic feature throughout the area. Formerly thought by Sollas (1880, pp 437–61) to be sponges (*Poterion cretaceum*) they are now regarded as of inorganic origin. The term paramoudra was first introduced from this district by Buckland (1817, p. 413) and is pre-sumably, according to Arkell and Tomkeieff (1953, p. 81), a corruption of an Erse name, probably padhramoudras, 'ugly Paddies' (Irish Naturalist, January 1927, quoted in *The Naturalist,* March 1927, 68) or, less likely, peura muireach 'sea pears' (*Proc. Norwich Geol. Soc.* **1**, 1881, 132). In Norfolk where similar flint forms have been described (Peake and Hancock 1961, p. 318) they are known as potstones. P.I.M.

LITHOSTRATIGRAPHICAL DIVISIONS

Until comparatively recently knowledge of the White Limestone was confined to broad correlations with the macrofossil zones of the English Chalk (Hancock 1961; Reid 1959, 1962, 1963, 1964), and no attempt was made at a detailed lithostratigraphical subdivision. Early in the resurvey of the Moira district, however, it was realized that the White Limestone could be subdivided utilizing bedding, flint courses and green-stained intraformational erosion surfaces (Manning *in* Fowler 1955, p. 52). This technique was later confirmed during the resurvey of the Giant's Causeway (7) Sheet (Manning *in* Fowler 1958, p. 45), and it was clear that detailed palaeontological study would produce an effective regional subdivision and zonation of the White Limestone. Systematic examination of the faunas was started in the Ballycastle (8) and Giant's Causeway (7) sheets by C. J. Wood and T. P. Fletcher, this work being later extended to other areas. Simultaneously Mr. Fletcher (unpublished M.Sc. thesis 1967) undertook a study of the White Limestone succession of the coastal outcrops in East Antrim and established a stratigraphy applicable in general to the whole of the Northern Ireland White Limestone. Unpublished palaeontological work by Wood based on this stratigraphy shows that, in broad outline, the stratigraphical palaeontology parallels that of much of the higher part of the English Upper Chalk.

Fletcher's succession comprises a number of lithostratigraphical units characterized by differences in the type and extent of flint development, and bounded in many cases by laterally persistent bedding-planes. The higher units are more variable than the lower ones, and are in consequence less easy to define in the field. In the Belfast Sheet it is these variable units which comprise the successions exposed in the numerous quarries along the outcrop while the lower units are either missing at non-sequences or else developed in an arenaceous chalk facies. The problem is further complicated by the fact that the diagnostic echinoids and belemnites are far less common in the Belfast area than in the coastal areas where the palaeontological sequence is known in great detail. At the time of writing it has not proved possible to establish an unequivocal correlation between the White Limestone of the Befast area and that of the coastal outcrops, although the broad zonal sequence is clear. It is also important to appreciate that while the Upper Cretaceous elsewhere in Northern Ireland ranges up to the Lower Maestrichtian, extensive erosion prior to the deposition of the basalts has removed much of the former Cretaceous cover in Sheet 36.

As it would be inappropriate to introduce Fletcher's stratigraphy, in whole or in part, in advance of publication, the following lithostratigraphical classification is adopted for Sheet 36.

	Top Flinty Beds	Bedded chalk with tabular and nodular flints.
Upper White Limestone	Paramoudra Beds	Bedded chalk with flint bands and paramoudras.
	Lower Flinty Beds	Bedded chalk with characteristic flints.

Green Beds	Practically flint-free chalk with, usually, 3 glauconitic bedding planes and a relatively abundant fauna.
Lower White Limestone	Bedded chalk with flints. Basal 1 ft of yellow chalk with glauconite and sand grains W. of Groganstown.
Glauconitic Chalk . . . . . . *non-sequence*	Glauconitic chalk with small quartz pebbles.
Conglomerate Bed . . . . . . *non-sequence*	Sandy glauconitic chalk or calcareous sandstone.

Conglomerate Bed. This bed, which was described by Hancock (1961) as the 'Cobble Bed', is 1 ft to 2¼ ft in thickness (0·3–0·8 m), and rests with major non-sequence on the eroded top of the Hibernian Greensand (Plate 4). Lithologically it varies from a sandy glauconitic chalk to a calcareous sandstone, and invariably contains rounded pebbles of quartz. Pebbles and small cobbles of various Hibernian Greensand lithologies, with Yellow Sandstone predominating, are present, together with a varied fauna of phosphatized hexactinellid sponges (Reid 1958, p. 263) and other fossils including frequent belemnites (*Gonioteuthis spp.*) and poorly preserved echinoids. The upper part of the bed is indurated and glauconitized with a brown phosphatic skin, the whole complex, known locally as the Mulatto, being terminated by a layer of algal stromatolites. The presence of these stromatolites shows that deposition at times took place in very shallow water—a depth of less than 6 ft (1·8 m) being quoted by Logan (1964) for Recent examples—and does not exclude actual emergence. The matrix of the Conglomerate Bed can be assigned to the lower part of the *Echinocorys depressula* Subzone of the *Offaster pilula* Zone on the basis of the occurrence of *Echinocorys tectiformis* Brydone, and is the lateral equivalent of White Limestone of *pilula* Zone age in the coastal outcrops. In the absence of palaeontological evidence for the age of the uppermost beds of the Hibernian Greensand the extent of the basal non-sequence cannot be determined. The non-sequence between the Conglomerate Bed and the Glauconitic Chalk is considerable, and involves most of the *pilula* Zone, and much if not all of the overlying *Gonioteuthis quadrata* Zone. The palaeontological evidence for this is taken from excavations for the Bellevue road-cutting, Glengormley, immediately to the north-east of the present Sheet.

Glauconitic Chalk. To the east of Groganstown the lowest few feet of the White Limestone consist of a variable thickness of glauconitic chalk. There is a gradual upward decrease in the amount of glauconite present, and in the lower part small quartz pebbles may occur. The Glauconitic Chalk is superficially similar to the upper part of the Hibernian Greensand, but the occurrence of *Echinocorys conica* Lambert and large belemnites of the group of *Belemnitella praecursor* Stolley establishes its equivalence with the basal *Belemnitella mucronata* Zone White Limestone of the coastal successions. C.J.W., P.I.M.

Lower White Limestone—This is the lowest member of the White Limestone proper (Chalk) over the area to the west of Groganstown and has a basal bed of 6 in (0·15 m) to 1 ft (0·3 m) thick. This consists of highly glauconitic chalk with a few phosphatized quartz pebbles and glauconite-stained and unaltered fragments of Keuper Marl up to 9 in (0·2 m) across. Where clearly seen, as at Laurel Hill Quarry [190 646] or at Hull's Quarry [187 642], the junction with the under-lying Keuper Marl is irregular and the latter may contain lenses of calcareous material several inches below the surface.

In the 23 ft (7 m) of Lower White Limestone above the basal bed, flints in nodules and tabular bands are common, as are also irregular discontinuity surfaces though these are not prominent. Minor impersistent green-stained planes of discontinuity are common at Knocknadona and there is at least one at Moira.

Green Beds. These comprise three glauconite-stained intraformational erosion surfaces with associated bioturbation (Manning *in* Fowler 1955, p. 52; 1958, p. 45; Manning *in* Hancock 1961, pp. 34–5; Manning *in* Bromley 1967, pp. 181–2) which have proved to be a useful marker group in the White Limestone (see Plate 5B) of the Sheet. Apart from a few nodules in the lower third of the intra-green-bed chalk the sediment is flint-free in the Moira area where it is 3 ft 6 in (1·1 m) thick. From Collin Glen eastwards, only two glauconite-stained erosion levels are usually present and the intervening chalk sediment contains tabular and nodular flints.

Upper White Limestone

Lower Flinty Beds These, the lowest beds of the Upper White Limestone, consist of bedded chalk with nodules of flint or characteristic branching flints known to quarrymen as the 'Og–Knee' flint. A little glauconite is present in the lowest 1 ft and the fauna is comparatively rich. The top is drawn at a poorly developed bedding plane immediately below the lowest paramoudra. The 'Book' flint also occurs within these beds.

Paramoudra Beds—These are characterized by a large increase in the amount of flint with massive nodules and nodular zones and with fairly numerous paramoudras, although the latter are not invariably present. The top of this subdivision is marked by a prominent and laterally continuous dirty bedding plane, below which *Inoceramus* debris is very abundant. Apart from *Inoceramus*, fossils are infrequent and difficult to extract. This unit is from 25 ft to 60 ft (7·6–18 m) thick.

Top Flinty Beds—These are very fossiliferous, and are characterized by slabby rather than massive chalk with courses of tabulate flints, which in places exhibit complex silicification phenomena. This unit is sometimes entirely cut out by pre-basaltic weathering but when present it is up to 35 ft (11 m) thick.   P.I.M.

STRATIGRAPHICAL PALAEONTOLOGY OF THE WHITE LIMESTONE

The White Limestone proper of the Belfast (36) Sheet—i.e., Senonian deposits excluding the Conglomerate Bed and the Glauconitic Chalk—falls entirely within the *Belemnitella mucronata* Zone of the standard English classification, and is of Upper Campanian age. Elsewhere in Northern Ireland the White Limestone embraces a zonal succession ranging from the Santonian *Uintacrinus socialis* Zone to the Lower Maestrichtian Zone of *Belemnella occidentalis*; Maestrichtian sediments were probably deposited in the area of the Belfast (36) Sheet, but have subsequently been removed by erosion prior to the extrusion of the Tertiary Basalts.

On the Continent beds equivalent in age to the Belfast White Limestone have been subdivided into three belemnite zones (Jeletzky 1951) and the resulting zonal scheme has proved applicable to the Chalk of England and Northern Ireland (Peake and Hancock 1961). The three zones are characterized by distinctive belemnite assemblages, the lowermost *Belemnitella senior* Zone containing corpulent and strongly ornamented guards, the middle *Belemnitella minor* Zone slim elongated guards, and the uppermost *Belemnitella langei* Zone predominantly small waisted guards. All three assemblages are present in the complete succession seen on the north Antrim coast.

In the Belfast area belemnites occur throughout the White Limestone, but are too scarce to be of more than broad use in zonal correlation. The lowermost belemnite assemblage has not been proved in the White Limestone, but is probably represented in part at least in the underlying Glauconitic Chalk. The Green Beds appear to mark the junction between the *minor* and *langei* Zones, but further work on the belemnite succession in other areas needs to be carried out before this zonal boundary can be drawn with confidence. Unpublished work on the palaeontology of the coastal successions has shown that it will prove possible to subdivide the *mucronata* Zone part of the White Limestone using a sequence of echinoid and brachiopod faunas, but in the Belfast area the comparative scarcity of echinoids and brachiopods, except at certain levels, has so far precluded a refined subdivision of this kind.

The macrofauna of the Lower White Limestone is restricted. Collections from Richardson's Quarry during the resurvey yielded *Cretirhynchia woodwardi* Pettitt, some poorly preserved *Belemnitella sp.*, and large ammonites including *Pachydiscus* cf. *oldhami* (Sharpe) and a coarsely ribbed pachydiscid; *Cretirhynchia* cf. *arcuata* Pettitt was recorded from this level in other localities.

The Green Beds can be described as a complex of hardgrounds, pointing to comparatively shallow-water conditions with reduced and, at times, arrested sedimentation. In common with other hardground complexes in the Chalk (e.g., the Chalk Rock and the Catton Sponge Bed), the Green Beds and the lower part of the overlying Lower Flinty Beds contain, in addition to hexactinellid sponges, a rich fauna of aragonitic-shelled molluscs preserved as moulds and steinkerns, coated with a green skin of glauconite. In some instances the form of the original shell is preserved as a pseudomorph in calcite after aragonite. The Green Beds and related deposits are best developed and most fossiliferous in the western part of the sheet, in the area around Moira and Lisburn, but do not compare in richness of fauna with the White Limestone basement beds in Derry and Tyrone. The relationship between the Belfast Green Beds and 'green beds' in other areas is complex, and requires further elucidation. It is likely however, that the Green Beds are not the correlative of the green beds in the Giant's Causeway (7) Sheet, but are developed at a higher horizon approximately corresponding to the basement beds of Derry.

The fauna of the Green Beds and the immediately overlying chalk is extensive, but tends to be monotonous in any one locality. The Green Beds in the pit at Kilcorig near Lisburn, from which Tate (1865) described many new species, are no longer exposed. Apart from poorly preserved sponges, haploid corals (*Desmophyllum spp.*) and the aragonitic-shelled molluscs, brachiopods are conspicuous, particularly *Carneithyris spp.* and *Cretirhynchia norvicensis* Pettitt, with subordinate *C. woodwardi*; the large *Neoliothyrina obesa* Sahni (synonym

N. abrupta Tate sp., occurs not uncommonly, usually as a rolled and glauconitized steinkern. Echinoids are represented by *Echinocorys* cf. *humilis* Lambert and *E.* cf. *meudonensis* Lambert, together with *Galerites spp.*, also preserved as rolled steinkerns. The molluscan fauna is dominated by *Margarites (Periaulax ?) striata* (S. Woodward), *Turritella unicarinata* S. Woodward, *Meiocardia? trapezoidale* (Roemer) and several species of *Pholadomya*. Ammonites are commonly represented by poorly preserved *Pachydiscus spp.*, but include *Baculites* cf. *vertebralis* Lamarck, *Desmophyllites larteti* (Seunes), *Gaudryceras?* cf. *jukesi* (Sharpe) and *Parapuzosia icenica* (Sharpe), in addition to three species originally described from the Belfast Sheet: *Glyptoxoceras hibernicum* (Tate) *Phyllopachyceras occlusum* (Tate) and *Scaphites elegans* (Tate). Of the other cephalopods, belemnites occur in large numbers, but are usually extensively attacked by the boring sponge *Cliona* and other organisms, and rarely capable of determination beyond *Belemnitella sp.* Nautiloids are represented by *Cymatoceras bayfieldi* (Foord and Crick), which is comparatively common.

The Paramoudra Beds are sparsely fossiliferous, yielding *Carneithyris spp.*, *Cretirhynchia* cf. *norvicensis*, crushed but thick-tested small *Echinocorys sp.*, and elongate belemnites of the *langei* assemblage.

The Top Flinty Beds yield a rich fauna dominated by brachiopods of Maestrichtian affinities. The absence of the diagnostic Maestrichtian belemnite *Belemnella*, however, precludes references to this stage, and it is likely that these uppermost beds of the Belfast succession can be correlated with pre-Maestrichtian brachiopod-rich chalk developed in the Giant's Causeway (7) Sheet near Portrush. Fossils recorded include: thick-tested *Echinocorys sp.*, *Micraster* (*Isomicraster*) *sp.*, *Carneithyris spp.*, *Cretirhynchia arcuata*, *C. lentiformis* (S. Woodward), *C. magna* Pettit, *C. sp.* (*norvicensis-triminghamensis* Pettitt group), *Neoliothyrina obesa*, *Trigonosemus sp.* (one specimen collected), and numerous *Belemnitella* cf. *langei* Jeletzky.

EXPLANATION OF PLATE 5

Representative Cretaceous fossils.

Specimens representative of the fossils occurring in the Cretaceous deposits in the Belfast area, though not all collected from this Sheet. Registered numbers are given in square brackets.

1. *Schloenbachia subvarians* Spath; Glauconitic Sands, Carr's Glen [NIF 348] (One-inch Sheet 28).
2. *Chlamys fissicosta* (Etheridge); Glauconitic Sands, shore-section, Cloghfin Port, Island Magee [FD 4656]. (One-inch Sheet 29).
3. *Turritella unicarinata* (Woodward); White Limestone 'Green Beds'. J. Richardson's Quarry (Maghaberry Limeworks), 750 yd N. 119 W. of Magheraberry village Crossroads [NIM 131].
4. *Scaphites elegans*, Tate, holotype; horizon and locality given as 'Upper Chalk, Lisburn'. [37263].
5. *Neoliothyrina obesa* Sahni; horizon and locality as for Fig. 4 [108870].
6. *Exogyra obliquata* (Pulteney) [synonym *E. conica* (J. Sowerby)]; Glauconitic Sands, Collin Glen, 240 yd upstream from Glen Bridge [FD 4092].
7. *Exogyra columba* (Lamarck); 'Serpula' Grit' (i.e., Upper Glauconitic Sands), Collin Glen, 300 yd W. of Glen Bridge [FD 4333].
8. *Entolium orbiculare* (J. Sowerby); Glauconitic Sands, horizon and locality as for Fig. 2 [FD 4668].

PLATE 5

(For explanation see p. 80)   Representative Cretaceous Fossils

A.—Chalk Quarry, Ballymakeonan, Moira   (NI 231)

(For explanation, see p. xiii)   PLATE 6

B.—White Limestone, Legmore Quarry, Moira   (NI 233)

The skin of a semi-silicified flint from this level yielded a rich silicified micro-fauna which provided further indication of Maestrichtian affinity. The foramini-feral fauna was dominated by long-ranged planktonic species such as *Globigerina cretacea* (d'Orbigny), *Heterohelix globulosa* (Ehrenberg) and *Planomalina aspera* (Ehrenberg), but included the following species indicative of a horizon near the Campanian-Maestrichtian boundary: *Bolivina incrassata* Reuss, *Bolivinoides decoratus* Jones, *Eponides beisseli* Schijfsma, *Gavelinella pertusa* (Marsson), *Gavelinopsis voltziana* (d'Orbigny), *Neoflabellina praereticulata* Hiltermann, *Osangularia* cf. *lens* Brotzen, *Pseudouvigerina cristata* (Marsson) and *Stensioina pommerana* Brotzen.   C.J.W.

DETAILS

The southernmost exposures within the Belfast (36) Sheet are near Mathersesfort [115 576] where there are three small, almost completely overgrown quarries. In the original survey the beds were recorded as dipping W.N.W. at 5°, but the quarries were disused even then and no ascertainable dip was observed during the present survey. The most southerly excavation shows only some 12 ft to 15 ft (3·6–4·5 m) of boulder clay and the remaining two are both flooded, but 4 ft (1·2 m) of chalk are still visible in the centre quarry and 5 ft (1·5 m) of creamy white chalk in the northern one.

The Chalk forms a prominent escarpment running north-west from these quarries, but there are no further exposures between here and the village of Magheralin.

In Ballymagin townland, 330 yd (274 m) N.W. of Magheralin church, there is a group of quarries, long disused, in which some 20 ft to 30 ft (6–9 m) of chalk were formerly visible (Berger 1816, p. 174; *Mem. Geol. Surv.* 1871, p. 34; Hume 1897, pp. 545–6) with a dip N.N.W. of 5°. The 35 ft (11 m) thick overburden of drift has now obscured much of the rock and a maximum of 11 ft (3·4 m) of chalk is now visible. A typical *mucronata* Zone fauna was obtained from this locality at the time of the original survey (*Mem. Geol. Surv.* 1871, p. 13, Location 13) but only fragmentary belemnites were observed in the present resurvey.

Ballymakeonan Quarry [135 595] was being worked at the time of the original survey and about 40 ft (12 m) of Chalk were seen. Now disused, the section exhibits a maximum of 22 ft (7 m) of Chalk of the Top Flinty Beds in the north-west corner and this is now being covered by tipping; 10 ft (3 m) of the overlying basalt can be seen on the north side, but the junction is not observable. On the east side of the quarry a basalt dyke is seen in contact with a pinnacle of sheared chalk left by quarrying. The over-burden of basalt and boulder clay at this locality is about 30 ft (9 m).

The best section hereabouts lies further east at Small and Hayes Quarry [138 596] in Ballymakeonan townland (Plate 6A). The total thickness of chalk seen in the main quarry face is 67 ft (20 m) and the chalk/basalt junction with the intervening clay with flints is visible though inaccessible in the north face. The uppermost 35 ft 6 in (11 m) are referable to the Top Flinty Beds and the lower part of the Paramoudra Beds. The base of the quarry is about 10 ft (3 m) above the Green Beds horizon which is known to be present by quarry workers. The base of the Paramoudra Beds has been noted in later and slightly deeper excavations at the main quarry face. In an access tunnel beneath the Moira–Lurgan road a green horizon can be seen, but correlation with the Green Beds is uncertain.

Further east in Rileystown Quarry [142 598], just south of the main Belfast–Lurgan road, 26 ft (8 m) of chalk are seen in a face close to the road. The basalt capping visible during the original survey (*Mem. Geol. Surv.* 1871, p. 53) in the north-east corner of the quarry was not seen. Since only the top 4½ ft (1·4 m) are referable to the Top Flinty Beds and the remainder of the exposure is in Paramoudra Beds it seems probable that the basalt formerly exposed was an intrusion. No identifiable fauna was obtained at this locality.

Immediately north-east of Waringfield House a disused quarry is now becoming gradually obscured by tipping; 40 ft (12 m) of chalk were visible during the original survey and 29 ft (9 m) of Paramoudra Beds are now exposed dipping at 5° N.W. One small dip fault with a downthrow of 2 ft 3 in (0·7 m) to the W. was noted in the western portion of the quarry face.

Some 350 yd (320 m) S. of St. John's Church, Moira, are the extensive Clarehill quarries. Of the 65-ft (20-m) section exposed in the western quarry, the uppermost 11 ft (3·4 m) belong to the Top Flinty Beds; 42 ft 6 in (13 m) to the Paramoudra Beds; 5 ft 6 in (1·7 m) to the Lower Flinty Beds; 3 ft 6 in (1 m) to Green Beds and the remainder is Lower White Limestone. The locality is referred to by Buckland (1816, p. 413), Tate (1865, p. 25), *Mem. Geol. Surv.* (1871, p. 32) and Hume (1897, p. 546). The original survey appears to have confused the Green Beds horizon with Mulatto Stone which they record from the eastern quarry. The quarry workers have no record of the basal beds or the 'Greensand' mentioned in the Memoir.

To the east and north of the Clarehill quarries there are no natural exposures. A small area at Potterstown shown on earlier editions of the map as an outlier is now thought to be continuous with the main mass of chalk. At a farm just opposite Berwick Hall [159 607] chalk was said to have been met with under 17 ft (5·2 m) of boulder clay. In the main street of Moira village, excavations for a pipe line showed the chalk *in situ* at the eastern end of the village and similar excavations nearer the Sewage Works [153 611] proved the chalk *in situ* in that locality. To the north of Moira the Lower Basalts form the main scarp feature, the Chalk cropping out on the lower ground. Two well records and a note on the original field-map 300 yd (274 m) N. of Drumbane House [148 615] show that the Chalk continues to near the main railway line. A fault trending N.50°E. with the downthrow on the northern side, displaces the base of the Chalk 750 yd (684 m) to the S.W. In Legmore townland a large quarry was operated by the Ulster Limestone Corporation Ltd. for the production of ground agricultural line. A borehole at the quarry (36/55) showed 36 ft (11 m) of Chalk; the basal 9 in (0·23 m) were of grey glauconitic chalk.

In Legmore Quarry the beds dip W.N.W. at 1° to 2° and a total thickness of 36 ft (11 m) was seen in 1955, the succession being as follows:

			ft	*m*
Upper White	⌠ Paramoudra Beds	. . . . . .	23	(7·0)
Limestone	⌡ Lower Flinty Beds	. . . . . .	4	(1·2)
Green Beds	. . .	. . . .	3	(0·9)
Lower White Limestone	. . .	. . . . seen	6	(1·8)
			36	(10·9)

Between Legmore Quarry and Soldierstown the only Chalk seen *in situ* was about 2 ft (0·6 m) in the north side of a well 180 yd (165 m) S.W. of Oldmill Bridge. North-east of Soldierstown on the southern flank of Quarry Hill is a group of quarries which have been abandoned since before the time of the original survey. The quarries appear to have been worked haphazardly and no more than 7 ft (2·1 m) of Chalk are visible within the quarry group. A fault running N.37°E. through Primrose Hill Farm has an easterly downthrow and is partially responsible for the preservation of the triangular outlier of Chalk to the north of Moira station.

The main outcrop crosses the Soldierstown–Maghaberry road where the Chalk was worked by J. Richardson & Sons for whiting and putty [167 630]. The eastern quarry has long been abandoned and recent working has been in the western quarry. The section shows:

			ft	*m*
Upper White	⌠ Paramoudra Beds	. . . . . .	35	(11·0)
Limestone	⌡ Lower Flinty Beds	. . . . . .	4	(1·2)
Green Beds	. . .	. . . .	3	(0·9)
Lower White Limestone	. . .	. . . . seen	10	(3·0)
			52	(16·1)

The Paramoudra Beds contain fine examples of paramoudras, one specimen extending vertically for about 12 ft (3·7 m) (Hume 1897, pp. 545, 547). The Lower Flinty Beds include the characteristic 'Book' and 'Og-Knee' flints and contain *Belemnitella sp.*

The Green Beds are exposed in the western quarry, in the central quarry and in a small exposure at the base of the north-west face of the eastern quarry.

About 200 yd (183 m) E. of the eastern quarry, a fault running parallel to that through Primrose Hill Farm displaces the base of the Chalk 1050 yd (960 m) to the south.

Just east of Trummery House [170 618] 22 ft 6 in (6·9 m) of Chalk are visible with a westerly dip of 14°. The section here is:

			ft	m
Upper White	Paramoudra Beds	. . . .	4½	(1·4)
Limestone	Lower Flinty Beds	. . . .	4½	(1·4)
Green Beds	. . . .		3½	(1·1)
Lower White Limestone	. . . .		22½	(6·9)
			35	(10·8)

From immediately above the top of the Green Beds, *Cretirhynchia sp.* and *Echinocorys sp.* are recorded.

A well at Trummery House penetrated 46 ft (14 m) of Chalk below a 4-ft (1·2 m) cover of drift. Beside the railway at Church Hill Farm, a well 30 ft (9·1 m) deep is said to penetrate the Chalk while 22 ft (6·7 m) of chalk could be seen in a roadside well [172 625]. Former small quarry scars immediately west of Chestnut Hill Farm are now overgrown.

At the Glen Quarry (Balmers' Glen) [182 631] a total of 47 ft 6 in (15 m) of Chalk is exposed. The base is no longer visible but was formerly seen in a cutting at the north-east end of the quarry. A composite section at this locality is as follows:

			ft	m
Upper White	Top Flinty Beds	. . . .	3	(0·9)
Limestone	Paramoudra Beds	. . . . . .	48	(14·6)
	Lower Flinty Beds	. . . . . .	4	(1·2)
Green Beds	. . . .		3½	(1·1)
Lower White Limestone	. . . .		7	(2·1)
			65½	(19·9)

The beds dip at 5° to the W. though just east of a N.–S. dyke there is a local steepening to 10°. The western face is composed of Paramoudra Beds and Top Flinty Beds, but the Green Beds appear on the eastern face on either side of the N.–S. dyke.

Just beyond Baytree Hill Farm about 550 yd (503 m) further north, a long-disused quarry exhibits 30 ft 3 in (9·2 m) of Chalk, mostly Paramoudra Beds. The top horizon of the Green Beds just appears at the base of the section seen in the north-west wall, but is more clearly visible on the north-east side of the quarry. The north-west quarry wall is affected by small faults.

North-east of the quarry the escarpment continues to just beyond Yew Tree Farm, where a former quarry is now completely grassed over. The Chalk is seen *in situ* 200 yd (183 m) N. of the last location in the roadway leading to a small farm and a few yards further north is well displayed in the Mullaghcarton Lime Quarries [187 642].

Only a very small exposure was noted at this locality at the time of the original survey, but later expansion of lime burning has given two large quarries, Hull's Quarry

North and Hull's Quarry South, which show an almost complete section from the
Keuper Marl to the Lower Basalts. Fifty-nine feet (18 m) of Chalk, out of an estimated
total of approximately 61 ft (19 m), are visible and is divisible as follows:

			ft	m
Upper White	{ Paramoudra Beds	. . . .	25½	(7·8)
Limestone	{ Lower Flinty Beds	. . . .	4½	(1·4)
Green Beds	. . . .	. . . .	3¼	(1·0)
Lower White Limestone		. . . .	26	(7·9)
			—	——
			59¼	(18·1)

This is the most complete section of Lower White Limestone seen. The base is visible in
a drainage tunnel between the two quarries and also, though inaccessible, in an air shaft
for the tunnel in the south quarry. Some 5 in to 6 in (0·14 m) of glauconitic sandstone
with quartz and lydite pebbles form the base and above this the Chalk is yellow-
stained for about 1 ft 6 in (0·46 m) and contains occasional quartz pebbles. Fragments of
Keuper Marl were incorporated in the basal bed, one specimen exhibiting mudcracks.
The glauconitic sandstone is sparsely fossiliferous and contains *Neithea* (*Neitheops*)
sexcostata (S. Woodward), some indeterminate bivalve internal moulds probably
referable to *Isoarca spp.* and a fish tooth, *Isurus* or *Lamna sp.*

A trial pit in the north quarry showed light olive-coloured sandy calcareous shales
immediately beneath the chalk, but they were not penetrated to the Keuper Marl
beneath.

About 350 yd (320 m) N.E. of Hull's quarries formerly defunct quarries known
locally as Anderson's and Clarke's quarries at Laurel Hill [191 645] run for several
hundred yards along the escarpment and have been re-opened.

In the southernmost quarry a good section up to the top of the Green Beds is
measurable with about 25 ft (7·6 m) of poorly exposed Paramoudra Beds above. A total
of 24 ft 3 in (7·5 m) is recorded giving:

		ft	m
Green Beds	. . . .	3¼	(1·0)
Lower White Limestone	. . . .	21	(6·5)
		—	— —
		24¼	(7·5)

Though the drainage sump was said to penetrate into Keuper Marl, the basal 3 ft to
4 ft (1 m) of the succession are obscured by debris. The quarry floor is here of yellowish
chalk immediately above the basal glauconitic bed.

In the northern quarry a full sequence of 58 ft 6 in (18 m) has been measured:

			ft	m
Upper White	{ Paramoudra Beds	. . . . . .	26	(7·9)
Limestone	{ Lower Flinty Beds	. . . . . .	5	(1·5)
Green Beds	. . . .	. . . . .	3	(0·9)
Lower White Limestone		. . . . . .	22½	(6·9)
Chalk with glauconite grains		. . . . . .	1–2	(0·6)
			—	— —
			58½	(17·8)

At the sump hole in the south-east corner of the quarry the glauconitic chalk rests
with irregular contact on Keuper Marl with a marked angular unconformity. Above
this the chalk has a greyish or yellowish tinge and the glauconite content decreased
rapidly upwards. The measured flint courses do not, with the exception of a pair of
tabular courses about 7 ft 6 in (2·3 m) and 8 ft 6 in (2·6 m) below the base of the Green
Beds, compare closely with those in the south quarry. A lower glauconite-stained

bedding plane in the latter, 13 ft 3 in (4 m) below the base of the Green Beds, has not been detected in the north quarry. This bedding plane is comparable in position with the uppermost of the two seen in the Lower White Limestone of the Legmore quarry. Though the quarry face is not fresh, the Green Beds can be clearly seen over more than half the exposure.

The top Paramoudra Beds are inaccessible in the upper part of the quarry, but the basal 10 ft (3 m) of rotten Lower Basalts, not formerly visible, can now be seen.

Small quarries west of Mullaghcarton House show a few feet of chalk much broken up by small faults.

P.I.M.

Between Mullaghcarton and Kilcorig House the outcrop of the Chalk is poorly exposed. Small exposures east of Mullaghcarton village and in a ditch near Fallow Hill [203 653] enable the outcrop to be drawn with some confidence, but the reliable report of chalk encountered in a well at Springvale [189 658] indicated an inlier of chalk in this area. A number of small quarries in the grounds of Brookhill show up to 18 ft (5·5 m) of chalk and flints, but drift cover is heavy to the north and only an abandoned quarry at Kilcorig [217 672] offers an exposure in this area. This quarry, now showing only about 6 ft (2 m) of chalk and flint, was presumably that from which Tate (1865, p. 26) described the 'Flinty Flag' of Kilcorig, 15 ft (4·6 m) from the base of the chalk, with its upper surface covered with sponge remains in a glauconite paste—the first mention of the Green Beds.

The extensive quarries at Knocknadona [225 667] and Rockville [230 671], both now abandoned, show good sections through the Chalk, the base being seen at Rockville and beds within a foot or two of it at Knocknadona. The succession is:

			ft	m
Clay with flints	. . . . . .			
Upper White { Paramoudra Beds	. . .		10	(3·0)
Limestone ⎨ Lower Flinty Beds	. .		5	(1·5)–7 ft (2·1 m) at
Green Beds .	. . . . .		3	(0·9) Rockville
Lower White Limestone	. . . .		15	(4·6)
Chalk with glauconite grains	. . . .		0½	(0·15)
			33½	(10·15)

In the uppermost 5 ft (1·5 m) of the Paramoudra Beds at Rockville patches of White Limestone are full of scattered fragments, 3 to 5 mm long and about 1 mm thick of dense and darker calcite usually accompanied by macro-crystalline calcite. The distribution of the crystalline material seems to show a preferential orientation or 'way up'. We are indebted to Dr. R. G. C. Bathurst and Dr. M. J. Wolfe for comments on these objects which appear to be fragments of aragonite shell. The mechanism of replacement is obscure but it is suggested that part of the aragonite was dissolved and replaced by drusy calcite crystals while the remainder inverted to calcite.

The Green Beds consist of two glauconite-stained erosion surfaces 9 in (0·2 m) apart, separated by 19 in (0·5 m) of chalk from a third, usually more distinctive, surface which is overlain by 6 in (0·15 m) of glauconitic chalk. Below the lowest of these surfaces three other less well-marked erosion planes with some glauconite are seen in the underlying 3 ft (0·9 m) of chalk.

H.E.W.

The Conglomerate Bed, 2 ft (0·6 m) thick is exposed in the side of the cutting leading to the now abandoned quarry [266 707] at Collinwell. The rock is glauconitic chalk with pebbles of quartz and sponges weathering out on the surface. In the main part of the quarry about 20 ft (6·0 m) of chalk with flint nodules are visible, all of which appears to be above the top 'Green Bed'.

The basal beds of the White Limestone are well exposed in John Crow's Pit [268 716] where the section is:

		ft	m
Glauconitic Chalk	Chalk with glauconite and some quartz grains seen	2	(0·61)
Conglomerate Bed	Calcareous pebbly grit with glauconite; harder, more calcareous and not so glauconitic in top 1 ft to 1¾ ft (0·3–0·5 m); some cobbles of cherty sandstone near the bottom. *Echinocorys sp.*, *Concinnithyris?*, *Neoliothyrina ?* juv., *Gonioteuthis* cf. *quadrata* (Blainville)	5½	(1·65)
		7½	(2·26)

The best section of chalk in the district is in the large active quarry [267 719] on the western flank of Collin Glen. The base of the limestone is not visible but is thought to lie about 4 ft (1·2 m) below the floor of the quarry. The west face is about 105 ft (32 m) high and as much of it is sheer it precludes accurate measurement. The lowest Green Bed occurs about 18 ft 8 in (5·7 m) above the floor of the quarry with the other two, poorly developed, 5 ft 5 in (1·6 m) and 3 ft 10 in (1·2 m) higher up. About 17 ft 5 in (5·3 m) higher are two prominent bedding planes 4 in to 6 in (0·15 m) apart.

In Collin Glen the basal beds were formerly well exposed on the west bank of the river (B, Fig. 10). Here the Conglomerate Bed is in places a glauconitic chalk and in others a grey medium-grained chalky sandstone. Small quartz pebbles are present and there is much glauconite; fragments of oysters have been observed. The Glauconitic Chalk is 1 ft 6 in (0·45 m) thick and flint nodules occur to within 6 in (0·15 m) of the bottom. On the east bank of the river (C, Fig. 10) the Conglomerate Bed 1 ft (0·3 m) thick is a calcareous, glauconitic sandstone containing sponges and some cobbles of glauconitic sandstone. In the gorge [267 723] upstream from the waterfall the basal beds of the White Limestone are well exposed when the river is low. Here the Conglomerate Bed is represented by 5 ft (1·5 m) of glauconitic sandstone with irregular cobbles of hard glauconitic sandstone. This bed yielded a limb of *Hoploparia longimana* (G. B. Sowerby).

In the main Hannahstown Quarry [278 720] part of the basal beds was visible in the south-east corner, but now only chalk with flints is exposed. The Green Beds are not so prominent as in the Moira district, but there is a full section including the Paramoudra Beds and part of the Top Flinty Beds. In a small quarry 150 yd (137 m) to the east the Conglomerate Bed, 2 ft 3 in (0·7 m) thick, is a sandy chalk grading to calcareous sandstone, highly glauconitic and with quartz pebbles. Fletcher has noted many fossils from the Glauconitic Chalk above this bed including *Echinocorys conica* and large belemnites of the group of *Belemnitella praecursor* Stolley.

In an abandoned quarry [289 729] on the eastern flank of Black Hill the chalk is horizontal. About 80 ft (24 m) are exposed and apart from the usual courses of flint nodules there are several prominent bedding planes. The Green Beds have not been identified with certainty.   J.A.R.

From the Langford Lodge borehole a small fauna was obtained (NIA 7384–7420): *Ditrupa* (*Ditrupa*) *sp.*, *D.* (*Pentaditrupa*) *sp.*, echinoid test fragments; *Carneithyris sp.*, *Cretirhynchia* cf. *arcuata* Pettit, *C. lentiformis* (S. Woodward), small kingenid, *Terebratulina sp.*, *Chlamys* cf. *cretosa* (Defrance), *Pseudolimea sp.*, *Belemnitella sp. juv.*, and bone fragments.

The fauna suggests a horizon above the Green Beds, and is most probably indicative of the Top Flinty Beds. It was collected between 2620 ft 4 in and 2630 ft (799–802 m). There is no information on the nature of the base of the formation which was drilled through by rock-bit.   H.E.W.

Chapter 7

TERTIARY: ANTRIM LAVA SERIES

INTRODUCTION

WITHIN THE BELFAST area representatives of the widespread igneous activity of early Tertiary times consist of a considerable thickness of basaltic lavas, a few small plugs and agglomerate-filled vents and many dykes and sills. The basalts occupy the north-western half of the Sheet and form the upper part of a scarp facing south-east across the Lagan Valley. The lava-surface falls gently from the mountains at the scarp — Divis 1574 ft (480 m), Black Mountain 1272 ft (388 m) and Collin 1085 ft (331 m) — towards Lough Neagh at 48 ft (14·6 m) O.D. This slope coincides in places with the dip of the flows and is largely covered with boulder clay so that the individual flows with the exception of a mugearite (Walker 1960, pp. 62–4) and the lavas on Collin Mountain form no mappable features, though a few characteristic flows in the north-east can be traced for some distance.

The mode of extrusion of the lavas is uncertain. In Iceland, fissure eruptions are known from historic times and much of that area is assumed to be of fissure origin, but the Mull lavas are thought to be the products of a central volcano. In Antrim both vent and fissure eruptions probably played a part though dykes are only very rarely seen to feed lava flows. The large plug at Slemish has been shown to have been active at four or more periods and to have fed a lava lake similar to those occurring today in the Hawaiian volcanoes (Preston 1963). It seems likely that Slemish acted as a shield volcano and that much of the Antrim plateau was in fact a basaltic shield. Within the Belfast area three volcanic plugs, two at Trummery and Kilcorig penetrating Triassic rocks, and one at Armstrong's Hill penetrating basalts, may have been feeders of the lavas at some stage. The small tuff pipes seen in Rockville Quarry, Kilcorig, and proved in a borehole at Tully-loob Bridge near Moira are only of local significance and no extensive pyroclastic deposits are known.

The age of the vulcanism is still uncertain. Only fragmental plant debris from inter-lava laterite beds is available for palaeontological dating. Early workers regarded the lavas as Miocene though Gardner (1885) stated categorically that they were very early Eocene. More recent palynological work has given ages ranging from Late Miocene or early Pliocene (Simpson *in* Eyles 1952, p. 4) to early Oligocene (Watts 1962, p. 600). Radioactive dating of the lavas now indicates that they are of early Eocene or Palaeocene age.

The dominant lava rock type throughout the North Altantic Province is basalt though some andesites and rhyolites do occur. In the Belfast area the lavas are almost exclusively of olivine-basalt, the only exceptions being a flow (or possibly flows) of mugearite which crops out west of Stonyford, and tholeiitic basalts which are known from low in the succession on Divis and at Aughrim Plantation.

87

The majority of the lava flows, which are only rarely well exposed in cliff or quarry sections, are of the Aa type, with a slaggy amygdaloidal top and base and a more compact, massive, commonly pseudocolumnar central part. Some of the flows are thin and amygdaloidal throughout. Near the top of the cliffs at White-rock on the west side of Belfast is an occurrence of Pahoehoe-type lavas. These lavas are thin, usually only a few inches thick, but they do not exhibit the ropy upper surface of the typical Pahoehoe flow. Weathering of the lavas after extrusion, before burial by subsequent outpourings, has produced a zone of kaolinized and in places lateritized rock at the top of most flows. This zone is generally thin but may reach a thickness of several feet. The most important of these weathered zones (the Interbasaltic Bed) is of much greater thickness than normal indicating a prolonged period of volcanic quiescence. During this time lateritization occurred in the topmost flow to a considerable depth and in a few districts penetrated to the flow below. Subsequent desilicification and concentration of iron resulted in bauxite and iron ore deposits, formerly of economic value.

In the northern part of County Antrim three groups of basalts have been recognized but in the south only two groups (the Lower Basalts and the Upper Basalts) are present, separated by the Interbasaltic Bed. Up to the present no diagnostic difference has been found between these two groups of lavas and this, coupled with the absence of any major outcrop of the Interbasaltic Bed in the Belfast area led to the belief that only Lower Basalts were present within the sheet. The only exception was a small outlier capping Divis Mountain which was shown on previous editions of the Belfast map as of Upper Basalts. This is now considered to be a flow of the Lower Basalts and what was regarded as the Inter-basaltic Bed is merely a better-developed lateritized flow-top in the Lower Basalts. However, as a result of the deep borehole at Langford Lodge it is now known that Upper Basalts are present near Lough Neagh and geophysical work suggests that the area west of the Glenavy Fault is underlain by this series. At Langford Lodge the Lower Basalts are 1743 ft (531 m) thick, the greatest recorded thickness for this Group.

No early writers appear to have considered the lavas in the Belfast area in any but the most general terms although zeolites were collected. The first major reference is in the Sheet (36) Memoir (Hull and others, 1871). The basalts of the north-east quadrant of Sheet 36 were described in the memoir on the Belfast District (Kilroe and others 1904). More recently Patterson (1951, pp. 286) analysed a basalt from a quarry 1 mile (1·6 km) north of Moira Station and G. P. L. Walker (1960, pp. 62–4) has described the mugearite of Y-Bridge.

H.E.W., P.I.M., J.A.R.

Lower Basalts

Where visible within the Belfast Sheet the Lower Basalts rest on an eroded surface of the Chalk usually with an intervening layer of reddish brown Clay-with-Flints. The presence of this layer and the sporadic occurrence of lignite at this level indicates that the lavas were extruded onto a land surface.

Apart from the main scarp and the many quarries opened up near it, exposures of the Lower Basalts are generally poor. West and south-west of Divis, however, exposure is reasonably good. In the western half of the area exposures are not good though in the ground south and west of the Lisburn–Glenavy railway there is a considerable number of small exposures, the drift cover being thin in places, and the Ballinderry River, in its lower reaches, provides a good section. On a line of disconnected hills forming a belt of higher ground extending north-eastwards from Upper Ballinderry there are a number of exposures.

With the exception of a flow or flows of mugearite and unusual flows of tholeiitic basalt near the base of the lava pile all the lavas are of olivine-basalt of varying degrees of vesicularity and coarseness. The lower flows, particularly in the neighbourhood of Belfast, were especially gaseous when erupted and so the rock is highly amygdaloidal and decomposed almost throughout each flow. Higher in the pile the rocks are fresh and massive with a tendency to be feldspar-phyric. In the neighbourhood of Collin Mountain the massive central part of several of the flows exhibits strong irregularly columnar jointing.

The mugearite is a conspicuous fine-grained rock showing flow-banding and the outcrop can be traced for about 4 miles (6·4 km). It appears to be about 700 ft or 800 ft (213–243 m) above the base of the lavas. It is not possible to say whether it extends far beyond the outcrop but it was not present in the Langford Lodge Borehole.

The thickness of Lower Basalts now remaining near Belfast is about 800 ft (243 m) and the thickness at Langford Lodge is 1743 ft (531 m).

INTERBASALTIC BED

This thick zone of laterite and lithomarge is not seen at outcrop within the limits of the sheet but was penetrated by the Langford Lodge Borehole in which it occurred from 792 ft to 845 ft (241–258 m). Most of this thickness was drilled by rock-bit and no core was obtained, but a short length of core was recovered from 820 ft to 828 ft (250–252·4 m) and proved to be mainly white-spotted lithomarge with concentric colour bands reminiscent of the residual basalt spheroids common in the lower part of the Interbasaltic Bed. There was a notable amount of crystalline siderite lining joints and in small cavities.

UPPER BASALTS

The occurrence of this series was first proved by the Langford Lodge borehole and its extent confirmed by the results of gravimetric survey. Apart from short sections in the Crumlin and Glenavy rivers the outcrop is mainly drift-covered and exposures are few. The exposures in the Crumlin River are mainly of amygdaloidal and very decomposed basalt but the others are of compact medium- and fine-grained olivine-basalt in no way distinguishable from the Lower Basalt lavas. The total thickness proved in the borehole was 762 ft (232 m) which is probably near the maximum thickness present in this Sheet.

P.I.M., J.A.R., H.E.W.

DETAILS

LOWER BASALTS

Lurgan District. Just beyond the Sheet margin at Creamline Products premises south of Lurgan, a borehole penetrated 360 ft (110 m) of basalt resting on Chalk. At the Boxmore Works [110 582] at Dollingstown 381 ft (116 m) of basalt overlie 10 ft (3 m) of fine sand containing mainly flint chips with some shell fragments. The site of this borehole is 700 ft (213 m) W. of the outcrop of the base of the basalt and if such a rapid thickening is maintained, well over 1000 ft (305 m) of basalt are probably present between the Chalk scarp and the Kinnegoe Fault near Lough Neagh.

North-west of the escarpment, exposures are very limited in number and small in extent. Basalt was formerly seen at Darganstown House [127 608] and at a farm [121 612] at Tullyloob. A few small exposures occur in a railway cutting [126 618] and, 700 yd (640 m) N.W. of the railway, basalt lies under a thin coating of soil over a limited area [120 622].

North-west of Kilmore School [103 617] a small overgrown quarry is the only exposure in an area thinly smeared with drift. Small stream exposures, probably *in situ*, occur at Heanys Bridge [098 628] and about ½ mile (540 m) S.E. of Bullocks Bridge [103 634]. East of Gilberts Bridge [111 640] basalt is visible in the bottom of the Goudy River for just over 200 yd (183 m) and an area of thin drift extends eastwards along Rock Lane for about 1000 yd (914 m). Only near Rockhouse [117 642] does the basalt come to the surface where it was formerly quarried on a small scale for local building stone.

Moira district. Along the escarpment the Lower Basalts were formerly visible in the old quarries at Magheralin [128 590], but exposures are now largely covered by slipped boulder clay though up to 10 ft (3 m) of basalt can still be seen. At Small and Hayes' Quarry [135 596], ½ mile (805 m) E. of Magheralin, 6 ft (2 m) (inaccessible) was seen and this thickness has increased as more extensive quarrying has taken place. At Fairmount Nurseries [141 602] 90 ft (27 m) of basalt were proved in a borehole about 400 yd (365 m) N.W. of the basalt margin. Near Moira the basalt boundary as formerly mapped has been altered, on well evidence, to extend north of Drumbane House. Near Moira Castle [149 607] 12 ft (3·6 m) of dark grey compact basalt are exposed in a small disused quarry.

About ¾ mile (1200 m) N.W. of Soldierstown up to 18 ft (5·5 m) of massive basalt are exposed in Gormans Wood [142 635] along the western edge of Broad Water and on the eastern bank a small stream near Broadwater House cuts through over 20 ft (6 m) of rudely columnar well-jointed basalt. North-east of Soldierstown, in a disused quarry at Quarry Hill, 22 ft (6·7 m) of massive basalt, amygdaloidal in the top 4 ft (1·2 m) and patchily amygdaloidal near the base, are exposed. A new quarry was opening immediately to the west of this site in 1955 and further extensive exposures should be revealed as the work progresses. Patterson (1952, pp. 283–99) records a chemical analysis of the Quarry Hill olivine-basalt which he stated was a typical representative of the Plateau Magma type of the Mull Memoir (Bailey and others, 1924). It exhibited no unusual features other than in the minor element content, where it was the only basalt containing lead in an amount above the sensitivity threshold of 20 p.p.m It was also the most cupriferous.

About 20 ft (6 m) of basalt are seen above the chalk at Maghaberry Lime Works [169 630]. Just east of this, wartime quarries show up to 20 ft (6 m) of massive, in part columnar, basalt at the same level as the chalk at the Lime Works and confirm the presence of the Trummery Fault. These quarries have since been extended to supply material for the M1 and now exhibit two flows separated by a reddened horizon and small scale faulting sub-parallel to the Trummery Fault. A well section at Aughnagarey exhibiting 11 ft (3·4 m) of columnar basalt above amygdaloidal basalt is probably in the second flow.   P.I.M.

In the Aghalee district the lavas are well exposed in a number of crags and disused quarries on Chareltons Folly [126 661] where they are massive, fresh, medium-grained basalts, sometimes feldsparphyric. At Lake View [134 664] a large disused quarry reveals a 15-ft (4·6 m) face in this rock. A quarry north-west of Friar's Glen [128 648] has a 20-ft (6-m) face in massive coarse-grained basalt, sparsely vesicular throughout.
                                                                                H.E.W.

Maghaberry–Mullaghcarton. At Rock Hill [173 641], ½ mile (805 m) N. of Maghaberry, a recently abandoned quarry shows a maximum of 30 ft (9·1 m) of massive basalt. At four places in the floor of the quarry the weathered top of a lower flow is visible in small dome-shaped exposures (probably tumuli). The main wall appears to be cut in the third flow and may correspond with that exposed at The Rocks Farm, 1 mile (1·6 km) to the west.

Part of the first flow (12 ft; 3[7 m) is clearly seen at Balmer's Glen Quarry [182 631] where the underlying Clay-with-Flints is well exposed and near Mullaghcarton Lime Works [187 642] where up to 25 ft (7·6 m) of massive basalt with an amygdaloidal top are visible. About 10 ft (3 m) of rudely columnar basalt are seen in a small exposure near Anderson's Quarry and up to 6 ft (1·8 m) of massive columnar basalt at the roadside [200 642] just north of Mullaghcarton House.   P.I.M.

At Mullaghcarton [199 648] the scarp above the Chalk outcrop has been quarried and an 8-ft (2·4-m) face of fairly coarse-grained vesicular and rudely columnar basalt is seen which is presumably part of the lowest flow of the series.

Between Mullaghcarton and Kilcorig the lower flows of the lavas are very rarely exposed. The railway cutting south-south-west of Brookhill [205 663] is in compact fine-grained basalt. Further cutting exposures along the railway for ½ mile (800 m) to the N.W., presumably in higher flows of the series, are in fine-grained compact porphyritic basalts with sparse olivine and feldspar phenocrysts, but the exposures are too discontinuous to permit any correlation. A small quarry 850 ft (259 m) S.W. of Kilcorig House [215 673] shows a few feet of massive, slightly porphyritic fine-grained basalt.

Knocknadona–White Mountain. The lowest flows of the series are seen in the numerous quarries from Knocknadona north-east to the southern flanks of White Mountain.

The Knocknadona quarries [225 667] show that the lowest flow here is over 50 ft (15 m) thick, this maximum being seen in the eastern end of the main quarry, where the lowest 20 ft (6 m) are of massive fine-grained porphyritic basalt passing up into 30 ft (9·1 m) of rubbly and vesicular material. In the quarry south-west of Rockville [230 671] the basalt face is 40 ft (12 m) high and the whole thickness seems to be decomposed and vesicular. In the old workings north-east of Rockville the lowest flow is seen to have thinned to 24 ft (7·3 m), of which the upper 9 ft (2·7 m) are rubbly and vesicular, and to be overlain by the lowest 6 ft (1·8 m) of a second massive flow.

On the south-west flank of White Mountain [239 678], above the road, this second flow is seen to be over 30 ft (9·1 m) thick while the basal flow has now thinned to 20 ft (6·1 m) of rubbly basalt. It may be this second flow which has been extensively worked in Castlerobin Quarries, on the other side of White Mountain, where it is over 30 ft (9·1 m) thick and shows rudely columnar jointing throughout its thickness. In these quarries the rock is broken by a series of small crushes parallel to the main Magheragall Fault. The large quarries in the upper part of White Mountain appear to work a flow or flows higher in the series and in both massive rock 35 ft to 40 ft (10–12 m) thick, with some sigmoid joints, is exposed.

On the downthrow side of the Magheragall Fault the large quarries [247 680] on the west side of the main road up from Boomer's Corner worked a massive flow over 50 ft (15 m) thick. The topmost 20 ft (6·1 m) or so show signs of kaolinization and a clear horizontal banding due to weathering is seen in the face. In a cutting at the top of the worked face this is seen to be overlain by 10 ft (3 m) of purple lithomarge and 2 ft (6·6 m) of red laterite and these beds, in the floor of the upper quarry, are overlain by very massive porphyritic basalt over 30 ft (9·1 m) thick, with strong jointing.

These quarry exposures, and numerous natural small exposures in the vicinity of White Mountain, where drift is often thin or absent, suggest that the lowest 200 ft (61 m) or so of the lava series hereabouts consist of a number of flows over 30 ft (9·1 m) thick interbedded with somewhat thinner lavas. It is particularly noticeable that in every flow about which information is available the upper part is deeply kaolinized and lateritized, often for 10 ft (3 m) or more. The basalts are generally of fairly fine grain, often porphyritic, and only a few exposures are of coarse basalt.

Portmore–Stonyford. In this drift-covered area exposures are few and no correlation of flows is possible. A pipe trench at Hallstown [209 678] revealed deeply weathered coarse-grained basalt and there are a few small exposures of decomposed porphyritic basalt to the north-east. Scattered exposures of medium-grained basalt, usually vesicular, occur south and east of Ballycarrickmaddy School [203 762], at Rock Mount [221 680], and near Coyles Bridge [220 687]. A flooded quarry at the south side of Stonyford Reservoir shows 15 ft (4·6 m) of massive medium-grained basalt above water-level.

The area of higher ground culminating in Crew Hill [180 702] and Cairn Hill [193 715] may be due to the occurrence of a particularly massive coarse-grained lava here. Immediately below this coarse flow occurs the very fine-grained conchoidal-fracturing mugearite flow which can be traced for some 4 miles (6·4 km) trending in a north-easterly direction.

The most northerly exposure of the mugearite is in the yard of Moatstown, [203 725] where very fine-grained flaggy-weathering mugearite is seen at the surface.

An abandoned quarry in Ballynacoy townland [201 719], ¼ mile (400 km) S.W. of the Y-Bridge, shows about 25 ft (7·6 m) of mugearite which is most irregularly jointed. It has a well-marked grain or flow-banding indicating a dip to the north-west of 10°–20°. The base is not seen but the top of the flow is indicated by an occurrence of red laterite at the top of the face, near the road, and overlying fine-grained basalt is seen in the road cutting.

In the townland of Ballypitmave a large quarry [189 710] has been worked and abandoned in the mugearite. Here an irregular thick band of purple lithomarge at the top of the face overlies 20 ft (6·1 m) of fine-grained mugearite. The full thickness is not seen as the quarry is flooded, but if the dip is as much as 20° to the N.W., as the flow-banding of the mugearite suggests, the full thickness of the flow is over 50 ft (15 m) here. The mugearite is exposed in a small quarry on the east flank of Crew Hill [183 703] where it is overlain by flaggy vesicular basalt and it is visible in the lane outside the house at Ivyhill [180 695].

The most southerly exposure is in a small quarry near the stream 200 yd (182 m) N.W. of Crooked Bridge [173 674] and in the stream nearby where faint flow-banding is seen in the rock.

The higher flow of coarse basalt is seen south-west of Y Bridge, on Cairn Hill and on Crew Hill, and what may be the same flow is seen in the river east of Templecormack [173 681].

West of these outcrops exposures of the basalts are somewhat more plentiful than to the east but they are still mainly so small and limited as to give little information about the lavas. In general, fine-and medium grained basalts predominate and only a few exposures show coarser material.

There are numerous exposures in the upper reaches of the Ballinderry River north of McPherson's Bridge [167 668], Upper Ballinderry, mainly of coarse and medium-grained basalt. One of the most extensive natural exposures is in the lower reaches of the river where rock is seen more or less continuously for over ½ mile (800 m) from Donnellys-town [136 681] to the edge of the alluvial flat round Portmore Lough. The flows have a gentle westerly dip, slightly greater than the inclination of the river bed, and about twelve flows can be distinguished by the occurrence of thin red laterites. Most of the lavas are thin, 10 ft to 15 ft (3–4·6 m) being usual, but conditions make accurate esti-mates of thickness very difficult. Most of the basalt is of medium or rather coarse grain, but a thick finer-grained flow is seen just down-stream from Drumart Bridge [132 680]. Amygdaloidal texture is common, many of the thinner flows being vesicular throughout, while the more massive lavas have kaolinized and vesicular tops.

In the area north of the Ballinderry River, rock is visible fairly frequently as knolls and in streams. As a rule only fairly compact basalt is seen in these exposures and much of it is medium or fine in grain.

Castle Robin–Collin Glen. About 200 yd (183 m) S.E. of Castle Robin two small quarries [250 685] worked a massive flow, probably the second of the series, and show 10 ft (3 m) of purple lithomarge on 35 ft (11 m) of massive fine-grained close-jointed basalt. The reddened lithomarge top of what is probably the basal flow was seen in a ditch between the quarries.

Small quarries on the scarp below the road north-east of Quinn's Bridge [251 689] expose the lowest flow here which is a coarse-grained basalt.

The quarry [245 689] north of Castle Robin, affected by small faults, is in the second and subsequent flows of the series, the basal flow, which appears in occasional small exposures, being nowhere worked. In a cutting between two of the fault-separated blocks of this quarry what is probably the fourth flow is seen to be over 25 ft (7·6 m) thick, massive and compact and to overlie a 22-ft (6·7 m) flow of which the upper 6 ft (1·8 m) is decomposed and vesicular. This in turn rests on a flow over 35 ft (11 m) thick, base unseen, of which the uppermost 15 ft (4·6 m) are kaolinized and vesicular.

In the small quarry south-east of Boomer's Hill [249 697] the purple vesicular top of what is probably the first flow is visible in the bottom of the working while the second flow is 37 ft (11 m) thick, the topmost 12 ft (3·7 m) being kaolinized and vesicular. Up to 15 ft (4·6 m) of massive basalt, the lower part of the third flow, is seen at the top of the face. H.E.W.

On Boomer's Hill there are three features suggesting the presence of three flows and these can be followed north-eastwards on to the high ground of Aughrim. The basalt here is fresh and medium- to slightly coarse-grained and the highest flow, which forms the top of Aughrim, above the Plantation, is variable in composition (p. 99). A number of small disused quarries, and one large working quarry, south-east of Aughrim Plantation, are in a massive flow near the base of the series. In the working face, about 60 ft (18 m) high, this flow is seen to be compact and medium-grained with signs of weathering to lithomarge in the uppermost 16 ft (5 m) or so.

On the lower eastern slopes of Collin [261 705] the lowest three flows are exposed in a few quarries east and north-east of Groganstown. The lowest flow rests on a bed of flints 1 ft to 3 ft (0·3–0·9 m) thick which are reddened and crushed at the top and in turn this Clay-with-Flints rests on the Chalk. This flow is about 100 ft (30 m) thick comprising 30 ft (9 m) of amygdaloidal slaggy basalt at the bottom, a massive central portion 40 ft (12 m) thick overlain by about 30 ft (9 m) of kaolinized and lithomarged top. The second flow is about 40 ft (12 m) thick and is not so well exposed but there are three extensive quarries in the third flow which is about 80 ft (24 m) thick with a massive central part about 60 ft (18 m) thick. This massive basalt shows poorly developed columnar structure. Continuing up the hill, exposures are rare for the next 200 ft (60 m) but towards the top the scarps of three flows can be traced part way round the hill. These flows total about 70 ft (21 m) in thickness and are composed of medium- to coarse-grained basalt with few amygdales.

A large quarry [265 712] to the north-north-east of Collin exposes about 80 ft (24 m) of rather coarse basalt showing faint columnar structure resting on 20 ft (6 m) of slaggy reddened, highly amygdaloidal basalt. From the general appearance of the rock and the thickness of massive basalt this would appear to be the third flow and if so then the combined thickness of the first and second flows has decreased from at least 140 ft (43 m) at Groganstown to less than 100 ft (30 m) or it is possible that the second flow is missing. About 440 yd (400 m) to the north, along the strike the massive flow has been extensively quarried but here the basalt does not appear to be so fresh and banding is present in places. The largest quarry in the district, in what is considered to be the third flow, but which may be the second in this area, lies across the Glen River [265 728] about 1200 yd (1097 m) N.W. of Glen Bridge. The rock is a fresh, medium to slightly coarse olivine-basalt, porphyritic in places and strongly jointed. The base of the flow is seen at the bottom of the working face which is 100 ft (30 m) high, and the total thickness of the flow appears to be about 120 ft (37 m).

Hannahstown–Divis. The best exposure in this district occurs in the extensive Black Mountain Quarry [280 724]. The working face is up to 80 ft (24 m) high and appears to be in one flow which, from evidence farther north along the escarpment, may be the fourth flow. The rock is a fairly fresh medium-grained basalt with strong joints trending east-north-east and a horizontal banding is apparent in the non-amygdaloidal parts. Just below the floor of the main quarry, towards the top of the tramway, up to 12 ft (3·7 m) are exposed of decomposed red amygdaloidal basalt. Some of the amygdales are large and crystals of analcime up to ¾ in (19 mm) across were found.

North-eastwards along the escarpment to Black Hill the basalt is on the whole not very fresh but the rock (fine- to coarse-grained basalt) exposed in Allens Ravine [289 735] is usually fresh.

A general picture of the lava sequence can be obtained from disconnected exposures in the escarpment [293 744] above Whiterock Road.

	ft	m
Alternations of coarse and rubbly and fine and dense basalt in bands from 1 in (25 mm) to about 1 ft (305 mm) thick (Pahoehoe-type basalt)	100	(30)
Basalt, red, decomposed with large amygdales in places .	5–15	(1·5–4·6)
Basalt, massive	5– 8	(1·5–2·4)
Basalt, decomposed, much of it red and some of it amygdaloidal	5–15	(1·5–4·6)
Obscured	30	(9·1)
Basalt, some amygdales, decomposing, weathers spheroidally	8	(2·4)
Basalt, massive with some irregular joints	6	(1·8)
Basalt, decomposing, weathers spheroidally	12	(3·7)
Basalt, amygdaloidal with flow structure at the top and elongated amygdales at bottom	15	(4·6)
Basalt, massive with vertical joints	10	(3·0)
Clay-with-Flints, burned	0½	(0·15)
Obscured	4	(1·2)

This exposure is notable for the fine development of the unusual Pahoehoe-type of flow, which is well displayed and easy of access.

Away from the escarpment there are many small isolated exposures of basalt. Basalt was also exposed in several places during the construction of the road leading to the B.B.C. television station [287 750].

More important exposures occur on and near the summit of Divis Mountain [281 754]. On the 1901 edition of Sheet 36 a restricted outcrop of the Interbasaltic Bed was shown with a capping of Upper Basalts and this interpretation was repeated on the colour-printed Drift Series map of Belfast, published in 1904. The Belfast Memoir (Lamplugh and others, 1904) makes only a brief reference (p. 39) to the Interbasaltic Bed at Divis. The evidence from the present-day exposures indicates that the bed merely represents an interflow pause although there are no exposures of the clay material to verify this. This weathered top of a flow makes a discontinuous platform round the summit of Divis. Close to the fence, 110 yd (101 m) E.N.E. of the triangulation station, this platform is free of peat and there is a fine debris of decomposed reddish, medium-grained amygdaloidal basalt. At the time of the original survey (about 1868) 'clayey debris with scattered fragments of decomposing ferruginous basalt, sometimes approaching a lithomarge' was noted. According to a note on the map 'on digging into this, quantities of lithomarge, with occasional pieces of hematite (brown) and decomposing basalt are found'. There seems to be no justification for regarding this material as representing the Interbasaltic Bed especially as a mile to the north (just beyond the northern limit of the Sheet) a few feet of similar material exposed on the eastern side of the summit of Wolf Hill was not regarded as Interbasaltic Bed. The lava overlying the decomposed basalt and forming the summit of Divis is a fresh, dark grey medium-grained porphyritic basalt.   J.A.R.

Tornaroy–Dundrod. The high ground behind the basalt escarpment is drift-free over wide areas on Priests Hill [247 715], White Hill [254 724], Rushy Hill [244 725], Standing Stones Hill [252 743], Armstrong Hill [265 758] and the high ground round Budore. Exposures, however, are usually very poor and only weathered knobs and floors of basalt can be seen, except in the few quarry sections.

In a disused quarry [254 713] 880 yd (804 m) E. of Priests Hill, up to about 12 ft (3·7 m) of medium-grained non-amygdaloidal basalt are visible and a similar rock is exposed in a quarry [248 721] 650 yd (600 m) N. of Priests Hill. About 20 ft (6·1 m) of basalt are visible and the rock is traversed by joints about 3 yd to 4 yd (3–4 m) apart. These joints are approximately vertical but are sinuous or sigmoid and in plan trend north-westerly.

In the drift-free district at Tornagrough the basalt is generally medium-grained and fairly fresh. In the floor of a disused quarry [251 734] the red amygdaloidal bottom part of a flow is exposed whilst near the top the basalt occurs in 3-in to 6-in (75–150 mm) bands. Some of these bands are amygdaloidal and some are not which would suggest pause, however slight, between the ejection of each band. In another quarry 300 yd (274 m) to the north the basalt is characterized by sporadic amygdales measuring up to 1 in (25 mm) across.

Exposures in the streams north and south of Galway's Hill [230 703] show reddened lithomarge which in both cases suggest that the dip is to the west, and the southern exposures also show over 12 ft (3·7 m) of flaggy-weathering massive medium-grained basalt.

Further north the extensive exposures at The Rock [226 717] are all in very massive fine-grained basalt, up to 12 ft (3·7 m) of which is seen in a quarry at Rock farm. One exposure, 450 yd (411 m) E.S.E. of St. Peter's Church, is of feldsparphyric basalt with phenocrysts up to ½ in (12 mm) long in a fine-grained matrix.

In the north face of a quarry on the south side of McGowans Hill [235 737] medium to slightly coarse, finely amygdaloidal basalt shows undulating flow banding. A quarry on the north side of the hill exposes about 25 ft (7·6 m) of brownish medium -to coarse-grained basalt. There are some scattered amygdales and many nearly vertical sigmoid joints. On the south face of another quarry, 200 yd (182 m) N.N.W. of the one just described the arrangement of the joints gives the appearance of a magmatic roll.

There are numerous small crags and quarries on Bo Hill, the largest quarry [222 736], 300 yd (274 m) W. of Rock View, showing a 14-ft (4·3 m) face in massive fine-grained basalt with remarkable curved joints.

The area round Dundrod, in the northern part of the Sheet, is heavily drift-covered and small exposures in streams and occasional knolls average about one per square mile (2·6 km^2). One of the more extensive is in the Crumlin River at Adjeystown [217 763] east of Dundrod Bridge, where fresh, sparsely porphyritic, fine-grained basalt shows occasional flow banding.

On the northern slopes of Budore Hill [241 752] there are several exposures of slightly coarse porphyritic basalt but no large quarry has been opened in this rock. On the high ground almost 1 mile (1·6 km) N.E. of Budore Bridge [248 760] there are several small exposures of decomposed basalt which weathers slabby. The rock, which dips north-west at a low angle, is scoriaceous in places. Just above this and forming a slight feature is a band of lithomarge. Capping this and forming a small knoll is medium-grained basalt.

Langford Lodge Borehole. Apart from one short core, the whole thickness of the Lower Basalts was drilled with a rock-bit. Therefore the only record available of the succession was derived from inspection of the cuttings collected at five-foot intervals from the drilling mud, and at the considerable depth involved this method can lead to inaccuracies. By recording the occurrence in the cuttings of red laterite from the weathered tops of lava flows it is possbile to get a general impression of the frequency with which these laterite beds occur in the 1743-ft (531 m) succession. It seems that the topmost four or five flows are thin, averaging about 20 ft (6 m), but for the next 900 ft (274 m) the lavas average over 40 ft (12 m) thick with exceptional flows reaching 80 ft (24 m) or more.

Below 1900 ft (579 m) it becomes almost impossible to distinguish separate flows, perhaps because of the increasing depth and increasing degree of mixing of cuttings during their long ascent up the mud column. It does seem, however, that between 1900 ft and 2000 ft (579 and 609 m) and from 2200 ft to 2500 ft (671–762 m) the flows are very thin. In the lowest 100 ft (30 m) or so, however, separate flows seem to be distinguishable and to be 40 ft (12 m) or more thick.

The only section cored was from 2015 ft to 2025 ft (614–617 m) and was all of highly kaolinized basalt with many zeolite-filled vesicles. Reddish brown at the top, it changed downward into a green-streaked brown rock and was probably approaching the more compact base of a flow. J.A.R., H.E.W.

UPPER BASALTS

Fairly extensive exposures of compact medium-grained basalt occur on the lake platform due south of Langford Lodge [090 748] but the total visible thickness is probably not more than a few feet. A mile to the north-east a disused and partly filled quarry behind Gartree House [106 757] reveals a few feet of similar compact medium-grained basalt.

The westernmost exposures in the Crumlin River, due south of Mourne View [128 765] are of decomposed vesicular fine-grained basalt, in places showing spheroidal weathering. Vesicles are often full of green chlorophaeite. A couple of hundred yards upstream a 20-ft (6·1 m) cliff on the north bank is entirely made up of lithomarge with residual spheroids of basalt. This lithomarge can be seen to dip downstream, and to pass laterally into decomposed but unlithomarged basalt.

There are small exposures in the bed of the river near Cidercourt Bridge [135 765] but the next extensive exposures are north of Glenfield [141 765] where 30 ft (9·1 m) of basalt are seen in the north bank. Here a thin lateritic bed is visible about 6 ft (2 m) above the water, dipping gently downstream. The underlying flow top and the overlying flow, at least 25 ft (7·6 m) thick, are both broken and kaolinized. West of Glendaragh [148 768], further exposures of kaolinized basalt are seen in the north bank though more compact medium-grained basalt occurred in the stream bed between here and the last exposure.

The extensive exposures in the turbulent section of the river at Crumlin Saw Mills are all in the same flow which dips gently downstream. Decomposed rubbly amygdaloidal basalt seen in the steep banks passes down into massive medium-grained basalt, very irregularly jointed.

A railway cutting ¼ mile (400 m) S. of Crumlin Railway Station shows a few small exposures of decomposed amygdaloidal basalt, but apart from this there are no exposures of the lavas between the Crumlin and Glenavy Rivers.

In the latter, at The Leap [136 726], there are extensive exposures of massive coarse feldsparphyric basalt at least 25 ft (7·6 m) thick, with strong vertical jointing giving a rudely columnar appearance. Just west of the Mill Pond this is cut off by a fault and upstream there are exposures of massive fine-grained basalt, sometimes with flaggy jointing. This jointing, inclined W. at 15°, probably indicates the dip of the flow which may therefore be 70 ft (21 m) or more in thickness. Vesicular basalt seen at the weir probably represents the base of this flow.

Upstream from Leap Bridge [139 725] there are a few small exposures, generally of fresh, medium and fine-grained basalt. Compact fine-grained basalt is also seen in small quarries 300 yd (275 m) E. of Mount Pleasant [139 719], south of the river.

The southernmost exposures seen of the Upper Basalt are in Ballyvanen townland where small quarries north-east and south-west of Posey Hill [128 711] show a few feet of feldsparphyric fine-grained basalt.

It is noteworthy that the exposures in the Glenavy River and southwards are generally of fresh compact rock, while those in the Crumlin River are mostly of decomposed and lithomarged basalt.

Langford Lodge Borehole. Examination of the cuttings from the upper part of this borehole [090 748] suggests that there are about 26 flows in the 762 ft (232 m) ascribed to the Upper Basalts. The average thickness of almost 30 ft (9·1 m) would be increased but for the occurrence in the upper 120 ft (37 m) of no less than 7 flows with an average thickness of only 17 ft (5·2 m). It may well be significant, however, that this concentration of thin flows is recognized where the hole was so shallow that the cuttings were not mixed during their ascent through the mud column, and it may be that the number of flows identified at lower depths is too small.

The thickest flows recognized measure about 65 ft (20 m) and have, apparently, lithomarged tops which are 25 ft (7·6 m) or so thick. Most of the chippings are of fine or medium-grained basalt and only one flow of even moderately coarse-grained basalt occurs at a depth of about 565 ft (172 m).

H.E.W.

PETROGRAPHY

The specimens have been examined by J. R. Hawkes, P. A. Sabine and H. E. Wilson

Olivine-Basalts. Lavas of this type are rarely seen in continuous section within the boundaries of the Sheet, but by analogy with other areas they are assumed to be laterally restricted flows, often of no great thickness, though they may reach 100 ft (30 m) or so. Many of the thinner flows are vesicular throughout but in the more massive lavas amygdales are confined to a foot or two at the base and an upper zone which is often many feet thick. Rocks selected for sectioning were generally taken from the freshest and most compact part of the lavas, and no attempt has been made to investigate vertical or lateral variation in individual flows.

The Lower Basalts are normally holocrystalline olivine-basalt, consisting essentially of labradorite, augite, olivine, and iron ore with accessory apatite, and analcime. Though few are macroscopically porphyritic the majority are microprophyritic with feldspar and olivine phenocrysts, the latter being ubiquitous. Both feldspar and olivine occur commonly in two generations. Augite is almost always ophitic to feldspar and olivine of the first generation but occasionally occurs in sub-ophitic or granular form. In rare instances it occurs as phenocrysts.

There is occasionally some interstitial chloritic and zeolitic material in the groundmass of these rocks between the idiomorphic or sub-idiomorphic plates of ophitic augite. A little glassy mesostasis is also present in rare instances.

The granularity of the rocks examined in thin section varies considerably, but the variation is less than might be expected from macroscopic appearance, the coarser textures commonly being due to micro-phenocrysts rather than to coarse-grained groundmass. The average linear intercepts of minerals in the groundmass varies from 0·23 mm in the case of an exceptionally coarse rock from the Ballinderry River, which may be intrusive, to 0·04 mm, but are mainly within the range 0·07 mm to 0·1 mm. The finest grained rocks are notable for the occurrence of well-marked fluxion texture, scarce olivine and augite and much iron ore. (NI 1297, 1236, 1330).

The most abundant constituent is plagioclase feldspar which forms about half the rock by volume, the actual percentage ranging from 41 to 69 and averaging 54. The laths of the groundmass are generally from 0·1 to 0·3 mm long but exceptionally range up to 1 mm or larger (NI 1325). Maximum symmetrical extinction angles, generally 30° to 35°, indicate that most of the groundmass feldspar is labradorite (An 55-60). Most of the feldspar is fresh in unweathered rocks but a few specimens have saussuritized and kaolinized laths and, rarely, marginal corrosion of feldspar is seen.

Though the laths are usually random in orientation there is occasionally a suggestion of flow-structure and several specimens show strong fluxion texture (NI 1297, 1326, 1330). In all, no less than 25 per cent of the specimens examined show some feldspar alignment.

Feldspar phenocrysts are present in about a quarter of the specimens examined. They are mainly labradorite in composition but bytownite occurs in the core of a zoned phenocryst in one specimen (NI 1323). The phenocrysts are usually lath-shaped, up to 5 mm long but generally smaller, and albite-twinned though some simple twinning is seen, and one specimen has a cruciform twin (NI 1298). Glomeroporphyritic aggregates (NI 1321) and stellate aggregates (NI 1294) occur.

The pyroxene is a pale brown or colourless titaniferous augite which forms from 12 to 38 per cent of the rock volume. It is almost always ophitic or subophitic to feldspar and olivine, forming plates which are sometimes of considerable size but usually less than 0·5 mm across. Some of these plates are so crowded with feldspar laths as to be indistinguishable in ordinary light (NI 1294–6). Rare specimens show the pyroxene to be granular and non-ophitic (NI 1321, 2039–40); granular and ophitic augite may also occur in the same rock (NI 1314). In two slides augite is seen to occur as phenocrysts (NI 1316–7).

Olivine, frequently partly or wholly pseudomorphed by serpentine, chlorite, or 'iddingsite', is ubiquitous and though in one specimen it made up only 4 per cent of the rock it usually ranges fron 10 to 20 per cent of volume. It occurs as small grains in virtually every specimen examined and also as phenocrysts of earlier generation in over half the rocks. The latter are generally 1 mm in diameter or less but some large phenocrysts up to 4 mm or more have been noted. Most of the phenocrysts have good idiomorphic form while the smaller grains are usually rounded. Extremely abundant porphyritic olivine makes up 25 to 30 per cent of the rock by volume in two specimens (NI 1310, 1312) both from flows very low in the lava succession—the lowest flow in the case of NI 1310—and these chrysophyric basalts approach picrite in composition. In some cases the phenocrysts show a glomeroporphyritic tendency. Picrite lavas are known from low in the lava succession in the Dungannon area (Fowler and Robbie 1961, p. 117)

Though in many instances wholly or partly replaced by secondary minerals some of the olivine is remarkably fresh. For example, a rock from Tullyrusk

(NI 1317) shows phenocrysts of clear and pale brown-tinted olivine which show only slight alteration on some cracks, as well as an abundance of fresh granular olivine.

Iron ore is present in all sections examined, averaging about 5 per cent of the rock by volume though in one case it is over 10 per cent (NI 1330). It generally occurs in very small grains but large sub-idiomorphic grains up to 1 mm are present in some specimens. Skeletal grains are common.

Though zeolite-filled vesicles are common in most flows few are seen in the thin sections. In many slides, however, zeolites are seen to occur as interstitial patches representing the last products of crystallization. In some cases, notably the coarse basalt from the Ballinderry River (NI 1325), it may make up from 5 to 10 per cent of the rock by volume. Much of the zeolitic material is chabazite. Fibrous aggregates (NI 1313, 2019) are probably natrolite. Levyne and thomsonite are also recognised (NI 1496, X-ray film X3176).

Chlorite, pale green non-pleochroic, is sometimes abundant as a mesostasis and is present in some quantity in most specimens. It occasionally partially replaces feldspar and pyroxene.

Calcite occurs rarely as a mesostasis (NI 1314) and a small amount of interstitial glass is occasionally seen. Acicular apatite is a rare accessory.

There appears to be no significant difference between the olivine-basalts of the Upper and Lower Basalts. Generalizations made in the above account apply with equal significance to both groups. On the limited number of grain-size measurements and modal analyses carried out there is a tendency for the Upper Basalt group to be slightly finer in grain and to have a smaller proportion of olivine, but these differences are too small to be statistically significant.

Tholeiitic Basalts. Though known from a group of lavas above the Lower Basalts in North Antrim, tholeiitic basalts have not hitherto been recorded from the predominantly olivine-basalt series. Two specimens from the lower part of the succession west of Belfast have now been found to be tholeiitic.

A specimen from the Bobby Stone [289 749] (NI 2018) consists of labradorite laths and scattered grains of olivine in a fine-grained matrix of labradorite, pale green pyroxene (2V positive, 30°– 40°, sub-calcic augite/pigeonite), magnetite, fibrous zeolite (levyne) and occasional patches of interstitial glass. Pyroxene crystals are concentrated round the margins of amygdales.

Four miles (6·4 km) to the south-west a specimen (NI 2046) from the northwest corner of Aughrim Plantation [250 698] consists of clots (up to 3 mm diameter) of plagioclase, pyroxene, magnetite, brown glass, and some zeolite, set in a fine-grained matrix of the same constituents. Two pyroxenes are present: hypersthene (2V negative, large) and pigeonite (2V positive, 20°–30°), and a few crystals with 2V large negative and inclined extinction may be hypersthene in a state of inversion to pigeonite. The flow from which this specimen came is very variable in appearance over short distances, and is mainly olivine–basalt. It is probable that the tholeiitic material is the result of segregative processes during the cooling of the lava.

Mugearite. This flow, or flows, has been collected from five exposures at widely separated points. With minor variations all the specimens are of a very fine-grained and fluxion-textured microporphyritic rock consisting essentially of oligoclase laths and granular pyroxene and iron ore.

The feldspar laths (oligoclase) average 0·02 m to 0·03 m long and show remarkable parallelism. This parallelism is not always aligned with the flow-banding; islands of well-aligned laths often show a direction at right angles to the flow-banding round them.

Some well-defined patches of feldspathic material between the tightly-packed laths of oligoclase are probably the interstitial alkali-feldspar referred to by Walker (1960, p. 62).

The feldspar phenocrysts—andesine-labradorite—are scarce but occur in all the specimens examined except one from near Ballinderry (NI 1328). They are up to 0·5 mm long and coincide with the parallelism of the groundmass feldspars.

Granular colourless or very pale brown augite is sparsely present and occasionally forms sub-ophitic clusters. Where most abundant (NI 1328) it makes up less than 10 per cent of the rock. Granular magnetite, in very small grains, is abundant. Olivine occurs rarely as small (0·5 mm) phenocrysts and small granules. It is usually completely pseudomorphed by greenish chlorite or red 'iddingsite', but can be fairly fresh (NI 1322). Rare needles of apatite are the only other visible accessory mineral but the rocks also contain some chlorite material dispersed among the feldspar laths.

Texturally, the specimen from Moatstown (NI 1318) differs from the rest in the occurrences of bands of extremely fine-grained material among the fluxion textured feldspar laths. Very finely divided magnetite and plagioclase are predominant in these areas, with some phenocrysts of altered olivine occasionally associated with poorly formed crystals of plagioclase (0·1 mm).

This rock differs from Harker's type mugearite from Skye in the absence of hornblende and biotite. The chemical analysis given by Walker (1960, p. 64) indicates that the Antrim mugearite lies in a field roughly midway between those occupied by the British Tertiary and the Skye mugearites.

Zeolites. Gas cavities in the lavas usually contain minerals of this group, in some instances complemented by the occurrence of calcite and montmorillonoids. Some of these crystals are of considerable beauty and the quarries on the hills above Belfast are an excellent source of good specimens.

The following varieties have been recognized as a result of X-ray powder photography by Mr. B. A. Young: apophyllite, chabazite, gyrolite, garronite, heulandite, levyne, mesolite, natrolite, stilbite.

The distribution of amygdale minerals in the Antrim lavas has been described by Walker (1960) and the only addition to his account is the record of garronite, which occurs in radiating aggregates round a core of gyrolite in a dyke in Black Mountain Quarry.   H.E.W.

Chapter 8

LOUGH NEAGH CLAYS

GENERAL ACCOUNT

THOUGH AN AREA of about 15 square miles (39 km²) on the eastern shore of Lough Neagh is underlain by this formation it is exceedingly poorly exposed at the surface and our scanty knowledge of the succession comes from occasional wells and bores, and the accounts of previous observers.

The area underlain by Lough Neagh Clays is flat and low-lying, most of it being only a few feet above the level of the lough, and it is covered by thin lacustrine and fluviatile deposits and peat through which 'islands' of boulder clay rise. Only in the area west of Crumlin and Glenavy does the surface of the formation rise to much above the lough level. At the Plantation it appears to reach a height of about 100 ft (30 m) O.D. or about 50 ft (15 m) above the lough, while near Bessfield, 1¾ miles (2·8 km) S.W. of Glenavy it reaches the same altitude.

The occurrence of white and grey clays, lignites, and silicified wood in this area was known locally from an early date, the latter being famous as material for hones, but the first scientific account of the formation was given by Barton (1757) who lived near the lake and was an assiduous collector of fossil wood. His account of the occurrence of the silicified material in the lignite is so clear and factual that subsequent speculation on it is difficult to understand. Griffith, in his 'Geological and Mining Surveys of the coal districts of Tyrone and Antrim' (1829), recognised the Lough Neagh Clays as Tertiary deposits and correlated them with the Bovey Tracey Beds in Devonshire. In 1837 Scouler reviewed the earlier writings on the Lough and its alleged petrifying qualities and described the 'beds of clay and bituminous wood' which surround the southern end of the Lough, comparing them to the 'beds of clay, lignite, and volcanic rocks which are found in the basin of the Rhine'. Scouler considered that the silicified wood occurred only on the east side of the Lough and suggested that it was in some way connected with the 'trap formation' or basalt. He claimed that it was not 'co-extensive with the deposits of lignite', noting in particular that it had not been found in any of the interbasaltic lignites of Antrim.

Portlock (1843) reviewed the previous authors' work and suggested that Scouler's conclusions on the distribution of the silicified wood were doubtful. He considered that the Lough Neagh Clays were of lacustrine origin but seems (p. 165) to have considered them to be of fairly recent origin. In a footnote (p. 82) he quotes the journal of Bishop Nicholson who, on August 9, 1718, visited Antrim where he saw Lough Neagh—"Here are ye famous hones for razors".

The subject of the Lough Neagh Clays was discussed by E. T. Hardman of the Geological Survey in a series of papers written after he had surveyed the Dungannon (35) Sheet. In 1875 he described sections and clay pits seen by him near Sandy Bay, and discussed fully the geological relationships of the formation, concluding that the clays are of Pliocene age. He casts doubts on Barton's accounts of the occurrence of silicified wood in the lignite and concluded that it must have been derived from the basalt country to the north of the lough and have been formed in the occasional thin lignites which occur between flows of lava. In an appendix to the paper he describes the occurrence in the Crumlin River of fossiliferous Pliocene clays overlying basalt, and a note on this exposure was published in the Geological Magazine (1876) and is also added as an appendix to the 1877 Memoir on Sheet 35 of the one-inch map.

Swanston (1879) re-examined the Crumlin River and differed vigorously from Hardman on the interpretation of the section, denying the existence of Pliocene clays at this locality and ascribing the deposits to a recent date. He doubted Hardman's conclusions on the Pliocene Clays as a whole and also disagreed with his conclusions on the silicified wood.

Hardman (1879) replied to this attack and while apparently not very confident of his fossiliferous clays at Crumlin defended vigorously his thesis on the age of the main deposit.

Gardner (1885) considered the clays to be of Interbasaltic age, and from the plants obtained in ironstone nodules described them as Eocene. Like Swanston, with whom he visited the deposits, he had no doubt that the silicified wood occurred in the lignites.

Geikie (1897, 2, pp. 448–52) briefly reviewed the subject and dismissed the idea of a Pliocene age, suggesting that the clays were deposited in a depression during Interbasaltic times. He apparently based this conclusion on the observation by Clement Reid that the clays were overlain by drift, though this fact was known to Hardman and Swanston, and properly disregarded by them.

Cole and Seymour (*in* Cole 1912, pp. 99–101 and 121–6) gave some useful information about contemporary exploratory work in the area. Cole, in his summing-up, stated that the clays are post-Basaltic though their precise age was uncertain, but was categorical that the silicified wood and the ironstone nodules came from some inter-lava horizon in the Antrim Basalts. These authors were the first to use the term Lough Neagh Clays to describe the formation.

Wright (1924) gave a detailed account of the whole Lough Neagh Clays succession as revealed by the Washing Bay borehole in County Tyrone, though the area in Sheet 36 is only mentioned indirectly and briefly in the subsequent discussion. Wright's contention, based on the evidence of a single analysis, that the Clays were largely derived from the weathering of the Tertiary basalts was demolished by Fowler and Robbie (1960, pp. 127–30) who, in a comprehensive review of all the evidence, give the most complete account of the age and origin of the formation.

Robbie (p. 127), in reviewing the age of the series, pointed out that recent boreholes show that the beds are older than the faulting in the Lough Neagh area which has been thought to be of early Miocene age, but the occurrence in the clays of *Sphagnum* is regarded by botanical authority as evidence of later age, as this species is not found in European deposits older than the Upper Pliocene. This is in agreement with Simpson's view (Eyles 1952, p. 4) that the Upper Interbasaltic Bed of North Antrim is Late Miocene or Early Pliocene in age.

Watts (1962, p. 600; 1963, pp. 117–8), on the evidence of pollen in the lignites and lignitic clays, has concluded that the clays are of late Eocene or early Oligocene age, though he remarks that the precision of this method leaves something to be desired until more work has been done on Tertiary pollen in north-west Europe.

Fowler (p. 128) pointed out that Wright's specimen of Lough Neagh Clays, which on analysis contained 10 per cent titanium dioxide, is anomalous and that the source of the detritus which formed this deposit was not largely lithomarge but was probably the whole catchment of the basin and that the contributing formations included the Carboniferous rocks of Tyrone, the Newry and Mourne granites, the Cretaceous, and the basalts and rhyolites of Antrim.

The little recent information on the formation in this area indicates that the beds are mainly grey clays, some sandy. Sandy bands, sometimes with siderite concretions, are much less important. Particularly in the area around Sandy Bay, all the indications are that there are thick beds of lignite and lignitic clay in the succession and it is to this district that most of the accounts of fossil wood refer.

An account of recent investigations of the lignite deposits in this area is given in Chapter 13 (p. 166).

The only information about the thickness of the formation on the east side of the Lough is in the account of an unlocated borehole 250 ft or so deep, over a century ago, which failed to penetrate them. They are probably thickest in the area west of Portmore Lough, and thin to the north where, between Portmore Lough and the Crumlin River, they are deposited against the basalt spur of Darachrean.

Silicified Wood. As shown in the notes on the history of research on the Lough Neagh Clays the origin of the fossilized wood has been the subject of prolonged controversy. Barton (1757), Scouler (1837) and Gardner (1885) give factual accounts of its occurrence and recovery but Hardman (1875) and Cole (1912) insist that there is no evidence that it has been formed *in situ* and regard it as derived from interbasaltic horizons in the Antrim Lavas.

Barton's original observation referred to an area near the mouth of the Glenavy River. He describes the section found by the digging at the foot of the boulder-clay cliff as follows (p. 97)—"The upper stratum of matter is red clay, three feet deep; the second stratum is stiff blue clay, four feet deep; the third stratum is a black wood lying in flakes, four feet deep; the next stratum is clay . . . . . this stratum of wood . . . . is one uniform mass capable of being cut any way with a spade.

". . . . . It has evident marks of violent collision or pressure . . . . where the fibres of some are squeezed up into a complicated cluster, as if a man should squeeze end ways in a vice a short piece of hempen rope . . . .

". . . . . Sometimes this wood will not easily break, and in that case, requires the aid of some other tool to separate it from the mass which, if carefully done, may afford a block of two, three, or four hundred pounds; which when examined, are found to consist more or less of stone."

Gardner quotes Swanston and Stewart who in 1884 visited the area and talked to a farmer who was working the lignite. ". . . . on asking the farmer whether he had found any petrified wood in it he told us that he had carted some of the lignite to his house for burning, and that the heart of the largest piece turned out to be stone, which he kept . . . . It was a piece of veritably silicified wood."

No new information on the occurrence of the silicified wood has been obtained during the resurvey of the area and the material is now very scarce, but the occupation of the Langford Lodge estate by the Air Ministry during the 1939–45 War has revealed several large blocks of it used as garden ornaments. One about 2 feet (0·6 m) in diameter and 2 ft (0·6 m) high with a polished top is now in the Department of Geology, Queen's University, Belfast and a massive piece, 6ft (1·8 m) or more in length by up to 4 ft (1·2 m) in diameter and weighing well over half a ton, (500 kg) is now in the Ulster Museum. This piece is mentioned by Scouler (1837, p. 240) and Kelly (1868, p. 323). It is inconceivable that masses of this size could be found, as Hardman and Cole allege, in the boulder clay or river gravels, transported from some unknown source in the Antrim Lavas, and there is no doubt that, as Barton said over two centuries ago, the silicification has taken the form of a replacement of the more massive timber in the lignites.

This process can be well seen in the specimen at the Ulster Museum where the transition from lignite into completely silicified wood can be traced. In part of the block, sinuous bands of pale-weathering silica, apparently replacing particular layers of the heart-wood, can be seen standing out from the surrounding lignite. The first stages of the replacement appear to consist of the formation of small discrete crystals of quartz, and when the material in this state is weathered clear of the lignitic remains between grains the result is a porous saccharoidal texture which bears only a vague resemblance to the original wood. On the other hand, when the replacement is complete, the fossil wood is hard and compact, takes a high polish and completely pseudomorphs the original timber.

The date of the silicification of the timber, and the conditions which brought it about are uncertain and speculative. It seems possible that the silicification may have actually preceded the humification of the timber, and it is at least probable that the whole process took place soon after the deposition of the lignite beds and while the Lough Neagh Clays were still accumulating. At this time, soon after the end of the extrusive volcanic activity, there were probably widespread geysers and fumaroles in the area and the ground waters may well have been more heavily charged with silica than is usual.   H.E.W.

DETAILS

The northern boundary of the Lough Neagh Clays outcrop is formed by the north-east trending Lennymore Fault which, with downthrow to the south, brings the Clays against the Upper Basalts of the Langford Lodge promontory. The course of this fault, as established on geophysical evidence, runs well to the south of the locality on the Crumlin River, where Hardman (1875 etc.) recorded the presence of fossiliferous Pliocene clays and Swanston's (1879) denial of this is thus borne out by recent investigations.

The whole of the wide flat area south of the Crumlin River must be underlain by Lough Neagh Clays but is masked by superficial deposits. The eastern extent of the outcrop is, however, suggested by reports of wells at The Plantation [138 753] and Strandfield [144 758]. The former is said to have been sunk through 38 ft (11 m) of lignite below 20 ft (6 m) of red 'till' while the latter entered 'rotten rock', presumably basalt, under 5 ft (1·5 m) of boulder clay and 20 ft (6 m) of 'black gravel'. A recent trial borehole near The Plantation failed to find any lignite and was in boulder clay to 50 ft (15 m).

An exposure [133 740] in a small stream 650 yd (594 m) S. of Thistleborough shows a few feet of drift overlying 4 ft (1·2 m), unbottomed, of lignite which appears to be *in situ*. An unidentified note on the old geological field map of the area states that a

20-ft (6 m) square pit was opened near here and that under 5 ft (1·5 m) of surface deposits the following section was uncovered:

	ft	m
Blue clay (Lough Neagh Clays) . . .	2	(0·6)
Fine lignite (broken)	2½	(0·8)
Solid lignite. Base not reached . . .	12	(3·7)
	16½	(5·1)

A recent borehole 100 yd (91 m) N. of this exposure failed to find lignite in 100 ft (30 m) drilled.

Seymour (in Cole 1912, p. 150) stated that a pit at Bellbrook was sunk through 18 ft (5·5 m) of Lough Neagh Clays without reaching bottom or finding lignite. A recent borehole here reached the Clays under 19 ft (5·8 m) of drift and penetrated 81 ft (25 m) of grey and olive clay with a siltstone layer containing lignified plant debris. Half a mile to the south-east a well at Glenville was reported to have penetrated 6 ft (1·8 m) of sand and clay on the alluvial flat of the Glenavy River and then 8 ft (2·4 m) of bluish grey silty clay with some carbonized plant debris, occasional quartz and basalt pebbles, and rare blocks of basalt. This material was seen on the spoil heap and is probably either Lough Neagh Clays or the basal boulder clay which is presumably largely composed of local material. The section corresponds well with an old record of a trial pit which penetrated 3 yd (2·7 m) of 'Pliocene Clays' below 2 yd (1·8 m) of alluvium. If the clays are *in situ* there must be a limited area of Lough Neagh Clays south-east of the Crumlin Fault.

On the lake shore north of Sandy Bay, Seymour recorded a pit which showed the following section:

	ft	m
Lough Neagh Clays	30	(9·1)
Lignite	14	(4·3)
White Clay	6	(1·8)
Lignite, unbottomed	14	(4·3)
	64	(19·5)

There are a number of swampy hollows near the lake shore about here and this is clearly an area where lignite has been worked or explored over a long period. Griffith (1838, Appendix p. 22) gave the journal of a borehole made 'at Sandy Bay'. :

	ft	m
Blue Clay	10	(3·0)
Black lignite mixed with clay	25	(7·6)
Clay	2½	(0·8)
Black lignite	20	(6·0)
Clay	4	(1·2)
Black lignite	15	(4·6)
	76½	(23·2)

Gardner (1885) described the observation of Swanston and Stewart in 1884 who saw a pit in the lignite hereabouts which was being worked at that time. They observed silicified material in the lignite.

Cole (1912, p. 122) also recounted a visit to the area where he saw old pits and arranged a shallow trial digging in which he saw lignite masses embedded irregularly in Lough Neagh Clays. Both Gardner and Cole recount the occurrence here of nodules of ironstone which contain well-preserved plant debris. These were found lying on the shore and in the boulder clay but none were seen during the resurvey.

About ¼ mile (0·8 km) S. of Sandy Bay Point, Seymour (*in* Cole 1912) recorded a pit 65 ft (20 m) deep which reached the basalt. No lignite was encountered. A well sunk more recently near this point is said to have been sunk for 90 ft (27 m), all in blue clays. Some distance south-east of this a well 500 yd (457 m) W.N.W. of Lodge Hill is reported to have been sunk for 35 ft (11 m) in grey clays with some lignite and silicified wood, a specimen of which was produced.

On Tunny Island Seymour recorded a pit on the north side of the houses south of the church which was sunk for 27 ft (8·3 m) in Lough Neagh Clays. As the island is a boulder clay hill it seems probable that most if not all of this pit was actually in drift deposits.

Scouler (1837) quoted Stewart, who examined the area on behalf of the Royal Dublin Society, on a boring in search of coal at Portmore: "They bored through two beds of coal, or what is called black wood, 25 ft (7·6 m) thick each, and a third stratum 9 ft (2·7 m) thick and 80 yd (73 m) deep; they bored 18 inches (0·5 m) deep into a fourth stratum, having no more rods to go deeper". There is no indication of the location of this bore and the term Portmore seems at that time to have covered the whole area from Ballinderry to the shore of Lough Neagh. There is no more recent information about this district.

About 2 miles (3·2 km) S. of Portmore Lough the spoil heap of a well [103 655] 300 yd (274 m) W. of Cairnhall was seen during the resurvey. Apart from some red boulder clay at the top, most of the spoil showed the well, 55 ft (17 m) deep, to have been sunk in drab brown clays with occasional sandy films giving the clay a banded appearance at times. Half a mile further south, on Courtney's Island, a well [102 648] 350 yd (320 m) N.E. of Island Hill entered grey clays below 23 ft (7 m) of red boulder clay. The grey clay, seen in the spoil heap, was penetrated for only 6 ft (1·8 m) but the material dug out was partially pale grey and partly dark bluish grey in colour, the latter type showing occasional polished shear-planes.

Some distance to the east of the known outcrop of the Lough Neagh Clays reports of the material encountered in wells in and near the townland of Aghadolgan suggested that there was an outlying area of this formation east of the basalt spur of Darachrean. To check this, two shallow boreholes were put down in 1959. The first (36/594) was sited [143 715], 700 yd (640 m) W.S.W. of Ballymacricket School and entered decomposed basalt under about 9 ft (2·7 m) of drift. The second (36/595) was sited [135 708] beside the stream 100 yd (91 m) W. of Bessfield and entered Lough Neagh Clays under 13 ft (4 m) of alluvium. The beds, penetrated to a depth of 50 ft (15 m), were mainly bluish grey clays with polished shear-planes and occasional sandy bands, paler grey in colour. The sandier bands are often irony and there are occasional small concretions in the clay.

The occurrence of Lough Neagh Clays in this borehole confirms the existence of a tongue of this formation along the west side of the Glenavy Fault which probably extends as far north as the southern end of Devlin's Hill.   H.E.W.

There are no exposures at the southern end of the outcrop but a well in Derryhirk [087 641] was sunk to a depth of 48 ft (15 m) into what was described as blue clay, the lowest 2 ft (0·6 m) being 'bouldery'. None of the spoil was seen at the time of survey.

Geophysical evidence suggests that the southern margin of the Lough Neagh Clays outcrop is faulted. The Kinnegoe Fault, which is a major feature in the area to the west, runs eastwards towards Aghalee and forms the boundary between the Clays and the Lower Basalts.   P.I.M

Chapter 9

INTRUSIVE IGNEOUS ROCKS

INTRODUCTION

INTRUSIVE ROCKS of two generations occur in the Belfast area. Caledonian igneous rocks are represented by two lamprophyre dykes in the Lower Palaeozoic rocks of County Down and Tertiary intrusives by a quartz–porphyry boss, a few small agglomerate-filled vents, three small plugs, a sill and a large number of dykes, all of basalt or dolerite.

Minor intrusions of Caledonian age, mainly in the form of lamprophyre dykes, are abundant in east Down where they have been described by Reynolds (1931), but only two decomposed dykes are known to occur in this area. One is over 8 ft (2·4 m) wide and is exposed in Purdys Burn a few yards west of Charity Bridge [358 671] and in a knoll 300 yd (274 m) to the south. The other is seen in Drumbo Glen [327 652] and is over 4 ft (1·2 m) wide.

The Tertiary quartz-porphyry, which was erroneously included in the Caledonian igneous rocks on the One-inch Map, lies mainly in the adjacent Sheet (48) of the One-inch Map and only a small marginal area lies in Sheet 36. Though poorly exposed in the drift-covered country west of Hillsborough, it is thought to be an intrusive boss. It may be coeval with the Tardree rhyolites of County Antrim and the granites of the Mourne Mountains.

Some thirty dolerite plugs are known to be associated with the Antrim Lava Series in Northern Ireland, mainly intruded into the basalts and ranging from a few yards to 860 yd (787 m) in diameter. The three examples known in the Belfast area are all of dolerite or gabbro and do not form the strong features characteristic of some of the other plugs. The largest is intrusive into Bunter Sandstone 2 miles (3·2 km) E. of Moira; a small plug penetrates the Keuper Marl at Limestone Lodge 3 miles (4·8 m) N.W. of Lisburn; and the third is on the basalt plateau at Craig just over a mile (1·6 km) N.W. of Divis.

The only sill known in the district is intruded into the Keuper Marl and is exposed in the Ballygomartin area and in the Forth River where it is in two leaves 6 ft 9 in (2 m) and 40 ft (12 m) thick, separated by a band of mudstone.

The most impressive intrusive feature in the district is the very large number of Tertiary basalt dykes — over three hundred are known any many more must exist beneath the drift cover. The dykes are more numerous in the Mesozoic rocks in the Belfast area, and eastwards towards Carrickfergus (One-inch Sheet 29). To the west they seem to become rather less abundant, but how far this is due to absence of good exposures, particularly of the Trias, is not clear, The frequency is, however, much reduced in the Lower Palaeozoic rocks, presumably because of the greater resistance to lateral expansion in this massive formation. The apparent paucity of dykes in the basalts is partly due to the lack of fresh exposures as the similiarity of the dyke rock to the country rock makes detection less easy

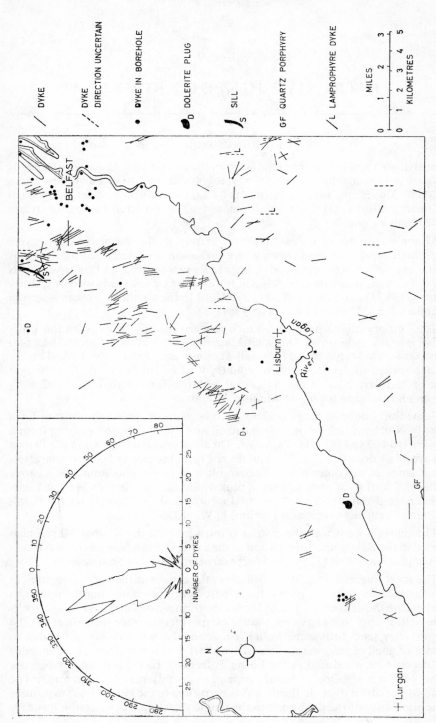

Fig. 13. *The distribution of Tertiary intrusive igneous rocks, and orientation of dykes in the Belfast area*

in weathered sections. It does appear, nevertheless, that they are less numerous in the basalts than in the underlying formations.

Charlesworth and Hartley (1935, pp. 193–6) have differentiated between two dyke 'swarms' in this area. The group between Carrickfergus and Lisburn is said to be part of the Tardree swarm, and the swarm running through Hillsborough and designated by that name is related to the group extending to the County Down coast. By analogy with similar Scottish and Irish dykes swarms, which converge upon granite bosses as foci, Charlesworth and Hartley suggest that concealed granite bosses exist near Tardree and Hillsborough. In the latter case, the quartz-porphyry lies eccentric to the swarm and it is suggested that the controlling granite (if present) lies at depth beneath the town of Hillsborough.

The orientation diagram for 250 dykes over the whole area (Fig. 13) shows that there are two maxima: (a) between 10° and 20° west of north and (b) 30° west of north, the predominant direction being the first mentioned. These maxima differ from the single predominantly north-west direction of the Mourne and Ardglass swarms (Tomkeieff and Marshall 1935, p. 253) and the similar direction which prevails among the dykes of the Isle of Man and Anglesey. There is no further evidence to suggest an earlier phase of intrusion prior to the intrusion of the main swarm (Walker 1959, p. 195). In the Belfast area the north-north-west direction is predominant, the trend becoming more northerly west of the city. In the Hillsborough area orientation is more random.

The thicknesses of the dykes vary from a few inches up to 60 ft (18 m) but the majority do not exceed 10 ft (3·1 m). Walker (1959, fig. 1) has attempted a calculation of the crustal expansion due to the intrusion of the dykes. In the western half of the Sheet he indicates a stretch of 2 to 4 per cent and in the eastern half a stretch of over 4 per cent.

Most of the dykes are vertical or within a few degrees of vertical, and they are usually reasonably regular in thickness and direction over their exposed courses. Exceptions to the latter characteristics occur in the Chalk and the Keuper Marl. In the former, dykes are locally exceedingly irregular in thickness and, to a lesser extent, in direction. In the quarries at Knocknadona dykes can be seen swelling, ramifying and re-joining in the Chalk but maintaining constant direction and thickness in the overlying basalt. (Plate 1A). In the Keuper Marl brickyard exposures, now mainly built over, some dykes of fairly constant thickness followed extremely sinuous and irregular directions with offshoots and protuberances.

The country rocks are little affected by the intrusions and any baking or metamorphism is normally confined to a marginal selvage a few inches thick. The effect is least in the Lower Palaeozoic rocks, Bunter Sandstone, and basalt lavas, while Keuper mudstones are slightly baked and bleached. Effects are greatest in the chalk and recrystallization to blue or greenish marble can be seen for distances up to a foot or more from the dyke in some cases. The marble is largely composed of calcite granules and crystals but apatite is sometimes seen in thin sections [NI 1500]. In a few cases the dyke-rock itself is altered to 'White Trap' where it intrudes marl or sandstone. This phenomenon, in which the rock minerals are kaolinized or calcified, has been recorded by Hartley (1928, p. 75) and Cleland (1932, pp. 93–4).

Two agglomerate-filled pipes, possibly due to phreatic activity, are seen penetrating the first flow of the Lower Basalts at Rockville quarry, Kilcorig, and associated beds of tuff are seen above and below the lava. A borehole on the site of a road bridge at Tullyloob, near Moira, penetrated 81 ft (25 m) of what appeared to be a vent agglomerate.   P.I.M., H.E.W.

DETAILS

Hillsborough Quartz-Porphyry 'Intrusion'. A small outcrop of quartz-porphyry, part of a larger mass which crops out in One-Inch Sheet 48, was recorded [191 575] 3½ miles (5·6 km) W.S.W. of Hillsborough at the southern edge of the Sheet. The old quarry in which this was formerly exposed is completely flooded and there were no exposures *in situ* at the time of mapping. A well record at Kilwarlin Cottage gave 39 ft (12 m) of drift on a 'black slaty' rock surface and the northern limit of the outcrop, therefore, lies south of that previously mapped. The field relations of the quartz-porphyry with the country rock of Lower Palaeozoic grits and shales are not apparent within the area of One-Inch Sheet 36. Patterson (1951, pp. 281, 283) records it as a 'Rhyolite ? lava'. A full analysis given by Patterson (1952, p. 286) shows it to be closely similar in composition to the Tardree rhyolite lavas.

Plugs. The best exposure of the Hollymount intrusion [182 615] is in an old quarry about 3 miles (4·8 km) E. of Moira, just north of the road from Trummery crossroads to Halfpenny Gate, where a well-jointed massive dolerite appears at the surface. Further small exposures are to be seen south of the road, one of which, 30 yd (27 m) S. of the road, exhibited fairly close sheet jointing with a westerly inclination.

Wells immediately to the east of the intrusion enter Bunter Sandstone and provide a close delimitation of the western margin of the dolerite. Similarly, a well to the west of the exposures enters sandstone beneath the drift while a well at Trummery Cottage cuts through 10 ft to 15 ft (3–4·6 m) of dolerite before penetrating about 60 ft (18 m) of Bunter Sandstone. The western part of the dolerite seems, therefore, to be a sill-like extension of the main dolerite mass giving the intrusion the appearance of an asymmetrical laccolite. It is difficult to establish any significant criteria for calling the intrusion a plug, except that the width is much greater than the average dyke and a steeply plunging junction with the country rock is indicated on the western side, but as a topographic feature it is not prominent. Patterson and Swaine (1957) omit it from their survey of the Tertiary Dolerite Plugs of north-east Ireland.   P.I.M.

About 100 yd (91 m) S.W. of Limestone Lodge a small circular hill [221 664], some 50 yd (46 m) in diameter, lying well south of the basalt outcrop, is shown by a small quarry on its crest to be composed of massive spheroidal-weathering coarse-grained dolerite. There seems no doubt that this is a small intrusive plug, but no contacts with the surrounding Keuper Marls are exposed.

On the high ground north of Armstrong's Hill a crag exposure [264 764] at Craig shows 8 ft (2·4 m) of coarse-grained gabbroic rock. The near-horizontal jointing seen appears to be domed and the form of the adjacent hill-side suggests that this is an intrusive plug about 100 yd (91 m) in diameter.   H.E.W.

Explosion Vents. In the Rockville quarry, Kilcorig [229 671], the first flow of the Lower Basalts is penetrated by two vents filled with a rubbly mixture of vesicular basalt and some indurated and calcite-veined Triassic marl. In the face of the working 25 yd (23 m) N. of the southern basalt dyke, which intersects the chalk and overlying lava, the massive lava is penetrated by a funnel-shaped pipe about 10 ft (3 m) in diameter at the lowest point seen and widening irregularly upwards. The rubbly material in the pipe appears to pass into an inaccessible bed of similar material which overlies the lava and extends for some 75 yd (69 m) along the face, its full extent being concealed by the overlying drift.

Thirty yards (27 m) N. of the first pipe a second equally irregular vent above 25 ft (7·6 m) across cuts the lava and passes up into the overlying bed of tuff which reaches a thickness of about 15 ft (4·6 m) in the face. Immediately over the second and larger pipe the bed of tuff is overlain for a few yards by a mass of compact basalt, which has apparently sunk down into it, probably due to compaction of the vent-filling by the weight of the succeeding lavas.

The base of the first lava appears to rest, as a rule, directly on the top of the Chalk with only a thin band of flint rubble, but for a distance of about 10 yd (9 m) between the vents it is underlain by a bed over 4 ft (1·2 m) thick of rubbly tuff with marl fragments, identical to that in the vents. Exposures are partially clouded by scree and it is not clear whether this underlying tuff is intrusive or whether it preceded the lava.

An unusual feature of the tuff is the virtual absence of any Chalk or flint debris which might be expected in this situation. Some small flints were seen among the scree derived from the pipes but none was seen *in situ* in the tuff. The marl debris is occasionally in lumps up to 9 in (0.23 m) long and appears to be of true Triassic type. It is possible that these vents and the bedded tuff are the results of phreatic explosions — the release of pressure built up in the water-bearing Triassic rocks by the injection of super-heated magma and formation of steam.   H.E.W.

In a series of borings put down in 1963 for a road bridge [120 616] at Tullyloob north of Moira, one borehole penetrated 81 ft (25 m) into what appears to be vent material. A broken core was examined on the site and the following section recorded:

<table>
<tr><td colspan="2" align="center">*Borehole at Tullyloob Bridge* (36/866a)</td><td>ft</td><td>in</td><td>m</td></tr>
<tr><td>Basalt, medium-grained</td><td>⎫</td><td></td><td></td><td></td></tr>
<tr><td>Basalt, amygdaloidal</td><td>⎬ · · about</td><td>5</td><td>4</td><td>(1·60)</td></tr>
<tr><td>Basaltic rubble with some chalk fragments</td><td>⎭</td><td></td><td></td><td></td></tr>
<tr><td>Basalt, medium-grained</td><td></td><td>0</td><td>6</td><td>(0·15)</td></tr>
<tr><td>Basalt, rather soft, decomposed</td><td></td><td>5</td><td>0</td><td>(1·52)</td></tr>
<tr><td>Basalt, amygdaloidal, lateritized</td><td></td><td>2</td><td>0</td><td>(0·61)</td></tr>
<tr><td>Chalk with a little reddish clay</td><td></td><td>1</td><td>6</td><td>(0·45)</td></tr>
<tr><td>Basalt, medium-grained with few vesicles</td><td></td><td>2</td><td>2</td><td>(0·66)</td></tr>
<tr><td>Basalt, amygdaloidal, reddened</td><td></td><td>1</td><td>10</td><td>(0·56)</td></tr>
<tr><td>Basalt, fine- to medium-grained, amygdaloidal</td><td></td><td>3</td><td>6</td><td>(1·07)</td></tr>
<tr><td>Basalt and Chalk fragments in brownish clay, slickensided in part [? Agglomerate]</td><td></td><td>22</td><td>6</td><td>(6·85)</td></tr>
<tr><td>Basalt, rotten, veined</td><td></td><td>1</td><td>0</td><td>(0·30)</td></tr>
<tr><td>Agglomerate: subangular fragments up to 2 in across of basalt (medium- to coarse-grained) and Chalk in a fine-grained matrix</td><td></td><td>0</td><td>6</td><td>(0·15)</td></tr>
<tr><td>[Core broken] Fragments of agglomerate as above with some rotten basalt</td><td></td><td>4</td><td>6</td><td>(1·37)</td></tr>
<tr><td>Basalt, medium-grained, a few amygdales</td><td></td><td>2</td><td>0</td><td>(0·61)</td></tr>
<tr><td>Chalk</td><td></td><td>0</td><td>4</td><td>(0·10)</td></tr>
<tr><td>Basalt, rotten, amygdaloidal</td><td></td><td>5</td><td>0</td><td>(1·52)</td></tr>
<tr><td>Basalt, fine- to medium-grained</td><td></td><td>2</td><td>6</td><td>(0·76)</td></tr>
<tr><td>Basalt, fine- to medium-grained, chrysophyric</td><td></td><td>3</td><td>2</td><td>(0·96)</td></tr>
<tr><td></td><td></td><td>63</td><td>4</td><td>(19·24)</td></tr>
</table>

No contacts were seen between the Chalk, agglomerate or basalt and no further boreholes encountered agglomeratic material.

No evidence was seen to substantiate the outlier of basalt postulated by Hull (1871, p. 32) as existing near the Presbyterian Church at Moira and coloured erroneously as an intrusion on the One-inch Map (1901 edition).   P.I.M.

Sill. The Ballygomartin Sill, the only sill within the Belfast Sheet, strikes a few degrees E. of N. and has been traced for about 1 mile (1·6 km). Almost half of this distance lies beyond the northern boundary of the Sheet.

The sill protrudes through the boulder clay [295 757] on the north side of a tributary of the Forth River where it forms a well-defined dip-slope. It is here cut off by a north-easterly trending fault and has not been found to the south. The top of the sill dips N.N.W. at 12° and the rock is a medium-grained dark grey basalt showing strong horizontal joints and a rude columnar structure. It has been quarried on a limited scale in the past and would make an excellent roadstone.

Close to the stream the sill is cut by a thin basalt dyke indicating that the sill was intruded into position before the end of the Tertiary dyke activity.

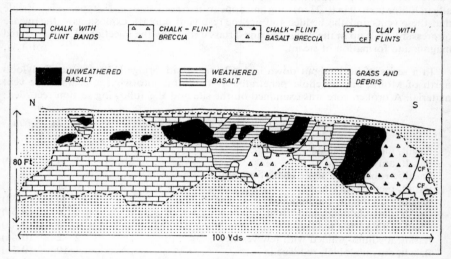

FIG. 14. *Section of Balmer's Glen Quarry showing disturbance of the Chalk by intrusive basalt*

The sill is also exposed [297 769] in the steep banks of the Forth River at the north margin of the Sheet. It is split into two parts here, the top part 6 ft 9 in (2·0 m) thick is of dark grey, medium-grained, slightly columnar basalt whilst the lower part, at least 40 ft (12 m) thick, is of similar rock displaying rudely columnar jointing.   J.A.R.

Dykes. The widths, location, and orientations of all known dykes are given in the following table while their positions are also given on the map (Fig. 13). In many cases the dyke has been exposed only in an inadequate section or recorded in a borehole, and details of width and direction are not availalbe.

One dyke which merits special mention is at Balmer's Glen quarry [182 631] east of Maghaberry where, in the eastern wall of the quarry, the intrusion has caused breccia-tion of the White Limestone. In the overlying Clay-with-Flints, fragments and small blocks of the basalt are incorporated in this rubbly material (Fig. 14).

A.—Striated Pavement in Greywacke, Carryduff Quarry   (NI 238)

PLATE 7

B.—Drumlin Topography near the Temple, County Down   (NI 240)

A.—Black Hill and Belfast   (NI 216)

(For explanation see p. xiii)   PLATE 8

B.—The Lisburn Esker   (NI 435)

No. of Dyke	Direction	Thickness		Grid Reference	No. of Dyke	Direction	Thickness		Grid Reference
		Feet	Metres				Feet	Metres	
1	333°	?		126591	43	167°	1	0·30	224667
2	14°	?		126591	44	163°	4–8	1·22–2·44	224667
3	337·5°	?		135595	45	170°	7	2·13	225668
4	16·5°	18	5·49	138596	45a	2°			226667
5	282°	8	2·44	134610	45b	290°			227667
6	322°	1	0·30	134610	46	140°	6	1·83	229670
7	322°	19½	5·94	134610	47	134°	4 +	1·22	229671
8	322°	8	2·44	134610	48	121°	?		231666
8a		5¼	1·60	136614	49	130°	?		234662
8b		32½	9·90	137613	50	110°	6	1·83	234662
8c		44½	13·56	138614	51	32°	3	0·91	238660
8d		36	10·92	135616	52	152°	?		239663
8e		16¾	5·09	138617	53	160°	6 +	1·83	233672
9	32°	12 +	3·66	155602	53a	308°	—		233672
10	341°	7	2·13	157632	53b	160°	—		233672
11	?	?		157632	54	162°	12	3·66	235674
12	301°	45	13·72	169629	54a	14°	—		235674
12a	336°			169629	55	165°	9	2·74	238673
12b	336°			169629	56	12°	10 +	3·05	238673
13	5°	?		182631	57	140°	4 +	1·22	238673
14	?90°	?		180581	58	140°	12	3·66	239674
15	275°	3	0·91	205579	59	140°	9	2·74	239674
16	31°	18 +	5·49	206582	60	3°	6 +	1·83	239670
16a	90°			214580	61	160°	c.4 +	1·22	239670
17	83°	?		217588	62	?	72 +	21·95	252656
18	315°	?		237591	63	?	64	19·51	259660
19	148°	?		239594	64	?	10	3·05	259660
20	140°	?		239594	65	?	29	8·84	249669
21	168°	8	2·44	273603	66	?	89	27·13	249669
22	15°	6 +	1·83	288577	67	93°	?		267670
23	20°	1½	0·46	298601	68	134°	5	1·52	270674
24	?	½ +	0·15	310602	69	163°	?		270674
25	172°	15 +	4·60	309593	70	160°	1	0·30	268678
26	15°	7 +	2·13	322586	70a	160°	—		270679
27	115°	?		333595	71	177°	22	6·71	270679
28				341596	72	94°	10	3·05	264681
29	136°	9	2·74	323579	73	158°	7	2·13	263683
30	150°	6 +	1·83	327576	74	54°	?		261683
31	15°	?		340575	75	150°	4	1·22	262686
32	—	2	0·61	249621	76	47°	2 +	0·61	253686
33	?46°	?		258623	77	172°	?		253683
34	?	?		269644	78	43°	4 +	1·22	253685
35	?	12 +	3·66	245639	79	156°	2–5	0·61–1·52	247687
36	155°	2½	0·76	235645					
37	160°			232654	80	45°	5 +	1·52	242682
38	94°				81	155°	4	1·22	241681
39	175°	5	1·52	226658	82	126°	3 +	0·91	239679
40	116°	3 +	0·91	227663	83	137°	9–10	2·74–3·05	239678
41	170°	4	1·22	223667					
42	170°	10	3·05	224667					

No. of Dyke	Direction	Thickness Feet	Thickness Metres	Grid Reference	No. of Dyke	Direction	Thickness Feet	Thickness Metres	Grid Reference
84	188°	?		262692	127	168°	?		295682
85	180°	?		262692	128	177°	8	2·44	295686
86	10°–164°	35	10·67	262695	129	177°	4	1·22	295686
87		?		265692	130	173°	12½	3·81	295686
88		?		266691	131	173°	30	9·14	295686
89				271695	132	173°	1	0·30	295686
90	120°	?		272693	133	105°	1½	0·46	295686
91	76°	?2	0·61	274697	134	175°	3	0·91	298690
92	—	6	1·83	276678	135	10°	10+	3·05	278705
93	151°			284673	136	160°	12+	3·66	268708
94	3°	?		284669	137	160°	3–4	0·91–1·22	267708
95	127°	4	1·22	299662					
96	24°	?		302638	138	173°	12+	3·66	266708
97	166°	13	3·96	306624	139	170°	3	0·91	264718
98	175°	2	0·61	316653	140	160°	3	0·91	264720
99	142°	c.3	0·91	316649	141	175°	¼–1	0·08–0·30	267719
100	155°	—		323657					
101	160°	—		323656	142	129°	?		270718
102	131°	9	2·74	344638	143	150°	4	1·22	263720
103	?34°	2+	0·61	349643	144	170°	3	0·91	262722
104	78°	10	3·05	355646	145	122°	1–1½	0·30–0·46	235736
105	53°	2+	0·61	356642					
106	150°	8–12	2·44–3·66	357642	146	170°	2	0·61	263722
					147	157°	5	1·52	264721
107	20°	1–6	0·30–1·83	364637	148	110°	69	21·03	269724
					149	170°	?		272728
108	175°	½	0·15	365636	150	160°	52	15·85	281711
109	180°	4	1·22	366640	151	?119°	?		281711
110	160°	¾–2	0·23–0·61	366647	152	163°	6+	1·83	281711
					153	169°	6	1·83	281711
111		1–1½	0·30–0·46	366647	154	180°	60	18·29	281711
					155	169°	4+	1·22	281711
112					156	161°	12	3·66	281711
113	130°	6	1·83	366647	157	?165°	?		283710
114	65°	?		336659	158	171°	1⅓	0·41	297706
115	64°	6+	1·83+	357753	159	171°	4½	1·37	297706
116	167°	3½	1·07	357753	160	171°	6	1·83	297706
117	132°	4–5	1·22–1·52	357753	161	181°	?		300706
					162	163°	6	1·83	300706
118	93°	?		357753	163	148°	9	2·74	305711
119	150°	9–10	2·74–3·05	347662	164	140°	18	5·49	305711
					165	?	15	4·57	293722
120	132°?	?		347666	166	160°	?		301719
121	110°?	?		346667	167	160°	4	1·22	301715
122	80°	6	1·83	347668	168	155°	2	0·61	274720
122a	345°	8+	2·44	358671	169	180°	?		274720
123	26°	6+	1·83	341681	170	135°	?		272714
124	147°			341681	171	145°	6	1·83	278720
125	147°			341681	172	160°	17	5·18	279723
126	103°			341681	173	163°	18	5·49	281723

No. of Dyke	Direction	Thickness Feet	Thickness Metres	Grid Reference
174	150°	3	0·91	282723
175	150°	?		285723
176	150°	?		285723
177	147°	?		286724
178	120°	7	2·13	287725
179	170°	14	4·27	288735
180	106°	8	2·44	269738
181	90°	5	1·52	293739
182	143°	45	13·72	294743
183	143°	60	18·29	294743
184	137°	?		296745
185	175°	6	1·83	291761
186	30°	?		298768
187	180°	?		299768
188	7°	?		300768
189	160°	6	1·83	302768
190	?	73½	22·40	302768
191	?	3	0·91	305767
192	7°	? (sm.)		309767
193	10°	?		307765
194	149°	3?	0·91	307763
195	145°	?		308761
196	180°	?		308760
197	85°	?		309758
198	160°	3	0·91	322761
199	160°	3	0·91	323761
200	164°	3	0·91	324761
201	165°	?		324671
202	165°	?		324760
203	?	78½	23·93	322756
203a		3 +	0·91	322756
204	80°	3	0·91	311750
205	104°	30 +	9·14	311749
206	155°	2	0·61	314750
207	150°	3 +	0·91	314750
208	151°	2	0·61	315747
209	165°	?		313747
210	151°	1½	0·46	315747
211	90°	?		314745
212	151°	2	0·61	315746
213	136°	1	0·30	316748
214	130°	2	0·61	316748
215	4°	10	3·05	307742
216	169°	4	1·22	308738
217	169°	2½	0·76	308738
218	169°	9	2·74	309739
219	150°	—		311740
220	150°	10	3·05	312740
221	142°	20	6·10	310730
222	140°	2	0·61	308725
223	150°	1½	0·46	308725
224	135°	15	4·57	309725
225	152°	?		310725
226	155°	2 +	0·61	312725
227	?	4½	1·37	339710
228	155°	?		365709
229	157°	3–3½	0·91–1·07	365706
230	165°	2½–3½	0·76–1·07	364706
231	?	3	0·91	330735
232	?	5	1·52	330735
233	?	66	20·12	329734
234	?	39	11·89	329734
235	?	33½	10·21	336734
236	?	4	1·22	337734
237	?	15	4·57	337734
238	?	2	0·61	340731
239	?	?		350736
240	?	37	11·28	350736
241	?	37	11·28	357740
241a	?	78	23·77	357740
242	?	14½	4·42	363738
243	?	2	0·61	363738
243a	?	10½	3·20	361739
244	?	c.3	0·91	347739
245	?	3	0·91	347739
246	?	1	0·30	343747
247	?	3½	1·07	343747
248	157°	60	18·29	336754
249	161°	3	0·91	337758
250	149°	13 +	3·96	338759
251	145°	?		337761
252	164°	?		338761
253	114°	?		338761
254	165°	27	8·23	337763
255	165°	?		338763
256	143°	20	6·10	339763
257	148°	10	3·05	339763
258	?			341764
259	?	?		357767
260	?	11½	3·51	362767
261	159°	12	3·66	307754
262	180°	6	1·83	307754
263	15°	6–8	1·83–2·44	307754
264	20°	3	0·91	307754
265	180°	0½	0·15	307754
266	22°	10	3·05	304755
267	70°	25	7·62	302757

No. of Dyke	Direction	Thickness		Grid Reference	No. of Dyke	Direction	Thickness		Grid Reference
		Feet	Metres				Feet	Metres	
268	150°	15	4·57	302757	284	?	7½ +	2·29	331740
269	120°	9	2·74	299758	285	?	6 +	1·83	319729
270	150°	15	4·57	299758	286	?	7½ +	2·29	335751
271	55°	?		297758	287	?	10½ +	3·20	338752
272	30°	?		294757	288	?	1	0·31	337751
273	180°	?		293758	289	?	0¾	0·23	321873
274	?	?		333738	290	?	0½ +	0·15	318722
275	?	7 +	2·13	343762	291	?	2 +	0·61	324735
276	?	0½	0·15	342753	292	?	4 +	1·22	324735
277	?	11½ +	3·51	338751	293	?	4 +	1·22	332738
278	?	6 +	1·83	337751	294	?	0½ +	0·15	332741
279	?	10 +	3·05	335751	295	?	?		334750
280	?	24½ +	7·47	333750	296	?	5 +	1·52	324735
281	?	8½ +	2·59	333750	297	?	37½	11·43	357754
282	?	9 +	2·74	333746	298	?	6½ +	1·98	358753
283	?	2 +	0·61	332745	299	?	8 +	2·44	338754

PETROGRAPHY

The specimens were examined by J. R. Hawkes and H. E. Wilson.

Quartz-porphyry. A specimen (NI 1490) from outside the Sheet area consists of rounded phenocrysts of quartz (0·04–6·0 mm) and a few oligoclase crystals (0·1–0·14 mm) in a matrix of anhedral quartz and alkali feldspar with average grain size 0·01 mm. Accessory constituents include muscovite, zircon, hematite, apatite, and possibly sphene. Patterson (1952, p. 286) gives its composition as : SiO_2 76·03 per cent, Al_2O_3 13·01 per cent, Fe_2O_3/FeO 0·15 per cent, CaO 0·36 per cent, Na_2O 2·25 per cent, K_2O 5·02 per cent.

Plugs. The Craig plug is coarse-grained hypersthene-olivine-gabbro (NI 2045). Clino-pyroxene (pale brown augite) and ortho-pyroxene (pale brownish green hypersthene) occur in approximately equal proportions and are ophitic to labradorite. Crystals are up to 5 mm long. The labradorite (An_{60}) ranges from 0·1 mm to 2·0 mm and in localized patches is replaced by fibrous levyne (X3290). Olivine accounts for 5 per cent of the rock and is fairly fresh. There is some accessory magnetite.

The small plug at Limestone Lodge (NI 1390) is an ophitic dolerite with large plates of pale brown augite enclosing feldspar laths, and glomero-porphyritic olivine partly altered to greenish chlorite and serpentine. The groundmass is partly replaced by vermicular chlorite and zeolite.

Sill. The Ballygomartin Sill is an olivine-basalt (NI 2009) with scattered phenocrysts of olivine and plagioclase (up to 2.5 mm long) in a matrix of labradorite laths which occasionally show zoning, sub-ophitic augite, olivine, accessory magnetite and interstitial chlorite. Olivine is altered to serpentine and calcite and patches of chloritic alteration centred on olivine phenocrysts affect the surrounding feldspars.

Dykes. The Caledonian lamprophyre dykes are extremely weathered and no fresh material was available for examination. One section (NI 2054) showed lamellar - twinned laths of andesine (Ab_{60}), granular quartz and feldspar, and abundant iron oxides, chlorite and calcite. The calcite is seen to replace feldspar in places and the larger patches of hematite and limonite may replace olivine or biotite.

The Tertiary dykes are all olivine-basalts or dolerites, and most show some signs of alteration. All consist of labradorite, titanaugite and olivine with accessory magnetite and occasionally interstitial chlorite or zeolite.

The labradorite (An_{55-60}) is often fresh but a saussuritization and replacement in part by zeolite, chlorite and serpentine is seen in some specimens. Two generations are rare, but phenocrysts up to 5 mm long occur in a few dykes (NI 1332, 1511, 2058) and are sometimes glomerophyric (NI 1333).

Pyroxene is pale brown slightly pleochroic titanaugite, usually ophitic, in plates up to 3 mm but also seen as tabular and granular crystals (NI 2049). It is normally fresh but one specimen shows it replaced by brown chloritic material (NI 1502).

Olivine is always rather decomposed, showing replacement by serpentine minerals and, rarely, chlorite. It usually makes up 5 to 10 per cent of the rocks but ranges up to 15 per cent. It sometimes occurs in two generations, larger grains up to 1·25 mm and smaller crystals, average 0·1 mm, appearing in the same specimen (NI 2013).

Vesicles with zeolite are less common than those filled by clay minerals, chlorite and calcite. The margins of a large dyke in Black Mountain Quarry, however, have amygdales containing fibrous garronite surrounding a core of gyrolite (X 3292, A, B, C).

The only specimen of 'White Trap' — a completely altered dyke rock which has lost all appearance of an igneous rock — to be examined was from [180 581] near Fortwilliam House, where the material (NI 1501) proved to consist of granular calcite and clay minerals replacing feldspar laths and set in a matrix of clay mineral, (probably kaolinite), sericite and a little chlorite. Limonite is abundant replacing olivine crystals and, with calcite, filling vesicles.　　　H.E.W.

Chapter 10

STRUCTURE

INTRODUCTION

APART FROM THE Lower Palaeozoic rocks, the structural details of which are described in Chapter 2, the area has suffered little from folding stresses. Mesozoic and later rocks have a north-westerly dip of about 5°. Faulting of some magnitude, however, affects rocks up to and including the Lough Neagh Clays which are of Eocene to Lower Oligocene age. Faults of several generations can be recognized and some relative ages discerned.

Though there are seven major unconformities within the area of the One-inch Sheet and it might therefore be expected that periods of earth movement could be accurately dated, the limited cover of the more recent formations reduces their value for chronological purposes. While the folding and faulting of the Lower Palaeozoics can be shown to be primarily of Caledonian age, the absence of Carboniferous rocks at outcrop makes it impossible to distinguish any Armorican effects. Pre-Tertiary down-warping movement may be indicated by the great thickness (2,600 ft: 792 m) of lavas preserved in the Langford Lodge Borehole but subsequent movements appear to be nearly all later than the Tertiary volcanic period, though not, perhaps, all later than the intrusive phases of Tertiary igneous activity. It is suggested that the main structural trends were established in Caledonian times and that the main faults inherit their directions from that tectonic framework.

An examination of the One-inch Geological Map will show that the most important structural lines are the faults trending in an east-north-east direction (Fig. 15). With the exception of those delineating Ordovician inliers in the Silurian rocks, these faults involve Tertiary strata and their final movements must be presumed to be of Tertiary age. Two further sets of faults trending north-west and north-east may be distinguished.

The sequence of movements which may be distinguished within the limits of One-inch Sheet 36 is as follows:

(a) Caledonian folding and faulting

(b) Pre- and intra-Cretaceous faulting

(c) Faulting affecting sediments up to late Oligocene in age (oldest episode first)

Group 1. N.W. trending faults

Group 2. E.N.E. trending faults

Group 3. N.E. tending faults

(d) Minor faulting along the escarpment of no provable relation to (c)

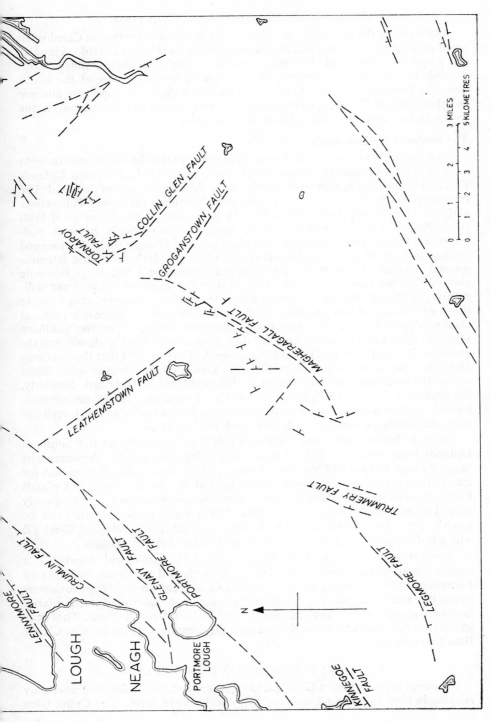

FIG. 15. *The principal faults in the Belfast area*

CALEDONIAN FOLDING AND FAULTING

The strike of the Lower Palaeozoic rocks is the normal north-east Caledonian trend and the major faults postulated to delineate four inliers of Ordovician run sub-parallel to this trend. The lack of palaeontological evidence renders it impossible to suggest the magnitude of these faults, but in the case of the inliers near The Temple they appear to be of very moderate throw. The apparent absence of other faults in this area, where they might be expected to be numerous, is due to the lack of exposure and palaeontological proof.

The Southern Uplands Fault

The northern margin of the Lower Palaeozoic massif has long been thought to be a close reflection of the supposed continuation of the Southern Uplands Fault which marks the south-east border of the Midland Valley of Scotland. The revision mapping has shown that the boundary between the Lower Palaeozoics and the Triassic sediments, formerly said to be almost rectilinear as if fault controlled, is in fact sinuous and far less straight than originally mapped. Furthermore, though a steep scarp with 'boiler-plates' of Silurian rocks was mapped in the Castlereagh area, the boundary has elsewhere little relief and is transgressed uncomformably by Permo-Trias sediments. The Long Kesh Borehole showed that the Lower Palaeozoic floor shelves gently under the Permian sediments. Anomalous records of breccia within the Bunter sandstones may indicate the continuation of this sediment to the west with Lower Palaeozoic rocks at shallow depths (see p. 35). Any major structure analogous to the Southern Uplands Fault must lie to the north of this. Anderson (1965) has shown that the claim by George (1960, p. 83) and Charlesworth (1963, p. 430) that the Southern Uplands Fault could be precisely located at Cultra (One-inch Geological Sheet 29) on the south-east shore of Belfast Lough could not be sustained. Similarly, the argument (George 1960, p. 39) that Belfast Lough may be the geomorphological expression of the fault is discounted by Anderson (p. 388). The Lough has been cut mainly in Mesozoic sediments and Tertiary lavas.

Both Anderson (1965) and Bullerwell (1961) pointed out that the Southern Uplands Fault cannot be identified on the gravity map, but the Aeromagnetic Map of Great Britain and Northern Ireland (1964) appears to offer support for the continuation of the Fault in Northern Ireland. Across the Southern Uplands Fault in south-west Scotland there is a considerable magnetic gradient, closely spaced isogams following the line of the Fault. This magnetic feature can be traced across the North Channel to Whitehead (One-inch Geological Sheet 29) where it disappears in the disturbance caused by the Tertiary Basalts.

South-westwards, beyond the basalts (One-inch Geological Sheet 47), a similar linear feature appears in County Armagh and continues into eastern Fermanagh. The Kinnegoe Fault (One-inch Geological Sheet 35) approximates to this line and within the present sheet the Legmore, Magheragall and Tornaroy faults are along a zone approximating to the Southern Uplands Fault. They may well have been instrumental in determining the formation of the present Chalk/Basalt escarpment.

PRE- AND INTRA-CRETACEOUS FAULTING

The area between Collin Glen and Groganstown shows evidence of instability ranging in time from post-Liassic to post-Antrim Lava Series. The Collin Glen

Fault was first active following the deposition of the Lias and the ensuing denudation removed the Lias, Rhaetic and part of the Keuper Marl west of Collin Glen.

This was followed by movement along the Groganstown Fault which probably gave rise to one of the most significant features in the history of the Cretaceous, namely the sudden disappearance of the Hibernian Greensands. Hancock (1961, p. 31) ascribed this disappearance to uplift in Senonian times. Furthermore, he stated that the general direction of this faulting was probably approximately parallel to the present outcrop, i.e. N.N.E. It may well be that the faulting was initiated in pre-Cenomanian times since there are several intra-Cenomanian unconformities. In the discussion on Hancock's paper, George (p. 34) said that the abrupt overstep of faulted Cenomanian [Hibernian Greensands] might reflect rift structure where the Southern Uplands Fault might be supposed at depth to lie beneath the Cretaceous rocks. The Cretaceous stratigraphy thus appeared to carry a mild impress of controls on sedimentation inherited from a tectonic framework of Caledonian and Hercynian establishment, despite the fact that the sediments post-date the latest known movements on the Southern Uplands Fault in Scotland (George 1960, p. 93 *et. seq.*). The limit of Senonian faulting postulated by Hancock (1961, Fig. 1) is hypothetical and a consideration of similar disappearances of the Hibernian Greensands at intervals along the Cretaceous outcrop (for example, at Woodburn near Carrickfergus) suggests that a series of *en échelon* faults similar in orientation to the Southern Uplands Fault may be the causal structures. The existing plexus of faults mapped in the Collin Glen area appears to be at variance with this solution.

Tertiary Faulting

North-West trending Fault Group

This group includes the Groganstown, Collin Glen and Leathamstown faults and an unnamed bifurcating fault under Belfast. In at least the first two cases Tertiary movement has taken place along earlier fault lines.

The Leathamstown Fault was postulated on geophysical evidence and is associated with a major positive aeromagnetic anomaly, which continues southwards into County Down. Its throw appears to be, however, only of the order of 150 ft (46 m). The geophysical evidence further suggests that it is terminated at its northern end by the Portmore Fault (possibly being displaced laterally north-eastwards for over half a mile) and is earlier than this east-north-east trending fault.

Similarly, the Groganstown Fault—50 ft (15 m) downthrow to the north-east—is cut off at its northwestern extremity by the Magheragall Fault. The branched plexus at the north-west limit of the Collin Glen Fault—100 ft (30 m) downthrow to the north-east—does not appear to cross the Magheragall–Tornaroy fault line. An unnamed fault from Brookhill to Brookmount has a similar orientation to the faults described above as has the unnamed bifurcating fault under Belfast. Throws of about 100 ft (30 m) in the directions indicated in Fig. 15 are typical of these faults.

East-North-East trending fault group

Faults in this group constitute the most important structural features of the One-inch Map. They fall into two main groups—a linear zone including the

Kinnegoe, Legmore, Magheragall and Tornaroy Faults which mark the Southern Uplands Fault at depth, and a belt of similar alignment to the north-west of the One-Inch Sheet which includes the Lennymore, Crumlin, Glenavy and Portmore Faults. There is borehole and geophysical evidence for the latter group. It is noteworthy that many of the large faults in the adjacent One-inch Dungannon (35) Sheet have a similar trend and there is some evidence for relating movements along them to a much earlier period, though only along the Low Cross Fault is the movement demonstrably pre-Permian (Fowler and Robbie 1961, p. 135).

The continuation of the Kinnegoe Fault from One-inch Sheet 35 cannot be proved or disproved either from exposure or from geophysical evidence but it does not trend to the north towards Langford Lodge as suggested by Fowler and Robbie (1961, p. 139) who thought that it was one of the most important movements contributing to the formation of Lough Neagh. The major fault line in Sheet 36, the Portmore Fault, lies over one and a half miles to the north and has a northerly displacement of about 1500 ft (457 m). It appears to be more in alignment with an unnamed fault in Sheet 35 between Lough Gullion and Lough Derryadd. North of the Portmore Fault the two faults named Glenavy — 75 ft (23 m) southerly downthrow — and Crumlin — 250 ft (76 m) northerly downthrow — are relatively minor compensatory structures while the Lennymore Fault — southerly downthrow of 400 ft (122 m) — is of larger dimensions. The possibility cannot be excluded that the Lough Neagh Clays overlap across these faults though it is unlikely. The final movement on the fault lines of this group may therefore be of Oligocene age.

The Legmore Fault has a southerly downthrow of about 150 ft (46 m) and is similar in orientation to the Magheragall — southerly downthrow of 100 ft + (30 m)—and the Tornaroy Faults. This structural line post-dates the north-west trending fault group, but is itself affected by the north-east trending group. The Legmore Fault is terminated at its eastern end by an unnamed fault at Magheramesk and the Magheragall Fault at its western end is cut by faults in the vicinity of Mullaghcarton. The cumulative effect of these faults is to displace the main east-north-east structure southwards by between half a mile and one mile.

North-East trending fault group

This group includes the two faults just described, the Trummery Fault—downthrow 150 ft (46 m) to the east—and possibly the north-trending unnnamed fault through Kilcorig—downthrow 250 ft (76 m) to the east.

Thus, in addition to the main east-north-east-trending 'Caledonoid' group, there are two complementary directional sets of faults which pre- and post-date the main structural dislocation. It may be of some significance that the minor faults accord in direction with the 130° dextral wrench faults and the 205° sinistral wrench faults mapped by Griffith in the Lower Palaeozoic rocks of the Carrickfergus (29) Sheet. The fault pattern may well be a reflection of the Lower Palaeozoic basement structures along which there has been posthumous movement.

Minor Faulting

There are a number of minor faults localized along the escarpment of no provable relation to the faults described above. Commonly north-east trending. they are not infrequently associated with dykes.   P.I.M.

Chapter 11

PLEISTOCENE

PRE-GLACIAL TOPOGRAPHY

FROM THE RESULTS of nearly 2000 boreholes in and around Belfast it has been possible to construct a contoured map of the rockhead (Plate 9) which is now buried by thick glacial and post-glacial deposits. Boreholes in the contiguous One-inch Sheets (28, 29, 37) have been used to facilitate drawing the contours on the map margins. The contours are given in feet above or below Belfast Mean Sea Level which is 8·9 ft (2·7 m) above the old Irish Datum (Poolbeg). Irish Datum is approximately 8 ft (2·4 m) below English Datum and thus the levels given are slightly lower than if related to the datum in Great Britain (Dixon 1949). Levels in the harbour area have been recalculated from the original figures related to Harbour Datum which is 2·4 ft (0·73 m) above old Irish Datum.

From Plate 9 it can be seen that there are under Belfast several narrow, deep, presumably pre-glacial valleys. Of the main examples, one extends south-west toward the Bog Meadows, the second lies between the Malone Road and the River Lagan which it crosses near Lagan Vale, the third runs south-eastwards from the Abercorn Basin (Belfast Harbour). The last, which Hartley (1940, Fig. 1) showed trending towards Knock, is thought to terminate within Sheet 36 and the channel at Knock connects with another channel extending through the Kinnegar area. The floors of the channels lie at below – 100 ft (–30·5 m) M.S.L. for much of their lengths while depths of over –150 ft (–45·7 m) are common in the city area. The lowest level of the rock surface actually determined is –191·9 ft (–58 m) M.S.L., though the channels probably extend to a depth of –200 ft (–61 m) or more, in Belfast Lough. Hartley suggested that the channels correspond approximately to the present day Blackstaff, Lagan, and Connswater rivers.

The narrow deep valleys appear to have steep or near vertical sides and to be, in effect, gorges incised in the soft Bunter Sandstone. They are filled almost exclusively with boulder clay, or in their narrower parts, with sand and gravel.

Howell (in Brindley 1967, pp. 246–7) has collated boreholes in the Cheshire–Shropshire region and has shown that the sub-drift rock surface is characterized by narrow deep valleys graded to levels of the order of one, two and even three hundred feet (30–90 m) below present sea level. They were, he suggested, drainage features in a fossil landscape, overdeepened in response to falling sea levels prior to the 'Newer Drift' glaciations. Buried valley rock-bottoms graded to similar levels are known all round the British Isles, though the maximum depths quoted disagree due to the chance relation of the borings to the talwegs and their distance from ancient shores (Charlesworth 1957, p. 1253).

It seems probable that subaerial rivers eroded the channels during a protracted period when sea level stood over 200 ft (61 m) lower than at present (possibly linked with glacial abstraction), though they may have been modified by direct glacial erosion or by subglacial streams. They have been shown to precede the infraglacial beach in south Ireland (Lamplugh 1905, p. 39).

GLACIAL DEPOSITS

Varied glacial drifts cover the greater part of the area of One-inch Sheet 36 particularly in the Lagan Valley where the mantle is most complete. Of the drift deposits, boulder clay is the most widespread, and extensive spreads of sand, either interbedded with or superimposed upon it, are a feature of the Lisburn–Belfast area. Laminated clays frequently occur at the base of the sand and at some levels within the boulder clay but are limited in areal extent. Despite the extensive cover of the drift, good natural sections in these deposits are rare and only where temporary sections have been excavated have the glacial deposits been adequately described (Fig. 17).

Though the area was completely covered by the ice sheet, no profound modification to the pre-glacial topography of the area was caused by ice; the main effect was depositional rather than erosional. During the latter phases of the main glaciation at least one glacial spillway and some marginal channels slightly altered the drainage system. The distribution of drifts and drift-features is shown in Fig. 16.

Early observers in the area, such as Bryce and Hyndman (1843, pp. 738–40) made observations on the shell content of the drift which led to the erroneous conclusion that the boulder clay was marine in origin. MacAdam (1848, pp. 36–41; 1850a, pp. 250–65; 1850b, pp. 265–8) described drift in the railway cuttings in the vicinity of Belfast and first recognized the overriding of east Ireland by Scottish Ice. In 1866, Close gave a classical general account of the glaciation of Ireland and, incidentally, brought the term drumlin into glacial literature. In the following year Gray (1868, p. 34) noted glacial 'markings' around Belfast.

A brief general account of the glacial deposits was given by Hull and others in the first memoir on the district (1871, pp. 37–8).

From 1894 onwards the Belfast Naturalists' Field Club published a number of reports on the distribution of erratic blocks, many of which refer to the Belfast area. The foraminiferal content of the glacial deposits also received attention from Wright (1881, pp. 149–63).

The most comprehensive report on the drifts of the north-eastern part of the One-inch Sheet is contained in the Special Memoir 'The Geology of the Country around Belfast' (1904, pp. 47–121). For the first time in this area the tripartite grouping of the drift into Lower Boulder Clay, 'Middle' or Malone Sands and Upper Boulder Clay was recognized. Previously this subdivision had been recognized in northern Ireland by Hardman (1877). An important paper on the glaciation of north-east Ireland by Dwerryhouse (1923, pp. 352–422) includes a short account of the Belfast valley which incorporates lists of erratics published by the Field Club.

Charlesworth (1939, pp. 255–95) amplified Dwerryhouse's work and correlated the deposits of the Belfast district with those of the other areas.

Whilst there is little doubt that the tripartite division of the drift can be applied in the vicinity of Belfast (as indeed in many glaciated areas of Britain) the relationship of the various members of the series can seldom be demonstrated. Outside the area where the sequence is clear and the Malone Sands are present

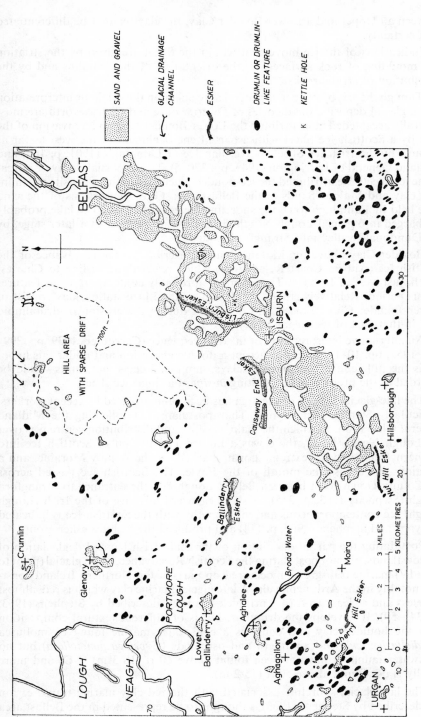

FIG. 16. *The distribution of sand and gravel and other glacial features in the Belfast area*

between an Upper and a Lower Boulder Clay, the clays cannot be differentiated with certainty.

Indications of the ice movement within the Sheet are given by the striation and moulding of rock surfaces, by the orientation of the drumlins and by the distribution of erratic rocks.

Though there is by no means general agreement on the origin or interpretation of the glacial deposits, the theories of Dwerryhouse and Charlesworth are most generally accepted. These attribute the Lower Boulder Clay to an invasion of the area by a Scottish Ice Sheet with a partial encroachment in the west as far as Lisburn by the Irish Ice Sheet — Ivernian Ice of Lamplugh (1901, p. 142) and Midland General of Farrington (1949, pp. 220–5). To the partial decay or retreat of the ice sheets are attributed the ice-ponded glacio-lacustrine deposits of laminated clay and Malone Sands in the Belfast area. Gravelly portions of the sand were held by Lamplugh to be the shoreline facies of the lake. The lake probably discharged into glacial Lough Neagh via the Broadwater or, at a later stage, by the Dundonald valley into Strangford Lough.

Boulder Clay above the Malone Sands is ascribed to the re-advance of the Scottish Ice (Antrim Coast Re-advance) which reached, according to Charlesworth, just west of Lisburn. The eskers are probably contemporaneous features of this period. Relatively stone-free clays near the Maze station may consist of Lower Boulder Clay resorted and deposited during the erosion of drumlinoid drift in that part of the Lagan valley.

An alternative interpretation of the drift sequence (Carruthers 1939, pp. 299–328; 1953, pp. 10–11) does not accept a double glaciation and under this hypothesis the full sequence of boulder clays, laminated clays and sands would be regarded as the product of the melting *in situ* of a single ice sheet.

The late-glacial history of the area has been re-interpreted by recent workers, particularly Stephens and Synge. They postulate that following the Midland General Glaciation, in which the main ice-shed and distribution centre was over the Lough Neagh Basin, there was a limited penetration of Scottish Ice into northern Ireland in the extreme north, marked by the Armoy Moraine, and a morainic line across the mouth of the Foyle. The Scottish Ice passed across the County Down coast between Belfast Lough and the entrance to Strangford Lough (Stephens 1963, p. 345). A short eastward advance of the Irish (Lough Neagh) Ice contemporaneous and in sympathy with the Scottish Ice is indicated by Synge and Stephens (1966, p. 114) and produced the Lisburn esker group.

Post-dating the partial dissolution of the main Irish Ice and deglaciation of the coastlands, but not post-dating the Scottish Re-advance, a late glacial (prior to Zone IV) marine transgression occurred on parts of the Northern Ireland coastline, notably in the Ards peninsula. A late glacial marine clay which is red, almost stoneless and at least 15 ft (4·6 m) thick has been described by Stephens (1963, pp. 345–59). It envelops the drumlins, but is of different textural composition from the boulder clay and contains a cold-water marine fauna of mollusca (*Yoldiella lenticula*), foraminifera and ostracods (*Cypridea punctellata*) but no freshwater fauna. It has not been found above 60 ft (18·3 m) O.D. and more usually is not found above 50 ft (15·2 m).

The late-glacial shoreline associated with the red clay marine transgression and described by Stephens (1963) is thought to be represented in the Belfast area by the wide abandoned flood-plain of the Lagan at 50 ft to 80 ft (15–24 m). The post-glacial shoreline is recorded at 25 ft to 33 ft (7·5–10 m) O.D.

Belfast is recorded by Stephens as an uncertain but probable locality for the red clay, but a re-examination by C. J. Wood of specimens described by Kennard from red clays in Belfast Harbour has revealed that a similar fauna in red clay occurs at the East Twin, Belfast below the Estuarine Clays. These coldwater forms include *Nuculana rostrata* (Gmelin) recorded by Kennard as *Nuculana pernula* (Müller) and *Portlandia* (*Yoldiella*) *philippiana* (Nyst) recorded by Kennard as *Yoldiella lenticula* (Möller).

The following table attempts to summarize the glacial deposits of the district and suggests their origin in upward chronological order:

Deposits	*Origin of deposits*
Eskers	Meltwater retreat deposits of final retreat
Upper Boulder Clay	Ground moraine of Antrim Coast Re–advance
Red Marine Clay	Marine transgression
(Belfast melt-water channels and Broadwater spillway)	
Malone Sands	Meltwater deposits of retreat stage of Main Glaciation in glacial lake
Laminated Clays	Lagan ponded by coastal ice
Lower Boulder Clay with drumlins	Ground moraine of (1) Scottish Ice (2) Scottish and Ivernian Ice of Main Glaciation (Midland General)

BOULDER CLAY

Over most of the area the main drift deposit is boulder clay which is essentially ground moraine of the ice sheets and may form a featureless spread. In the Aghagallon area and in much of the Lower Palaeozoic country drumlinoid topography is, however, more characteristic. The drumlins are low streamlined elliptical hills of boulder clay, sometimes mounded around an outcrop of grit which forms a rock-nucleus or core. They commonly show preferred orientation resulting from the interaction of the main ice movement and the sub-ice topography. In the County Down area they are packed in a close-set series and form the well known 'basket of eggs' topography typical of that county (Plate 7B).

The country west of the Divis range of hills is blanketed with a thick cover of boulder clay which is thickest on the low ground round Lough Neagh and on the gently rising slopes east of the main Lisburn–Crumlin railway and which thins out towards the hills on the east. Thicknesses of over 80 ft (24 m) of boulder clay have been reached at several places. The basalt scarp which rises above the Lough Neagh plain, west of the railway, is in general less thickly mantled. Scattered groups of drumlins, generally aligned west-north-west, occur over the western part of the area. They rise from the boulder clay blanket and occasionally appear to have rock cores.

Over the basalt plateau the boulder clay mantle is fairly complete. It thins towards the high ground north-west of Belfast, but even here pockets of drift or ice-borne erratics are present at high levels. The precipitous slopes of the escarpment are drift free in parts.

Over most of the ground underlain by the Antrim Lava Series the boulder clay is a reddish brown, somewhat friable clay with a very heavy load of basalt blocks and relatively few exotic boulders, mainly flint. The lower part of the clay is generally darker than the top and near Lough Neagh this darker clay contains appreciable amounts of schistose and Carboniferous rocks. Beds of sand, gravel and silt interbedded with the boulder clay are very common in the area north of a line from Glenavy to Dundrod and examples are seen in several river exposures. Elsewhere such intercalations are common.

The boulder clay on the Trias outcrop to the south-east differs in character from that on the basalts. It is redder in colour, more plastic and fine-grained where it overlies the Keuper marls, and contains fewer stones. In general, it is also thinner and it seems probable that the thickness of ground moraine may depend to some extent on the plasticity and shear-strength of the clay.

The boulder clay in the south-east corner of the Sheet is entirely in the form of drumlins which occur in groups and as isolated hills, often *en échelon* or in lines, and are aligned north-west over most of the district, though in the extreme eastern part of the sheet they have a more northerly trend. In this area there is virtually no drift on the rock-head between the drumlins but the hollows between the hills are commonly occupied by small lakes, as on the southern edge of the Sheet, or by the alluvium and peat which have succeeded them.

The boulder clay in this area is a tough brown clay, heavily charged with blocks of the local Palaeozoic rocks and with occasional basalt, chalk, flint and other exotic debris. There are no known intercalations of sand or gravel in this area.

Between Belfast and Lisburn an Upper Boulder Clay with a much lower stone content has been recognized, separated from a Lower Boulder Clay by the Malone, or 'Middle', Sands. Typically, this clay is a red or reddish brown unctuous plastic clay with occasional traces of laminations and containing very few pebbles. The Upper Boulder Clay can be recognized with certainty only where it is underlain by the sands and it occurs as sheets and disconnected patches overlying the sands.

No conclusions were drawn as to the origin of this Upper Boulder Clay by Lamplugh, and Charlesworth does not specifically refer to it as the ground moraine of his Antrim Costal Re-advance. Dwerryhouse regarded the Upper Boulder Clay as evidence for a re-advance during the general retreat of the ice.

The erratics of the boulder clays consist of both local and far travelled material. Two suites of the latter are recognized as emanating from Ailsa Craig and Arran. Other recognizable erratics include the Cushendun microgranite, the Tardree rhyolite, various members of the Tyrone igneous and metamorphic series and Carboniferous limestone. Marine shells have been recorded from some localities.

SANDS AND GRAVELS

Malone Sands. Extensive spreads of sand and gravel exist in the Belfast–Lisburn area and are partially overlain by Upper Boulder Clay (Fig 16). They were probably laid down in Glacial Lake Lagan during the glacial retreat which preceded the formation of the Upper Boulder Clay. The subsequent re-advance may have obscured their original shape though in part the surface is hummocky. Two kettle-holes have been mapped. Much of the material is clean stratified red or reddish brown sand of Triassic derivation with some gravelly layers and thin bands of brown clay, occasionally laminated. Beds of coarse gravel occur locally and Lamplugh regarded these gravelly portions as probably indicating a shore-line of the glacial lake which he considered responsible for the deposition of both the Malone Sands and laminated clays.

Upper Sands with Eskers. A few masses of sand and gravel of fluvioglacial type, all restricted in area, are free from any cover of boulder clay and retain their original form apparently unmodified by subsequent denudation. The most con-spicuous are the eskers between Dunmurry and Red Hill and their continuation westwards to near Aghalee, together with other eskers west of Moira and Hills-borough. They were probably laid down from melt-waters during the final retreat of the ice.

The Ballinderry Esker is a discontinuous sand and gravel ridge which can be traced for 4 miles (6·5 km) from near Brookhill to Ballinderry House. It is separated by a gap of about 1½ miles (3·9 km) from the much bigger Lisburn Esker.

Outcrops of sand and gravel in the western part of the sheet are located by the villages, most of which are sited on areas of these deposits. Aghalee rests on the gravel deposited at the outfall end of the Soldierstown overflow channel; Lower Ballinderry on the gravel spread at the western end of the Ballinderry Esker. Crumlin and Glenavy, too, are built on sands and gravels, the former perhaps on an intercalation in the boulder clay, the latter on sands probably connected with the small glacial overflow channel near the village. In addition to the larger areas of sand and gravel on which the villages stand, there are a number of small widely scattered patches, generally of gravel, which are probably in some cases intercalations in the boulder clay. Those west of Aghalee, however, are probably the remains of the formerly more extensive deposits of the Soldierstown overflow which doubtless poured into standing water when the level of Lough Neagh was higher than at present.

A few small areas of coarse gravel and sand occur in intimate association with drumlin boulder clay. The gravel occurs as patches on the south-eastern ends of drumlins in the Aghagallon area. Similar patches were mapped on the Dun-gannon (35) Sheet to the west.

LAMINATED CLAYS

At the base of the Malone Sands in the Belfast area, and at some levels within the Lower Boulder Clay, deposits of chocolate brown silty laminated clays have been noted. They were exposed principally in the now disused Stranmillis and Annadale brickpits and at Lagan Vale. These laminated clays have been usually interpreted as being the finer sediments deposited on the floor of Glacial Lake Lagan during the retreat stage of the main glaciation. Carruthers (1953, p. 6), however, regards clays similar to these as 'shear clays' or 'pressed clays' attribut-able to the bottom melting of the ice sheet.

GLACIAL DRAINAGE CHANNELS

The ponding of the drainage by ice during the retreat phase of the Main Glac-iation has left few overflow channels incised into solid rock. The largest overflow valley is that followed by the Lagan Navigation Canal at Soldierstown, near Moira. It drained Glacial Lake Lagan into Glacial Lough Neagh. The intake at Boyle's Bridge on the main Belfast–Portadown road just east of Moira is at about 140 ft (42·6 m) O.D. and the outfall at about 70 ft (21 m). The broad flat floor is now partially obscured by the disused Lagan Canal and by the Broad Water lake. The overflow may have been later partially re-excavated by melt waters from the Antrim Coast Re-advance ice.

Small marginal channels occur high on the basalt area to the north-west of Belfast. A group of three, falling southwards, is seen on the eastern face of Divis and two others are seen near Dandry Bridge [250 767].

Just south of the Dispensary at Glenavy a striking dry valley 50 ft (15 m) deep cuts a sinuous course across the boulder-clay spur between the Lisburn and Moira roads. This channel is about 400 yd (365 m) long and the water which cut it flowed from the north-east. The level of the intake is about 250 ft (76 m) O.D.

GLACIAL STRIAE

Clearance of drift from the area west of the working face at Carryduff Quarry, east of Lough Moss, revealed an excellent glaciated rock surface on the grey-wackes, with incipient roches moutonnées and well developed striae indicating ice movement from the north-north-west (Plate 7A).

A rock knoll in a field 1400 yd (1280 m) S. of the church at Upper Ballinderry shows less well defined east–west striations which are possibly glacial.

P.I.M., H.E.W.

DETAILS

Area North and West of the basalt escarpment. The Langford Lodge promontory and the area north of the Crumlin River are covered with thick boulder clay, gently undulating over most of the district but with a few well-marked scarp features parallel to the edge of the alluvial plain at the mouth of the river. These may be due to terracing by the lough when it stood at a higher level, but the northernmost, running east-north-east from Gartree House, corresponds to a known rock outcrop and may be in fact a solid feature.

The boulder clay is well exposed in the steep bank which backs the lake terrace from Lennymore Bay to the northern edge of the Sheet and reaches a height of 50 ft (15 m). It is generally a typical basaltic till with plentiful basalt blocks in a rather friable reddish brown clay matrix. Extraneous pebbles are rare and flint is the only common exotic. In the cliff, 1 mile (1·6 km) E. of Gartree Point the boulder clay seen in occasional exposures over a distance of some 400 yd (365 m) is intimately interbedded with sands and gravels. Two hundred yards W. of Nettletons Pad Plantation [103 748] 5 ft (1·5 m) of boulder clay are seen to overlie 3 ft (1 m) + of khaki brown fine-grained sand at the top of the bank, and 100 yd (91 m) to the east, 6 ft (1·8 m) of coarse gravel are seen in the over-grown slope. Just south of the Plantation the bank is largely obscured by slips but boulder clay with basalt blocks at the top overlies a bed of sand 2 ft (0·6 m) or more thick seen 12 ft (3·6 m) down, and 100 yd (91·4 m) further east the 40-ft (12 m) cliff shows only 5 ft (1·5 m) of sandy boulder clay over 20 ft (6 m) of iron-stained gravel in coarse and fine bands with some beds of sand resting on over 5 ft (1·5 m) of poorly bedded silt. The base of the cliff is obscured by slips. A further 70 yd (64 m) E. the height of the cliff has fallen to 30 ft (9·1 m) and only 15 ft (4·6 m) of boulder clay are seen, the base being hidden by falls. Exposures east of this are all in boulder clay but one shows thin bands of silt, apparently interbedded.

There are few exposures of the boulder clay in the area inland from the lake shore, but those few all show normal basaltic till. Known thickness of drift varies from 30 ft (9·1 m) at Langford Lodge to 60 ft (18·3 m) at Largy House. A well 100 yd (91 m) S.W. from Gortnagallon School and another at Gortnagallon House are both reported to have entered gravel from which plentiful water supplies are obtained.

The village of Crumlin is built squarely on top of a small spread of sand and gravel. The deposits are best seen in the bank of the river south of Glendaragh where 20 ft (6 m) of silty and laminated sands rest on the underlying boulder clay. Sand was seen in various excavations in and near the village and along the road which runs westwards to Mulleague House where up to 9 ft (2·7 m) of earthy gravel were seen in a trench. There is a small outlier of gravel on the north bank of the river above the Saw Mill, where over 6 ft (1·8 m) are exposed at the top of the steep bank. In view of the extensive inter-calculations of sand and gravel known in the boulder clay in this district it is possible that these Crumlin deposits may be of the same type.

The strip of country between the Crumlin and Glenavy rivers rises steadily from the shores of Lough Neagh to the hills beyond Dundrod and is covered by a thick blanket of boulder clay which only rarely allows rock to appear at the surface. The surface is generally undulating and only in occasional river sections are the deposits normally exposed.

In the area west of the railway the thickness of the boulder clay, as known from well sinkings, is generally fairly thin and apparently does not exceed 25 ft (7·5 m). Sections seen in streams at Cidercourt Mill [135 765] and east of Thistleborough [133 745] are of stiff brown clay with abundant basalt blocks, occasional chalk, flint, and quartz pebbles and rare lignite fragments. At the Plantation [138 753] the clay is reddish brown and contains some red laterite debris.

In the south bank of the Crumlin River at Glendaragh the boulder clay is over 50 ft (15 m) thick and is overlain by glacial sands. The upper part of the clay is reddish brown in colour, while the lower part is dark brown.

Just east of Crumlin, in the small stream which flows through Entwistlestown, reddish brown boulder clay with basalt blocks and some chalk and flint, about 20 ft (6 m) thick, rests on silty sand with poor laminations and cross-bedding which passes down into gravel with sand lenses at stream level. Rock is seen in the other bank of the Crumlin river a few yards away and is probably not far below this section, so the total sand and gravel thickness is probably not much more than the 6 ft (1·8 m) seen. Two hundred yards (183 m) away, due east of Glen Oak, a cliff in the left bank of the Crumlin River shows a section 30 ft (9 m) high through a variety of drift deposits. The uppermost 10 ft (3 m) which may possibly be high level river terrace deposits, are earthy gravels, with angular blocks of basalt up to 1 ft (0·3 m) long, and a bed 3 ft to 5 ft (0·9–1·5 m) thick of false-bedded reddish brown sand. The irregular but near horizontal base of the gravel rests on reddish brown boulder clay 7 ft to 13 ft (2–4 m) thick with its base dipping south at about 10°. This overlies a bed of gravelly sand, with occasional large stones, about 5 ft (1·5 m) thick which rests on a bed of coarse gravel about 3 ft (0·9 m) thick with cobbles up to 9 in (0·23 m) diameter of basalt and occasional quartz, flint and lithomarge pebbles. These beds dip south at 10°. The lowest bed seen is dark brown sandy silt with scattered basalt pebbles and blocks up to 15 in (0·37 m) long, mainly angular. This band is at least 5 ft (1·5 m) thick and its base is not seen.

In the area between Crumlin and Dundrod the undulating country affords few exposures of the drift save for occasional ditches and the banks of the Crumlin River and its tributaries. Information from well records indicates that the thickness of the drift varies considerably, up to 79 ft (24 m) being recorded in a well near Dundrod while depths of 30 ft to 40 ft (9–12 m) are common. Occasional mention of gravel bands and 'gravel bottoms' suggests that over much of this ground the boulder clay is probably intimately associated with deposits of this type. Near Crawfordstown [200 756] a small pit in earthy gravel suggests a band in the boulder clay and an old rath 300 yd (275 m) N.W. of the farm is built of earthy sand and gravel on a boulder clay knoll. In the Cochranstown [213 745] area the soil is light and sandy in places though exposures are all of boulder clay, and a field ½ mile (800 m) W. of the farm is known as the 'sandhole field' though exposures in ditches are of boulder clay. About 500 yd (457 m) W. of Bell's Hill old clay pits [193 751] in dark brown plastic clay with some stones are seen above an alluvial flat.

East of Dundrod there are a few well-formed small drumlins and beyond them the drift thins upwards on the rising ground towards Armstrong's Hill and Standing Stones Hill though the boulder clay is still fairly thick in the river valley. A well near Budore Bridge [248 760] is said to have penetrated over 20 ft (6 m) of till. There are a large number of small patches of sand and gravel in the valley near Budore Bridge and on the west flank of Budore Hill. These deposits are of very limited extent and have only been worked locally by small pits. A well [244 767] west of Dandry Bridge is reputed to have been sunk to 25 ft (7·6 m) in sand in one of these mounds.

Like Crumlin, the village of Glenavy is built on a spread of sand and gravel. Wells in the village are reported to have penetrated up to 30 ft (9·1 m) of these deposits, and a well in an abandoned sand pit just north of the railway station is reported to have met 20 ft (6 m) of sand resting on boulder clay. The sands extend to the south beyond Goremount and a small gravel pit is seen east of the railway 200 yd (183 m) from the house. To the north-east of the village the gravels extend over most of the area between the mill race and the river and 9 ft (2·7 m) of gravel are visible 400 yd (365 m) E. of St. Aidan's Church. On the west side of the river, old gravel pits occur north of Quigleys-hill Head [152 730] and on both sides of the railway to the south while the steep bank above the river terraces west of the Flock Mill [155 727] shows about 30 ft (9·1 m) of earthy gravel.

Small gravel pits on both sides of the road 300 yd (275 m) N.W. of the Vicarage indicate an outlying patch of gravels, and other small excavations in earthy gravels are seen on the south side of Devlin's Hill and 500 yd (457 m) W.N.W. of Ballymacricket School [149 717]. While the small outliers, particularly the last two, may be interbedded lenses in the boulder clay, it is probable that the main spread at Glenavy is of outwash gravels associated with the glacial spillway south of the village.

East of Glenavy, in an area of unbroken drift some 3 square miles (8 km^2) in extent, the boulder clay reaches a considerable thickness. Drumlin-like hills, probably with rock cores, and aligned north-west to south-east are seen on the southern edge of the area at Carnkilly Hill and Steels Hill but over the rest of the district the topography is of moderate relief with only a vague north-west grain. As in the area to the north, the boulder clay is commonly interbedded with sands and gravels and laminated clays. Five hundred yards (457 m) E.N.E. from St. Aidan's Church, Glenavy, beds or lenses of brown stoneless silt are seen interbedded with boulder clay in the south bank of the Glenavy River, and a well ¼ mile (400 m) N.W. of Beech Vale [181 720] is reported to have penetrated, below 8 ft (2·4 m) of red till, 10 ft (3 m) of blue and purple mottled clay, and 30 ft (9·1 m) of pink and purple laminated clay which came out 'like broken tiles'. It is possible, however, that this material may be an unusual weathering product of a flow-banded lava as the underlying rock is described as hard glassy or rosin-like and might be an obsidian, though nothing like this is known in the area. No specimens could be obtained.

Wells near Glenavy itself reached a depth of 83 ft (25 m) in unbottomed dark boulder clay and a depth of 75 ft (23 m) in clay, unbottomed, is recorded in Tullynew-bank. At Ivy Hill [180 718] rock-head is 44 ft (13 m) down and wells in Ballypitmave reach rock below 20 ft to 25 ft (6–7·6 m) of boulder clay. In general it is clear that the drift thins to the east and there is a well-marked ridge of rock running north-east across Cairn Hill and Crew Hill which is in places free of drift.

East of this ridge, in the area round Leathamstown and Stonyford reservoirs, the boulder clay cover is generally thinner and follows the rock-head contours more closely but there are some well-formed small drumlins, aligned west-north-west, just east of Garlandstown [191 730] and in the area between Stonyford and Forked Bridge. On the flanks of these, thicknesses of 65 ft (20 m) of boulder clay at Killultagh House [193 690] and 75 ft (23 m) unbottomed in Ballynadolly [209 690], are recorded. Two hundred yards (183 m) downstream from Leathamstown Reservoir 25 ft (7·6 m) of boulder clay are seen in a bank, while at the upper end of the reservoir a 30-ft (9·1 m) bank is all in clay. A little to the west of this on the south side of the reservoir a 4-ft (1·2 m) band of laminated clay, dipping west at about 20° is seen apparently overlain and underlain by boulder clay.

The ground south of Forked Bridge is thickly covered with drift and thicknesses of up to 70 ft (21 m) of boulder clay are reported in wells.

East of Leathamstown the drift thins against the flanks of Priest's Hill, Collin Mountain, and White Mountain, but recognizable boulder clay can be seen up to well over 1000 ft (305 m) O.D. in many areas. Few figures for thickness of drift are available in this district, but a well [242 724] on the south side of Rushy Hill recorded 24 ft (7·3 m) of boulder clay.

There are small areas of sand and earthy gravel east of Groves Bridge and east of Rushy Hill.

In the area west of the railway and south of the Glenavy River the boulder clay is less blanket-like and rock exposures are more frequent. In the district round Fruit Hill [147 704] and The Bleary [148 709] there are a few drumlin-like hills of boulder clay, trending north-west and further south the general grain of the undulating drift is in the same direction with occasional small drumlin mounds particularly round Upper Ballinderry. Still further south, in the townlands of Moygarriff, Aghadavey and Bally-lacky, there is a well-developed swarm of small drumlins trending west-north-west to east-south-east. Over the whole of this area the boulder clay is a homogeneous till, usually very heavily loaded with large basalt blocks. As elsewhere, the upper part is reddish brown while the lower part is darker in colour. There are very few records of bands or lenses of other material in the boulder clay in this district, but a well [118 666] is said to have finished in a bed of rounded gravel. The thickness of the boulder clay is rarely more than 30 ft (9 m) though a thickness of 68 ft (21 m) was reported in a well at Legatirriff and the thickness in some of the drumlins may exceed this.

It is very difficult to estimate the thickness of the drift on the low ground near the Lough where the underlying rock is Lough Neagh Clay, both from scarcity of informa-tion and the difficulty of deciding whether the 'clay' described by informants belonged to the latter formation. In general, the boulder clay appears to be less than 25 ft (7·6 m) thick but one well at Moss Vale apparently penetrated 45 ft (14 m). On Derryola Island wells dug just before the time of survey are said to have met 10 ft to 12 ft (3–3·6 m) of red boulder clay and then passed into dark brown sandy clay, very hard and compact, full of small rounded and subangular pebbles of basalt, chalk, quartz, and flint, with occasional larger cobbles of basalt, flint, Carboniferous sandstone and limestone, hornblende-schists and other schistose rocks. Thin streaks of lignite are said to occur though they were not seen in the spoil heaps. There can be no doubt that this is a boulder clay and it reaches an unbottomed thickness of 70 ft (21 m).

Isolated patches of sand and gravel occur at Portmore House [133 686] where a well penetrated 13 ft (4 m) of sand, and Irwinstown [152 692] where 4 ft (1·2 m) of bedded gravel and coarse sand were seen in a trench on the road. Both these occurrences are probably lenses in the boulder clay, as is the locality at Brackenhill [153 679] where old pits are said to have worked gravel. Fairly extensive spreads of sand and gravel round the village of Lower Ballinderry are clearly connected with the Ballinderry Esker which runs across country to the south-east. On the west side of the Lurgan–Crumlin road a crescentic area, ½ mile (800 m) long, south of Ballinderry Mill [124 679], is underlain by sands and gravel and a well 230 yd (210 m) W.N.W. of the Moravian Church is reported to have been sunk through 28 ft (8·5 m) of sand to rock-head. East of the main road an extensive area centred on Ballinderry House and a small patch at the Cottage [125 668] are also on gravels. Small excavations north of the road 250 yd (229 m) E. of Ballinderry House reveal cobbles and gravels, the coarse cobbles being all of basalt while the gravel is of basalt, both rounded and subangular, flint, quartz, Lower Palaeozoic grits and quartzite.

The Ballinderry Esker is, for much of its length, a discontinuous string of small mounds of earthy gravel and sand which lie on both sides of and athwart the Ballinderry–Lisburn road for about 4 miles (6·5 km) and follow the course of the valley of the Ballinderry River. The westernmost sections are contiguous with the Ballinderry gravels, rising from the river alluvium near Alma Farm, and followed by Bunker's Hill to the east. A gap at the 'Middle' church [151 672] is followed by a mound west of Glen Villa, and another gap at Upper Ballinderry is succeeded by the mound east of Rosevale. From here eastwards the esker is more typically formed and can be followed as a succession of elongate ridges through Lillyvale [172 664] and Bessvale [178 662] to Elm Mount [196 662]. The road runs along the ridge for $\frac{2}{3}$ mile (1·1 km) W. of Elm Mount. All the ridges are low and the greatest thickness of sand and gravel recorded is 15 ft (4·6 m) in a well at Lillyvale and 15 ft (4·5 m) in a well 400 yd (365 m) N. of Springvale [189 658]. In most wells the gravel overlies boulder clay but in the area north of Springvale this is absent and the esker rests on rock-head. An outlying sector of the long ridge is cut off by a small stream south of Elm Mount and is partly exposed in an old gravel pit south of the Orange Hall where 3 ft (1 m) of iron-pan-cemented gravel are seen.

Half a mile (800 m) E. of the end of the Elm Mount ridge a small outlier of gravel forms two small mounds on the north side of the road at Belvista [206 658] which are just 1 mile (1·6 km) from the western end of the Lisburn Esker (p. 140).

The village of Aghalee is sited on the eastern end of a large gravel spread which extends as far as The Cairn [112 649]. Wells in the village record up to 12 ft (3·7 m) of gravel and sand. The gravel consists mainly of basalt pebbles with some flint and Lower Palaeozoic grit. Small outlying patches of sand and gravel occur north-east of Cairnhall [109 654], at Thornleigh [105 660] where a well met 12 ft (3·7 m) of sand, and in the orchard west of Moss Vale [111 666]. These Aghalee gravels were probably deposited by the waters issuing from the Soldierstown spillway.

East of Aghalee and north of the Soldierstown channel there are a few small patches of sand and gravel. At Campbell's Hill [135 660] a gravel pit shows 5 ft (1·5 m) of unsorted gravel with rounded and subangular boulders, cobbles and pebbles of basalt, and some quartz, flint, and pink granite. A similar coarse gravel covers a small area at The Grove [142 656], and a mound north of Fortland House [137 647] is composed of earthy gravel. Small gravel pits in two other small mounds, at Friar's Hill, and on the south-east end of Fairies Hill, show earthy gravels. Some of these occurrences may represent intercalations in the boulder clay, but there is little other evidence of such interbedding in this district.

The district south of the railway and east of the Ballinderry–Moira road is covered with undulating boulder clay often in the form of small drumlins with the usual west-north-west orientation. Rock is exposed at several points but over much of the area the drift cover must be fairly thick; 40 ft (12 m) unbottomed at Hallsville [183 655] and 55ft (17 m) unbottomed at Cottage Hill [177 667] are the greatest depths recorded.

A quarter of a mile (400 m) N.E. of Oakville [188 650] a section in a small stream shows the boulder clay overlying and apparently transgressing across horizontally laminated sandy clay which contains grains of chalk, basalt, and lithomarge. Apart from this occurrence the boulder clay is a normal stiff reddish brown clay with abundant basalt stones, of the type common all over the basalt outcrop.   H.E.W.

The district between Aghalee and Lurgan is covered with undulating boulder clay, with well-formed drumlins in the area south of Aghagallon. Small patches of sand and gravel occur west and south-west of Aghagallon, with larger mounds at Islandhill [100 643] and near Killaghy Corner [087 617]. The former has a sand pit showing 18 ft (5·4 m) of coarse sand and pebbles.

Several of the drumlinoid mounds of boulder clay in this area have caps of sand and gravel on their southern ends. Wells sunk in the mounds suggest that the gravel is a surface capping and is not interbedded in the clay. Good examples of this phenomenon are seen in Round Hill [097 597] and Brocker Hill [094 610].

A sinuous esker follows the valley of a small stream north of Cherry Hill [122 603] for over 1 mile (1·6 km) and is continued to the west as patches of sand and gravel in the Kilmore area.

The urban area of Lurgan is underlain by boulder clay of moderate thickness— usually about 20 ft (6 m) in the rather scanty records available. The boulder clay was used for brickmaking in the area, now built over, of Mark Street and in the Bellevue area north of the town. Small areas of gravels occur in Brownlow Park and at Dollinstown.

The Lagan Valley. The drifts of this area are extremely diversified and include both Upper and Lower Boulder Clay and the Middle or Malone Sands.

North-west of Belfast the hummocky ground formed by the Keuper Marl immediately below the escarpment grades out to a more even slope covered by boulder clay— stiff reddish sticky clay with a matrix largely composed of the underlying mudstone. The drift under parts of the north-western suburbs of the city is thin. At Parkview Brickworks up to 10 ft (3 m) are seen resting on Marl and at the west end of Fifth Street just south of the Shankill Road only 3 ft (0·9 m) rest on sandstone. Along the lower course of the Forth River the drift is usually from 3 ft to 6 ft (0·9–1·8 m) thick but near Monk's Hill [300 759] 35 ft (10·7 m) of boulder clay are seen in the Ballygomartin River. Thicker deposits are recorded in trial holes and wells. At the Great Northern Mineral Water Company's premises the drift was recorded as 25 ft (7·6 m) thick and rested on sandstone. Twenty-five feet (7·6 m) of boulder clay are recorded at Luke Street south of the Shankill Road and 10 ft (3 m) of boulder clay resting on marl at the Crumlin Road Prison are mentioned by Young (1871, p. 33) though a recent bore at Court Street, south of the Crumlin Road and immediately opposite the prison, entered 30 ft (9·1 m) of drift. Several shallow holes show up to 3 ft (0·9 m) of sand intercalated in the boulder clay and at a site at Millfield up to 16 ft (4·9 m) of red sand were found under a thin cover of clay. In the excavation of the old Waterworks [329 766] shells were obtained from the boulder clay and accounts were published by Bryce and Hyndman (1843) and Bryce (1845), and a list of shells was given in a paper by Stewart (1881). Wright (1879) recorded six species of foraminifera. McAdam (1850) recorded further shells from an excavation at the Grove [c. 340 765].

Along the Forth River and at Old Park several brickyards, long disused and now built over, appear to have used boulder clay and laminated clays beneath it. Lamplugh (1904, p. 69) described stratified 'warp' clay beneath boulder clay in the 'Blackwater' brickyard (site not now identified but probably about [320 740]) and at the west end of Springfield Dam [311 741]. More recently a site investigation at the Royal Victoria Hospital revealed up to 18 ft (5·5 m) of laminated clay overlain by 32 ft (9·8 m) of boulder clay and resting on brown sand. These records indicate the north-western extent of laminated clay in the Sheet. About 10 ft (3 m) of laminated clay under 30 ft (9·1 m) of boulder clay are recorded in a bore at the Gasworks close to the Lagan [344 734] on approximately the same latitude.

Boreholes in the western part of the City Cemetery reveal a minimum of 13 ft (4 m) of boulder clay, whilst the eastern part is nearly drift-free. From boulder clay in the stream in the western edge of Falls Park, Stewart recorded several shells (Lamplugh 1904, p. 70).

On the eastern side of the alluvial plain of the River Blackstaff the western slope of the ridge between the Blackstaff and the Lagan is underlain by boulder clay, though in parts the clay is thin. At the Malone Training School, for example, sand occurs beneath an 8 ft (2·4 m) clay cover and 200 yd (182 m) N. of this, soft yellow sand was reached under 3 ft (0·9 m) of boulder clay. East of the Lisburn road the Malone Sands and Upper Boulder Clay underlie most of the ground to the Lagan.

In the excavation for the Keir Building of Queen's University [334 723] up to 35 ft (11 m) of reddish sand were recorded in boreholes. About 20 ft (6 m) of fine reddish sand with occasional thin clayey bands were seen during the foundation excavations. The overlying Upper Boulder Clay was seen nearby on the lower slopes of the hill leading down to Methodist College and College Gardens. In the excavation for the New Physics Block at Queen's [331 725] up to 14 ft (4·3 m) of Upper Boulder Clay—a reddish brown rather sandy clay with low stone content—were seen resting on Middle Sands (Fig. 17, 2). The top 8 ft (2·4 m) of sand showed thin bands of boulder clay and in parts the sand was full of clay streaks. A borehole adds a further 55 ft (17 m) to the thickness of sand at this locality. A pipe of Upper Boulder Clay about 3 ft (0·9 m) by 2 ft (0·6 m) extended downwards an unknown distance into the underlying sand.

The Upper Boulder Clay extends west and north to the margin of the Estuarine Clay. The former Windsor and Central Brickworks, on the Donegall Road, long built over, probably used this material. The boundary between the Upper Boulder Clay and Middle Sands between University Road and the Lagan is difficult to trace for lack of evidence. A bore near the Presbyterian College records an alternation of clay and sand.

Southwards to Stranmillis occasional patches of Upper Boulder Clay were recorded by Lamplugh (1904, p. 71). South of Stranmillis, the Lower Boulder Clay rises above alluvium level and is seen on the right bank of the Lagan.

At Lagan Vale Brickworks, now closed [338 708], Lamplugh (1904, p. 71) recorded 12 ft (3·7 m) of red boulder clay on 3 ft (0·9 m) of loamy stony clay, which an adjacent bore proved to be over 70 ft (21 m) thick. Sections seen in 1952 showed the Malone Sands resting on dark brown laminated clay with small stones and occasional cobbles. This bed is irregular in thickness, varying from 2 ft to 12 ft (0·6–3·7 m) and rests on gravel and sand of unknown thickness in one part of the workings and directly on boulder clay in another. The lower boulder clay was worked as well as the laminated clay, and the section seen by Lamplugh was presumably in this material.

On the east bank of the Lagan a series of brickyards between Annadale and Ormeau Bridge is now abandoned and built over. A section was seen in 1952 at Prospect Brickworks, 300 yd (273 m) S. of King's Bridge:

Prospect Brickworks [340 715]. *Section in working face*

	ft	m
Sand, yellowish brown thickening to the east; 6-in peat layer 4 ft from the base	6–10	(1·8–3)
Clay, brown, silty	0– 4	(0–1·2)
Laminated clay with well marked E–W channel cut in it 4 ft deep	0– 4	(0–1·2)
Fine plastic clay with occasional stones with irregular base	10	(3)
Laminated Clay	0– 3	(0–0·9)
Thin sand lenses on Boulder Clay	—	—
	16–31	(4·8–9·3)

A peaty layer similar to the above was noted by Lamplugh (1904, p. 73) in the Marquis Brickworks, immediately to the north of the Prospect Brickworks, from which Reid identified macrospores in the plant material. Chandler (1946) showed that most of Reid's identifications of *Isoetes* are wrong and that these spores are probably *Selaginella*.

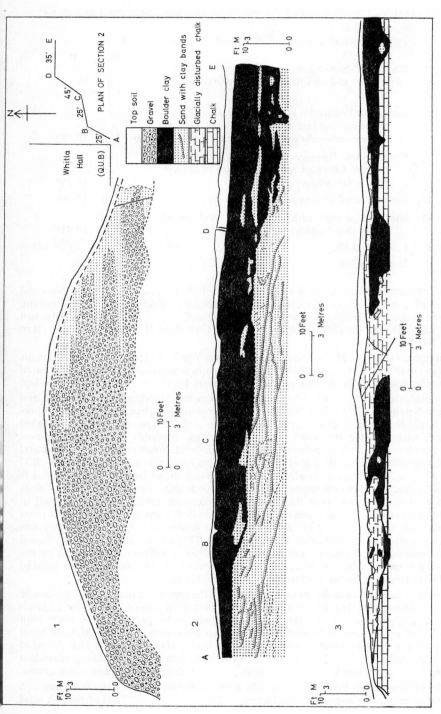

FIG. 17. Drift sections temporarily exposed in the Belfast area.

1. Section of part of the Lisburn esker 250 yd (228 m) S.E. of Clonmore House, Lisburn
2. Section in the Malone Sands and Upper Boulder Clay: Physics Building, Queen's University, Belfast
3. Trench section north-east from road junction [143 607], Moira

Mitchell (1951, p. 156) investigated the peat layer and gave the section as follows:

Section at Prospect Brickworks (*Mitchell*, 1951)

		ft	in	m
A	Sand, red-brown above, grey-brown below; coarser and finer bands occurred in the lower layers	3	11	(1·19)
B	Sand with horizontal layers of vegetable debris, including one leaf of *Salix herbacea*, nuts of *Rumex acetosella* and *Cenococcum* . .	0	3	(0·08)
C	Compressed *Parvo-caricetum* peat with many seeds of *Carex sp.* and megaspores of *Selaginella selaginoides*	0	4	(0·10)
D₁	Greyish sand with vegetable debris . . .	0	8	(0·20)
D₂	Sand becoming redder in colour and more clayey downwards	3	0	(0·91)
E	Laminated Clay	—		—
	Boulder Clay	—		—

Pollen samples from layers B and C yielded only a few grains of *Betula, Pinus* and *Salix*. In Layer C, pollen of *Cyperaceae* and microspores of *Selaginella* were common. Isolated grains of *Empetrum*, Caryophyllaceae and Compositae were seen. Mitchell allots Layer D (and possibly E) to Zone I, Layer C to Zone II and Layers A and B to Zone III (See Fig. 18).

Lamplugh (1904, p. 72) gave sections in Prospect Works and in the adjacent Marquis Works to the north and Annadale Works to the south, which showed that the laminated clay varies considerably in thickness and is overlain locally by Upper Boulder Clay.

An extensive area east of the Annadale embankment, extending east to Cregagh and south to Belvoir, is underlain by sand, but exposures in this urban area are poor. Boreholes in the shopping development at Breda Lodge [353 703] proved up to 53 ft (16 m) of sand, with variable laminated and gravelly clays—probably boulder clay—never more than 9 ft (2·7 m) thick between sand and rock-head. Small exposures of sand and gravel appear on Belvoir Park golf course and bores on Bowling Green Hill [341 701] proved up to 31 ft (9 m) of sand, 3 ft to 7 ft (0·9–2·1 m) of laminated clay, and 4 ft (1·2 m) of boulder clay unbottomed. Borehole records show that the thickness of drift increases from 50 ft (15 m) in the north at Templemore Avenue Baths [357 740] to 140 ft (43 m) at Castlereagh Laundry [358 730]. The drift thins slightly eastwards from Templemore Avenue to 37 ft (11 m) at the Belfast Ropeworks. The boulder clay was formerly worked at the Ravenhill Brickworks [352 727] which have long been disused and are now covered by streets and buildings. The area of Ballymacarret, shown on the published map as glacial sands, is apparently reworked sand in the form of a hooked spit dating from the higher sea level in post-glacial times.

South of Lagan Vale a line of seepage marking the top of the laminated clays can be traced for about 300 yd (273 m) but beyond this point the marginal bluff of the Lagan is featureless. Upper and Lower Boulder Clay appear to be present in the steep bluff 600 yd (549 m) N.E. of Newforge House with an intervening band of Malone Sand between them. About 300 yd (273 m) S. of Shaws Bridge about 6 ft to 8 ft (1·8–2·4 m) of sand and gravel overlie 15 ft (4·6 m) of Lower Boulder Clay containing abundant fragments of basalt, chalk, yellow sandstone, Lower Palaeozoic grits and some quartz.

The Malone Sands from the Stranmillis Road to Barnett's Park have been seen in a number of temporary exposures. At Danesfort 10 ft to 12 ft (3–3·7 m) of reddish sand were seen in foundation excavations and near White Lodge [325 702] a small section showed red sand interbedded with thin bands of boulder clay, the sand containing contorted streaks of clay. Up to 3 ft (0·9 m) of sand were seen in the foundations for new houses between Dorchester Park and Malone Hill Park.

The sand outcrop narrows towards Shaws Bridge and in Malone Upper townland becomes rather gravelly. In a bluff just south-east of Lismoyne House [316 680], about 4 ft (1·2 m) of coarse to medium bedded gravel containing occasional pebbles up to 6 in (0·15 m) in diameter, rests on up to 3 ft (0·9 m) of fine buff bedded sand with lenticles and thin beds of medium to fine gravel. In a lane 100 yd (91 m) N. of Ballydrain Lake, 10 ft to 12 ft (3·0–3·7 m) of red sand are seen in a cutting. Up to 20 ft (6·0 m) of reddish fine to medium sand are visible south of Ballydrain Lake in the road leading to Drum Bridge. Ballydrain Lake, a kettle hole, has apparently an underground outlet in the spring which issues about 350 yd (320 m) S.E. of the lake near the Lagan. The spring has formed a deep amphitheatre in the sands.

Westwards from Malone House, Upper Boulder Clay underlies most of the University playing fields and extends further west to merge with the wide area of boulder clay flanking the basalt escarpment. At Redhill [309 699], bricks were formerly made from this material; 800 yd (732 m) N.W. of Redhill, recent excavations showed up to 8 ft (2·4 m) of reddish marly clay with few boulders. At Black's Road, just south of the railway, boreholes showed up to 23 ft (7·0 m) of clay. Further north-westwards, in the absence of sand, it is not possible to distinguish between Upper and Lower Boulder Clay.

On the eastern side of the Lagan, near Shaws Bridge, reddish sand forms the bluffs margining Purdys Burn. A section at Crooked Bridge showed 4 ft (1·2 m) of sand and gravel wash on 6 ft (1·8 m) of sand resting on 6 ft (1·8 m) of boulder clay. A borehole just north of Shaws Bridge proved up to 43 ft (13 m) of sand on boulder clay and near Milltown 40 ft (12 m) of sand (unbottomed) were met. To the north-east the sand thins and lenses out. Above Corbie Wood Quarry [338 696] two sand horizons have been mapped, both lenticular in character and wedging out into boulder clay. A strip of sand runs eastward from Minnowburn Beeches through Milltown and 700 yd (640 m) beyond Milltown the sand becomes mounded. Small pits opened in this strip of sand have now become overgrown.

Similar pits in Ballynahatty townland were formerly opened for moulding sand. In this area, thicknesses of sand of up to 60 ft (18 m) are recorded in well sections. West of the Giant's Ring old gravel pit exposures show up to 6 ft (1·8 m) of gravel and sand with occasional cobbles. Boreholes immediately southwest of the Giant's Ring recorded up to 95 ft (28 m) of drift, though there is no differentiation in the logs between sand and clay. The underlying Boulder Clay is marked by a strong seepage line from Edenderry to Edenderry House [325 674]; 750 yd (686 m) E. of Edenderry a small marshy hollow is probably a kettle hole.

The south-eastern margin of the sand area runs south-westwards from Purdysburn Fever Hospital to Ballylesson and towards the margin the sand passes into gravel and becomes rather mounded. The total drift thickness in the neighbourhood of Ballylesson varies from 15 ft (4·5 m) to 65 ft (20 m) according to well records. South-west of Bally-lesson streams dissect the sand area and produce an irregular mosaic of boulder clay and sand as far west as Ballyskeagh. Part of the boulder clay mapped is clearly Upper Boulder Clay, but the tripartite division so well seen near Belfast is not so clearly defined.

Near Stanton Cottage [307 666] south of Drumbeg a 50-ft (15 m) well penetrated clay and sand and some laminated clay was seen in the debris. The earlier survey noted laminated clay beside a stream 700 yd (640 m) due E. of this locality. In both areas sand appears to overlie this lower clay and is succeeded by Upper Boulder Clay. At Gardners Loan End 3 ft (0·9 m) of boulder clay are seen over red sand. Sections 300 yd (274 m) S. of Drumbeg show very gravelly sand in small disused pits. Up to 15 ft (4·6 m) of gravel were proved in a nearby well.

At Ballyskeagh the sand forms an extensive spread and is slightly mounded. Wells in the village record up to 30 ft (9·1 m) of sand with some gravel. The spread of sand continues southwards on the right bank of the Lagan and forms a gently undulating area at about 120 ft (37 m) O.D. Immediately north-west of Green Hill [282 662] a well penetrated 10 ft (3·0 m) of sand over 70 ft (21 m) of boulder clay, and just south of Green Hill a well 55 ft (16·8 m) deep was entirely in boulder clay: 200 yd (182 m) S.E. of this, sand pits show up to about 15 ft (4·6 m) of sand whilst a well section near the pits gives 20 ft (6·0 m) of gravel and sand on boulder clay. Kilroosty Lough [286 660] is a good example of a kettle hole and a small hollow near Sycamore Hill [290 652] appears to be of similar origin. In this area the sand is reported to be up to 50 ft (15 m) thick.

South-west of Sycamore Hill the sand thins towards the southern limit of the sand and gravel deposit at Hillhall.

In the Dunmurry district sand underlies much of Dunmurry Golf Course and extends northwards to Brooklands farm. Two patches of boulder clay underlie The Park farm and the north-west part of Dunmurry. From Dunmurry to Lisburn an esker ridge, now removed to a large extent, can be traced along a sinuous course approximating to that taken by the railway which cuts across it at several points. A contiguous belt of sand covers the ground between the esker and the Lagan south to Lisburn and northwards towards Milltown. In parts of the Lisburn area the boulder clay is characteristically gravelly and is difficult to distinguish from associated gravel and sand. Elsewhere it is normal though still with a high stone content. On Lisnagarvey Hill just north of Wallace Park, Lisburn, brickmaking was attempted with it and a small pit shows a section of 10 ft (3·1 m) with occasional basalt boulders up to 2 ft (0·6 m) across.

Another esker extends in a sinuous course from Duncan's Reservoir [258 658] along the Causeway End Road to Magheragall. Most of the pits excavated in this ridge are now degraded and overgrown. Wells near Aikins Hill [253 646] record 40 ft (12 m) of sand and gravel. A mile (1·6 km) to the west, up to 12 ft (3·7 m) of sand and gravel can still be seen. At Red Hill [223 647] up to 30 ft (9·1 m) of reddish brown sand with thin gravelly layers are exposed in intermittently worked pits. Hartley (1937, p. 208) recorded in excavations on both sides of Red Hill, a 6 ft (1·8 m) layer near the base of the esker, of fairly coarse shingle composed mainly of subangular basalt pebbles associated with chalk and some (?) Scottish felsite and granite. At a higher level in the red sands he noted a 6-ft (1·8 m) band of basaltic gravel. At the base of the esker a few traces of peat and some apparently almost unrolled specimens of *Alectryonia carinata* [sic] were also seen. The esker appears to rest directly on boulder clay over much of its course though the junction was not seen. The general summit level of the ridge falls from about 250 ft (76 m) O.D. in the east to about 190 ft (58 m) O.D. at Red Hill. A section of the Lisburn Esker is illustrated in Fig. 17. 1 and Plate 8B.   P.I.M.

In the Magheragall area the boulder clay is gently undulating and, apart from a couple of small drumlins east of Mullaghcarton, has no distinctive grain. The thickness of clay in the relatively few recorded wells is not great and a thickness of 28 ft (8·5 m) at Magheragall rectory is exceptional. The frequent exposures of the underlying rocks in streams and ditches confirm that the drift cover is thin and it seems probable that the more plastic clay formed from these rocks gave rise to a thinner ground-moraine than the tougher sandy clay of the basalts.

The steep slopes below the Chalk and basalt scarp round White Mountain are covered with thick boulder clay much of which shows signs of slipping or solifluxion sliding. The only indication of the thickness of these deposits is given by the log of the well at Old Park [250 669] where J. J. Hartley recorded 72 ft (21·9 m) of drift.   H.E.W.

South of the esker the sand spread continues south and west in a belt on either side of the Lagan about 2 miles (3·2 km) wide. Up to 15 ft (4·6 m) of sand are seen in a cutting on the Hillsborough road just north of Moore's Bridge. Behind Grove Cottage

[250 641] a 10 ft (3 m) face of red sand was noted and west of this site an old sand and gravel pit shows coarse gravel of basalt, chalk, flint, red and yellow sandstone pebbles in a fine gravel or sand matrix. The original face is 20 ft (6·1 m) high. A half mile (800 km) W. of this pit a disused pit in the prominent mound of Knockmore Hill showed some 20 ft (6·1 m) of gravel and sand. Between the two pits the sand is thin and a small patch of boulder clay is exposed in the disused cutting just south-east of Knockmore Station.

On the northern side of the Lagan the sand appears to thin out round boulder clay mounds and the belt narrows towards Mazetown. Just north of Youngs Bridge and at Englishtown, however, thicknesses of up to 30 ft (9·1 m) are recorded in well sections. Beyond Mazetown the sand belt forms a narrow margin to the Lagan extending to about ½ mile (800 m) W. of Halfpenny Gate. Well sections record up to 20 ft (6·1 m) of sand in the latter area.

Between the sand belt and the basalt escarpment the valley is covered almost entirely by boulder clay. North of Halfpenny Gate thicknesses of up to 68 ft (21 m) are recorded in irregularly mounded boulder clay. In a well, 200 yd (182 m) N.W. of Halfpenny Gate crossroads, some 14 ft (4·3 m) of laminated clay were met beneath 6 ft (1·8 m) of soil and red clay. A well record [196 627] near Robinson's Hill gives laminated clay under 6 ft (1·8 m) of red clay and resting on Bunter Sandstone. North of the railway the boulder clay mounds have a roughly E.–W. trend. From 30 ft to 60 ft (9–18 m) of boulder clay are recorded in hilltop wells.

In the area west of Maze Station clay has been worked for at least a century. Old works between the main road and the railway have long been abandoned and appear to have wrought a shallow surface layer. To the north of the railway a small brickyard was producing drainage tiles until recently. Up to 6 ft (1·8 m) of brown plastic clay with few pebbles were worked, the depth being controlled by drainage problems. The Company ceased production because of the difficulty in making satisfactory bricks from this material.

South of the Lagan, near Lisburn, the main sand area runs southwards along the Ravernet River to just over a mile (1·6 km) N. of Hillsborough and then westwards, becoming less continuous, to approximately 2½ miles (4·1 km) E. of Moira.

In the Blaris–Magherageer area up to 46 ft (14·0 m) of sand are recorded, but 24 ft to 30 ft (7·3–9·1 m) is a more characteristic thickness. Along the north-western side of Long Kesh aerodrome the sands become very gravelly and form a north-north-east trending ridge or mound running for about 1 mile (1·6 km) and terminating at the Maze racecourse. Several pits have been opened along this ridge, showing up to 35 ft (11 m) of sand with pebbly layers.

North of the Bog Road the underlying clay is exposed over an area of about 50 acres (20 ha). Several pits, disused since 1936, have been opened in relatively stone-free boulder clay or laminated clay. Up to 12 ft (3·6 m) of clay have been removed but drainage difficulties in the area, which is only a few feet above the Lagan, seem to have discouraged any further working. Laminated clay is seen [215 611] 350 yd (320 m) W. of The Long Kesh where 4 ft (1·2 m) of reddish brown plastic clay rest on 1 ft (0·3 m) of laminated clay.

West of Long Kesh the sand is dissected by small tributaries of the Lagan into three north-west trending belts each about 400 yd (365 m) in width. Maximum recorded thicknesses of sand in well sections are 18 ft (5·5 m) in the north-eastern and central belts, and 28 ft (8·5 m) in the south-western belt. The sand terminates to the south against mounded boulder clay with tongues of sand lapping round the mounds.

In the area between Lisadian House [222 593] and the disused railway an irregular patch of sand and gravel has been exploited in parts for gravel. From Beechfield [213 591] a fairly continuous esker can be traced in a sinuous course westwards to Nut Hill [190 588]. Pits of poorly bedded coarse gravel up to 12 ft (3·7 m) have been opened at Beechfield, Corcreeny House and Nut Hill. There is little difference in level between the eastern or southern ends which are about 50 ft (15 m) below this level. Locally it forms well defined ridges and in other sections of its course appears to be cut into the underlying boulder clay.

West of Nut Hill the Lower Boulder Clay forms a fairly thick blanket with up to 60 ft (18 m) recorded in well sections.

North of the Lagan, in the quadrant bounded by Moira, Soldierstown, Mullagh-carton and Halfpenny Gate, the boulder clay cover thins against the Chalk escarpment but up to 40 ft (12 m) of boulder clay were recorded in a well at Lisnabilla [165 623] and 60 ft (18 m) of red clay near Brookfield. Half a mile (800 m) S. of Trummery House well records give up to 47 ft (14 m) of boulder clay.

Near Neworchard [175 608] shallow depressions mark a small area of sand. In a well at the house 9 ft (2·7 m) of sand were met with under 8 ft (2·4 m) of clay. To the east, wells show up to 26 ft (7·9 m) of boulder clay. Northwards the boulder clay thins over the Trummery Cottage intrusion and then thickens north-eastwards to 34 ft (10 m) or more before thinning at the chalk escarpment. A trench section near Moira showing glacially disturbed chalk is illustrated in Fig. 17. 3.   P.I.M.

Hillsborough–Newtownbreda area. In the south-east corner of the Sheet the boulder clay occurs in the form of low elliptical hills or drumlins. Much of the intervening ground between these hills is drift-free and the Lower Palaeozoic rocks are exposed on the low ground. The boulder clay is a stiff reddish brown till containing plentiful stones and boulders, rarely larger than 1 ft (0·3 m) or so in diameter. The vast majority of the stones are of local rock types, generally Lower Palaeozoic grits and greywackes, with some shales. A well at Lough View [342 645] showed the clay to contain subordinate but considerable amounts of basalt, dolerite, chalk, and flint, and occasional quartz, quartzite, and grey sandstone, probably Carboniferous. Exposures farther south indicate a similar assemblage, with a smaller proportion of exotic material. One roadside cutting [317 598], about 1 mile (1·6 km) N. of Larchfield, yielded a boulder of lampro-phyre though intrusions of this rock are not known in the immediate vicinity.

The drumlins, which are exceptionally well developed on the southern and eastern edges of the Sheet, vary in length from 200 yd to ½ mile (182 m to 800 m). They have an average length–breadth ratio of about 2: 1 but vary from extremes of about 3: 1 to 1: 1, some being virtually circular in plan. The long axes of the drumlins trend, on the whole, north-west, though on the east edge of the Sheet there is a tendency to swing towards north–south.

In section along the longer axis most of the drumlins have a pronounced streamlined form with the blunt end to the north-west or north and the tail to south-east or south This tendency to a tapered outline is also seen sometimes in plan and suggests that the hills were moulded by ice flowing from the north-west. They are up to about 100 ft (30·5 m) in height above the rock floor, and no evidence has been found to suggest that they reflect in any way the form of the floor. In particular, no indication of rock cores has been found, and a drumlin [365 647] east of Lough Moss which has been partially removed in quarrying operations rested on an irregular glaciated rock floor. At Fruit Hill [348 589] a well sunk in boulder clay reached rock head at a level below that of the drift-free inter-drumlin rock surface. Though often found as isolated hills, most of the drumlins occur in small groups, often *en-échelon* such as the group north-east of the Temple, but frequently in line ahead or abreast. Double forms also occur—generally large drumlins with typical tapered or streamlined form in plan but with two crests, that towards the broad end the higher, such as those north-east of Thorndale [346 608] and west of Lakeview [312 577].

The characteristic form of this drumlin country, often described as 'basket-of-eggs topography', has given rise to ill-drained hollows between the hills and the rock floor is often obscured by small lakes, particularly in the extreme south-east corner of the Sheet, and by alluvial and peat deposits which have accumulated in lake hollows.

The boulder clay of the drumlins generally appears to rest directly on rock-head as well records and observations show, but the hill east of Lough Moss which has been partly removed revealed an interesting section where the normal reddish brown boulder clay, 20 ft (6 m) thick, overlies 6 ft (1·8 m) of darker brown and more stony boulder clay which in turn rests on a bed of laminated clay. A note on this occurrence was published in 1958 (*Sum. Prog. geol. Surv. Gt. Brit.* for 1957, p. 45). The laminated clay is a plastic reddish brown clay with fine laminae and occasional small stones. It reaches a maximum thickness of over 9 ft (2·75 m) at the south end of the exposure where it was underlain by two or three feet of stony clay on rock-head. The laminated clay passes laterally into discontinuous streaks and lenses of laminated clay, up to 5 ft (1·5 m) long, interspersed with gravelly material with a clayey matrix and finally into occasional wisps of laminated clay in stony clay, apparently a boulder clay. In the gravel zone the laminae are often curved and appear to bend round gravel lenses but elsewhere they are straight and near horizontal. The lateral extent of the laminated clay occurrences is about 15 yd (14 m).

About ten yards south of these exposures the massive boulder clay was seen to rest on over 6 ft (1·8 m) of reddish brown plastic clay with stones. The stones are generally small and random in distribution but there are local concentrations of pebbles. The base of this deposit was obscured by fallen debris and it is not known what lies below it but it is at the same level as the laminated clay.

Farm houses are frequently located on the flanks or tops of the drumlins and wells sunk near the dwellings have been useful in determining drift thicknesses. Well depths up to 78 ft (21 m) are recorded.

North of a line from the Mental Hospital to Hillsborough the boulder clay forms a more continuous sheet and drumlins are less well developed. Drift thicknesses known near Drumbo range from 20 ft (6 m) to 24 ft (8 m) but near Newtownbreda 80 ft (24 m) of boulder clay were penetrated in a borehole near Cameron Lodge [358 686]. Further west, thicknesses in the Strawberry Hill–Ravernet area range up to 40 ft (12 m).

Throughout this district sand and gravel is virtually unknown, but approaching the Lagan Valley, small patches occur on Cairns Hill [357 687] (wrongly indicated on the one-inch map), in Ballycowan [338 671], near Duneight House [283 607], south of Beechmount [287 628], north-east of Ravernet village, and north of Hillsborough. None is of economic significance. H.E.W.

Chapter 12

RECENT

ESTUARINE CLAYS AND PEATS

THE DISAPPEARANCE of the ice in the late-glacial period was followed by a period of fluctuating sea levels during which the effects of the rising sea, due to the release of vast quantities of water by melting ice, were later matched and finally reversed by the isostatic uplift of the land, which gradually rose after the removal of the great weight of the ice sheets.

In the earliest stage, in late-glacial times, the land may have been as much as 120 ft (37 m) higher with respect to the sea than at present (Movius 1953b, p. 11) but the earliest deposits known in this area date from the Boreal period, around 6000 B.C., and are the peats found in the harbour area resting on the surface of the glacial drifts and as much as 40 ft (12 m) below present sea level.

The peat bed described at the Milewater Dock, Belfast, by Charlesworth and Erdtman in 1935 is assigned in age to Sub-Zone VIa (i.e. Early Boreal) in Ireland and roughly corresponds with similar horizons at Island Magee, Cushendun and Willsborough. Later work (Morrison in Stephens 1957, p. 139) on the peat from the site of the Victoria Power Station (West) in Belfast Harbour Estate indicates that the submergence which allowed the covering of the peat by estuarine clay had probably begun in Zone VIc.

Hartley (1940) considered that this peat was up to 3 ft (9 m) thick and was confined to those areas where the 'sleech' or Estuarine Clay reaches its maximum thickness. Further boreholes have shown that the peat is more widespread in its occurrence than Hartley supposed, whilst it is certain that in many shallow bores the peat layer has been missed or has not been recorded, particularly where it is thin. Its preservation under the thicker parts of the sleech may be ascribed to rapid covering when subsidence began, when at higher levels it may have been attacked by wave action and eroded away. Morrison suggests that the peat formed at about tide level and was intersected by streams and channels of open water.

An analysis by Charlesworth and Erdtman (1935, p. 234) gives the following proportions of pollen in the peat: Birch (*Betula*) 50; Pine (*Pinus*) 40; Oak (*Quercus*) 8; Elm (*Ulmus*) 2; Hazel (*Corylus*) 200. This analysis demonstrates the overwhelming preponderance of hazel pollen. Mitchell (1951, p. 118) remarks that the peat accumulated in Zone VI. Praeger (1887) recorded seed vessels of Alder (*Alnus*).

Morrison asserts that the analysis from Victoria Dock showed a much higher percentage of hazel pollen that that from Milewater Dock, and has full claim to be placed in Zone VIa (*Ann. Rep. Nuffield Quaternary Research Unit* Queen's University, Belfast, 1955–6, p. 3). This agrees with Jessen (1949, p. 136) who considered that the Milewater analysis might "belong to Subzone VIa and is possibly older than the Boreal hazel maximum".

144

Early Post Glacial Climatic Periods	Period	Palaeobotanical zone sequence in Ireland	Sub Zones	Land Movements and Temperature	Sequence of Deposits in Belfast Area
Sub-Boreal	Alder-Oak Period	VII	b	Emergence — Fauna as in present seas. Cool dry conditions. Regression of sea	Recent submergence - small depression. 2' yellow sand with 6'6" mud at Alexandra Dock (Praeger 1888). Raised storm beaches?
Atlantic	Alder-Oak-Pine Period	VII	a	Early post-glacial marine transgression — Maximum of Subsidence. Cooler at end. Warm and moist (Oceanic) conditions. Southern forms rare	UPPER ESTUARINE CLAY (Thracia Zone). Laid down in water c. 5 fathoms deep
	Hazel-Pine Period	VI	c	Submergence — Post-Glacial Climatic optimum (Maximum frequency of southern forms)	INTERMEDIATE ZONE of the ESTUARINE CLAY. Laid down in water c. 2-3 fathoms deep
			b		LOWER ESTUARINE CLAY (Scrobicularia Zone). Laid down under littoral conditions
Boreal	Hazel-Birch Period	V	a	Emergence — Warm and moist conditions. Fauna slightly warmer than that of the present seas. Emergence of the Irish Sea Basin to approximately the 20 fathom (=120 ft) line	Peat Bed (e.g. Milewater Dock). Weathering and erosion of the Late-Glacial land surface. Re-deposition of Boulder Clay
Pre-Boreal	Post-glacial Birch Period	IV		Late-Glacial Sea	Outwash Sands and Gravels. Final stages of ice recession
Late-Glacial		III Younger Salix herbera			Layer A and B
		II Late-Glacial Betula or Allerød			Layer C — Former Marquis Brickyards. Annadale Embankment
		I Older Salix herbera			Layer D and ? E (Lam. Clay)

Fig. 18. *Late-glacial and post-glacial chronology and deposits in the Belfast area*

Praeger (1892) records finds of the remains of wild boar and red deer from the peat layer during excavations for the Alexandra Dock. Derived remains of *Megaceros* have also been recorded (Charlesworth and Erdtman 1935; Mitchell and Parkes 1949) from the sands below the peat.

Peat from a depth of 49 ft (15 m) in a borehole at Castle Arcade, Belfast [339 742] has been determined by radio-carbon (C^{14}) analysis to be 9130 ± 120 years old. Wood from a depth of 38 ft (12 m) in the same borehole gave a date of 8715 ± 200 y. B.P.

After the accumulation of the peat horizon a period of relative submergence ensued during which the Estuarine Clay or 'sleech' was deposited. This deposit is the most widely distributed of the Post-Glacial marine transgression sediments. Movius claims that the term 'Estuarine Clay' is a misnomer (Baden–Powell 1937a, p. 95; Movius 1953b, p. 12) and that the foraminifera and the mollusca indicate marine conditions of sedimentation. Nevertheless, physiographically the beds are estuarine clays and the term, well established in the literature, is retained, using its synonym 'sleech' where it has been so described in drillers' records.

Stewart (1871) and Praeger (1887) made a basic twofold division of Upper Estuarine Clay and Lower Estuarine Clay, with a suggested intermediate stage, on basis of the contained molluscan fauna. Later workers (Movius 1953a, pp. 697–740 and McMillan 1957, p. 188) have confirmed a threefold division outside the Belfast area (see also Fig. 18):

3.	Upper Estuarine Clay	*Thracia convexa* Zone
2.	Intermediate Estuarine Clay	Characterized by species of *Pholas*
1.	Lower Estuarine Clay	*Scrobicularia plana* Zone

The greatest known thickness of the Estuarine Clay is 55 ft (17 m) but as it has been sub-divided palaeontologically only in areas where it is much thinner the proved thicknesses of the Upper and Lower divisions total only 26 ft (8 m).

Lower Estuarine Clay. The *Scrobicularia* clay is typically a brownish blue some-what sandy clay containing remains of *Zostera marina,* the grass wrack, and a large molluscan fauna indicative of intertidal conditions. The characteristic fauna includes: *Mytilus edulis* Linné, *Cerastoderma edule* Linné, *Venerupis decussata* (Linné), *Macoma balthica* (Linné), *Scrobicularia plana* (da Costa); *Hydrobia ulvae* (Pennant).

The Lower Estuarine Clay is not seen at the surface and is known only from occasional deep sections and boreholes. Recorded thicknesses vary from 6 in (150 mm) to 14 ft (4·3 m) and collections have been made from Victoria Square, Millfield and other localities. The faunas collected are listed in Appendix 4.

Sus scrofa has been recorded from a sandy layer just above the peat (Savage 1964).

Intermediate Estuarine Clay. Stewart (1871, p. 28) and Praeger (1888, p. 31) suggested the existence between the Upper and Lower Estuarine Clays in the Belfast area of an Intermediate zone characterized by an abundance of species of *Pholas*. Three species, *Zirfaea crispata, Barnea candida* and *Pholas dactylus* were noted, the first named attaining a large size and all still in their natural vertical positions. No thickness of this zone was given, and it has not been recognized in recent micro-faunal work.

Elsewhere in the north of Ireland a zone intermediate in character showing a transition from the intertidal *Scrobicularia* clays to those of the *Thracia* Zone has been recorded. In the Larne region, for example, the Intermediate zone has been identified at Island Magee (Burchell 1934, p. 370; Baden-Powell 1937a, p. 86). Jessen (1949, p. 139) assigned the zone to the lower portion of palaeobotanical subzone VIIa. The sediment was probably laid down in water up to 3 fathoms (5·5 m) deep.

Upper Estuarine Clay. The upper or *Thracia* clay is a tenacious blue-grey clay with fewer shells of greater variety. It represents the maximum transgression of the sea and the littoral zone fauna is replaced by faunas typical of the coralline and laminarian zones. Characteristic species include: *Mysella bidentata* (Montagu), *Acanthocardia echinata* (Linné), *Mysia undata* (Pennant), *Abra alba* (W. Wood), *Thracia convexa* (W. Wood), *Turritella communis* Risso. Recorded thicknesses of this clay vary from 1½ ft (0·45 m) to 12 ft (3·7 m).

A further period of submergence is indicated by this zone which accumulated in approximately 5 fathoms (9·1 m) of water.

Numerous workers have given details of faunas from the Estuarine Clay, especially Grainger (1853, 1859), Stewart (1871) and Praeger (1888, 1892, 1896). A modified correlation chart following Movius and others is given in Fig. 18 and a revised list of the Mollusca, compiled by C. J. Wood, is given in Appendix 4. An account of the foraminifera by M. J. Hughes is given in Appendix 5.

DETAILS

The succession of the Estuarine deposits has been given in the general account. The 1904 Belfast Memoir gave some details of the results of shallow boreholes in the Harbour area. Since that time many more boreholes have been put down in the Harbour area and a considerable number in the town area. The physical nature and engineering properties of the sleech have made it desirable to investigate each site before piling. As a result of these numerous borings it has been found possible to construct an isopachyte map of the Estuarine Clay and Recent Deposits (Plate 10).

It can be seen that the sleech reaches its apparent maximum thickness of about 50 ft (15 m) near the Twin Islands and under central Belfast. Hartley (1940) remarked that this thickness is unlikely to be exceeded seaward (beyond the limits of One-inch Sheet 36) since firm clay has been reached in dredging in the Musgrave Channel and beyond. This has been confirmed by boreholes put down in the Harbour area lying within One-inch Sheets 28 and 29. The deposits thin out to the south-west and die out where the change of slope occurs at about +25 ft (7·6 m) O.D. This change of slope has been mapped as the edge of the Estuarine Clay, though in some areas the deposit is very thin or absent on the wave-cut platform of boulder clay, marl or sandstone. Only a thin veneer, for example, is present at the Park Parade School [350 733], McClure Street [340 730] and at Gallaher's factory, York Street [341 753]. At York Road (R.U.C. Station) [341 763] and at Cullingtree Road [328 739] no deposits were found in excavations.

In the present Lagan Valley the Estuarine Clay outcrop narrows rapidly above Ormeau Bridge and passes into normal river alluvium. Similarly, in the Connswater valley the sleech passes into alluvium about Beers Bridge. The Blackstaff is margined by a 200-yd to 300-yd (182–273-m) wide flat underlain by Estuarine Clay which expands at Donegall Road into the extensive and formerly swampy area (now artifically drained) known as the Bog Meadows. An attempt has been made to delineate the area underlain by Estuarine Clay as distinct from the sandy river alluvium into which it passes laterally. The Estuarine Clay here extends about 3 miles (4·8 km) inland from the present shore-line.

For the purposes of detailed description, the Estuarine Clay outcrop may be divided into the following areas:

        (1) Bog Meadows and Blackstaff
        (2) Main Lagan Valley
        (3) Connswater area
        (4) Central area
        (5) Harbour area

1. **Bog Meadows and Blackstaff.** A large number of shallow bores has been put down in this area for the M1 Motorway. Much of this area is now covered by fill, particularly near the eastern boundary which has been obscured by railway sidings and buildings and allotments since the 1902 resurvey. Just north-west of Malone Training School [318 716], wide areas of land are covered to a depth of several feet by dumped rubbish. Though the drains show sections of brown or reddish alluvial clay, thicknesses of 4 ft (1·2 m) of sleech are recorded in bores under 4 ft to 5 ft (1·2–1·5 m) of alluvial clay as far south as Malone Training School. Hartley (1940) records the maximum thickness of alluvium in the Bog Meadows area as 15 ft (4·6 m). East of Andersonstown a low mound appears from the records to be boulder clay. Five bores penetrated a maximum of 33 ft 9 in (10 m) in red clay. A boulder of red sandstone 4 ft 6 in (1·4 m) in diameter is noted in one record. Just north of this the ground is flat, rushy and poorly drained and sleech up to 20 ft (6 m) thick has been recorded in recent boreholes. Five feet (1·5 m) of sleech are recorded at the western end of Windsor Park Football Ground. In the neighbourhood of the Clarence Engineering Co. Works [319 739], about 14 ft (4·3 m) of sleech are present under a cover of fill and reddish clay up to 10 ft (3·1 m) thick. The red clay is probably the equivalent of the yellow sands and other supra-sleech deposits noted later. Near the Motorway–Donegall Road junction, 14 ft (4·3 m) of sleech have been proved and in the same area, 60 yd (55 m) from the Blackstaff, the sleech thickens to 24 ft (7·3 m). In the Celtic Park carpark over 30 ft (9·1 m) are present but 100 yd (91 m) N.E. of this site only about 13 ft (4 m) are recorded. Twenty-eight feet (8·5 m) are recorded near the Electricity sub-station at Milner Street [325 734], though this figure decreases rapidly north-westwards in the Royal Victoria Hospital grounds. Boreholes put down for the first phase of the proposed Belfast Ring Road have added further records in this area.

A lobe of Estuarine Clay probably extends a short distance up the Clowney River, though no bore records are available.

2. **Main Lagan Valley.** Lamplugh (1904), records a series of eighteen borings from 9ft to 16 ft (2·7–4·9 m) in depth between Lagan Vale and Ormeau Bridge but the logs of these boreholes have not been traced at the Harbour Office nor at the Geological Survey Office, Dublin. Sleech from 7 ft to 14 ft (2·1–4·3 m) thick rests on gravel or clay. A bore on each embankment 40 yd (46 m) S. of the King's Bridge gave 4 ft and 8 ft (1·2 and 2·4 m) of sleech, though fill up to 19 ft (5·8 m) thick is recorded and the sleech may have been originally much thicker. Farther south, though thick alluvium has been proved, there appears to be no sleech recorded in the logs. One hundred yards (91 m) N. of King's Bridge, on the left bank of the Lagan, 17 ft (5·2 m) of sleech under 8 ft (2·4 m) of fill are recorded. Right bank bores extending 300 yd (274 m) S.W. of Ormeau Bridge average 14 ft (4·3 m) of sleech; at a site near the bridge [343 726] 20 ft (6·1 m) are recorded.

At McConnel Lock about 22 shallow bores have shown sleech varying from 10 ft to 15 ft (3–4·6 m) in thickness overlying thin drift resting directly on Bunter Sandstone.

In half the bores peat was present below the sleech. Praeger reported (1888, p. 232) similar thicknesses in the excavations for the foundations for the Albert Bridge:

Section at Albert Bridge

	ft	m
River mud and coarse sand	10	3·0
Lower Estuarine Clay	14	4·3
Gravelly bed	1	0·3
New Red Sandstone [Bunter]	—	—

A group of shallow bores at Park Parade School showed practically no deposits referable to Estuarine Clay and the northern portion of Ormeau Park is probably a wave-cut platform in Boulder Clay.

3. **The Connswater Area.** At Inglis & Co., Biscuit Factory [364 742] the maximum thickness of sleech recorded is 9 ft (2·7 m). Part of the sleech in two other bores on the same site may be represented by grey sand. Sandy sleech 8 ft (2·4 m) thick resting on 8 ft (2·4 m) of fine grey sand is reported at the northern end of the Irish Distillery ground [364 745]. On the Connswater bank 100 yd (91 m) E. of this site only 3 ft (1 m) are present. There is a rapid thickening north of East Road. One bore at the Esso Petroleum Company's site [363 750] showed 20 ft (6·1 m) of sleech underlain by 1 ft (0·3 m) of peat.

Excavations along the Lower Newtownards Road have shown that where Estuarine Clay was mapped in the earlier revision, the area is underlain by reddish or yellowish sand. This sand has only been penetrated by bores in the Sirocco Works East Yard [352 742] where up to 11 ft (3·4 m) rest directly on Boulder Clay. Present information suggests that the configuration of the sand area approximates to that of a hooked spit and that the deposit is partially reworked glacial sand. The sand overlies Boulder Clay in the Templemore Avenue Baths borehole where it was considered to be glacial sand (Lamplugh 1904, p. 147).

4. **The Central Area.** In the area to the south of a line running along the Great Northern Railway and Ormeau Avenue the sleech is thin and may in parts be absent or represented by a sandy wash. Along the lower half of McClure Street [340 730] only 1 ft 6 in to 2 ft (0·46–0·61 m) of sleech were seen in temporary excavations resting directly on Boulder Clay. In Brigg's borehole at Pine Street [340 732] no sleech was recorded, and immediately to the north in Donegall Pass, buff sand or boulder clay was revealed in road excavations. At the Pure Ice Company bore in Great Victoria Street [336 734] only 4 ft (1·2 m) are recorded.

In Ventry Street and Lane some 2 ft (0·6 m) of reddish yellow sand rested on 1 ft to 2 ft (0·3–0·6 m) of sandy sleech with shells, and in other temporary exposures along the Dublin Road similar sand was seen on sleech or bluish grey silty sand. At the B.B.C., Ormeau Avenue, 10 ft (3 m) of sleech are recorded. Greyish yellow sand to 4 ft (1·2 m) was seen in excavations along Sandy Row, no sleech being recorded at Miller, Sons and Torrence's borehole [334 733]. Some 9 ft to 12 ft (2·7–3·7 m) are recorded further north at Murray's Tobacco Factory.

In the area south and east of the City Hall the sleech is thin. A section in Linenhall Street records 4 ft (1·2 m), overlain by 4 ft (1·2 m) of grey sand. In May Street only about 6 ft (1·8 m) are present.

There is a rapid thickening to Donegall Square North. At the Belfast Water Commissioners' Building 32 ft (9·7 m) of sleech overlie 2 ft (0·6 m) of peat. In Donegall Place the following section was recorded:

Section at Donegall Place/Fountain Lane corner

						ft	m
Fill .	.	.	.	.	.	6	(1·8)
Estuarine Clay	.	.	.	.	28	(8·5)
Peat	.	.	.	.	.	1	(0·3)
Sandy clay	.	.	.	.	3	(0·9)
Stony clay	.	.	.	.	7+	(2·1+)
						45	(13·6)

At Cornmarket, 55 ft (17 m) of sleech are developed, again underlain by up to 3 ft (0·9 m) of peat. Part of the top of the sleech in the Bridge Street area is replaced by up to 11 ft (3·4 m) of fine-grained sand. A thinner development of this upper sand continues north of Talbot Street, and at St Anne's Cathedral it is a fine yellow silty sand. At York Street/Great George's Street junction, 11 ft (3·4 m) of grey and reddish sand overlie some 4 ft (1·2 m) of sandy sleech. Shallow trenches along York Street, between Great George's Street and Henry Street, showed up to 3 ft 6 in (1·1 m) of yellowish and grey silty sand which may be regarded as a marginal facies of the sleech though in part it may be equivalent to the beds above the sleech recorded by Praeger at Alexandra and Milewater Docks.

Twenty-one feet (6·4 m) of sleech overlie 1 ft (0·3 m) of peat in the Victoria Street/Oxford Street area.

At Victoria Square borings put down in 1956 yielded an abundant fauna from the sleech which is listed in Appendices 4 and 5. It is noteworthy that the Upper Estuarine Clay is present only in one borehole which may indicate channelling of the Upper Estuarine Clay into the Lower.

5. Harbour Area. The results of over 300 borings are available in this area as a result of the activity of the Harbour Board on improvements carried out in the port. The Harbour Engineer has kindly made these records available and copies are now on file at the Geological Survey Office.

Details of a few of the borings will be given, but the results are best summarized in diagrams (Plates 9 and 10).

All bores save a group at the east side of the Herdman Channel were done before the present survey and few new faunal collections have therefore been made.

The section recorded by Praeger (1888, pp. 29–51) was used in the 1904 Belfast Memoir as the type section and a modified summary of this is as follows:

Section: Alexandra Graving Dock

	ft	m	
Blackish clay with sandy layers	6–7	2·0	*Mytilus arenaria* (v. abundant) *Cardium edule* (abundant) *Tellina balthica* *Mytilus edulis*
Yellow Sand . . .	2	0·6	*Pecten opercularis* *Littorina littorea* *Mytilus edulis* *Tapes pullastra* *T. decussatus* *Thracia convexa*
Upper Estuarine Clay .	6	1·8	*Thracia convexa* *Lucinopsis undata* *Cardium echinatum* *Scrobicularia (Abra) alba* *Ostrea hippopus* *Acera bullata*

'Pholas' Zone . . .	—	—	Pholas crispata
Intermediate Estuarine Clay	—	—	P. candida
Lower Estuarine Clay with	6–7	2·0	Scrobicularia piperata
Zostera marina			Littorina littorea
			Cardium edule
			Tapes decussatus
Grey Sand . . .	2	0·6	None
Peat	1–2	0·3–0·6	Corylus etc., and Vertebrate remains—Sus, Red Deer[1], etc.
Grey Sand (passes down into red sand below) .	2	0·3	Foraminifera and ostracods
	25–28	7·6–7·9	

At the Entrance Basin, 200 yd (182 m) N. of the dock, the Estuarine Clay thinned to 4 ft (1·2 m) and this rapid attenuation was accompanied by a mixed fauna of both Upper and Lower Estuarine Clay species.

In 1931 the excavation of the Milewater Dock revealed a similar succession to that above and Charlesworth and Erdtman (1935) record the following section:

Section: Milewater Dock

	ft	m
Surface Clays and Sands . .	16	(4·9)
Estuarine Clays	12	(3·7)
Red Sand	3	(0·9)
Peat with tree remains . .	3	(0·9)
Red Sand	7	(2·1)
Grey Sand	2	(0·6)
Red Sand with Irish Deer . .	3	(0·9)
Gravel resting on Red Sand (un-bottomed)	2	(0·6)
	48	(14·6)

The basal 8 ft (2·4 m) of the Estuarine Clay were said to be very shelly though no faunal lists were given.

The most comprehensive study of the physical properties of both the Glacial and Post-Glacial sequence is contained in an unpublished report on boreholes at the East Twin.

Generalized Section, East Twin Island (Average of 11 Bores)

		ft	m
Recent Estuarine deposits	Dredged Fill (Mixed) . . .	16	(4·9)
	Grey Sand	5	(1·5)
Estuarine Clay	Soft Blue Silty Clay with basal peat	8	(2·4)
Late Glacial Deposits of Lake Lagan	{ Reddish brown Sand . . .	4	(1·2)
	{ Firm Red Plastic Clay . .	27	(8·2)
Boulder Clay	{ Red Sandy Clay with stones .	11	(3·4)
	{ Red Sand	6	(1·8)
	{ Red Sandy Clay with stones on Red Sandstone	18	(5·5)
		95	(28·9)

---

[1] The 'Irish Elk' recorded from the peat was later said to come from beds below the peat. The revised names of these fossils are given in Appendix 4.

A small macroscopic fauna from each of five holes was submitted to A. S. Kennard and all were from the Lower Estuarine Clay. From one bore, however, shells were obtained from red sandy clay with stones beneath Estuarine Clay and peat. The shells included *Hydrobia ulvae*, *Cerastoderma edule* and *Venerupis decussata*. Kennard adds, "These beds have been claimed to be of glacial origin, but shells show quite clearly that they belong to the Estuarine Series with climatic conditions similar to those of today". This appears to imply some redistribution of the pre-existing glacial deposits prior to the emergence during which the peat was formed. Nevertheless, in a re-examination of the specimens C. J. Wood notes:

"The following species given in Kennard's list are cold-water forms and would be anomalous in an essentially 'southern' molluscan fauna; examination of the matrix shows that these species were not collected from the Estuarine Clays. *Nuculana rostrata* (Gmelin), recorded by Kennard as *Nuculana pernula* (Müller): matrix red clay. *Portlandia* (*Yoldiella*) *philippiana* (Nyst), recorded by Kennard as *Yoldiella lenticula* (Möller): matrix red clay.

Note that both the above species were probably collected from the weathered top of boulder clay underlying the peat deposits and Estuarine Clays".

The red clay clearly is similar to that described by Stephens (1963) and ascribed to a late glacial marine transgression.

Boreholes on the site of the new Dry Dock on the East Twin revealed a maximum of 47 ft 9 in (14 m) of Estuarine Clay, usually underlain by peat. Over part of this site, however, much of the recorded thickness of Estuarine Clay may represent dredged fill. On the site of the new Building Dock in Musgrave Channel, thicknesses of up to 26 ft (8 m) are known.

RAISED BEACH

During a period of post-glacial submergence, when the sea stood 10 ft to 20 ft (3–6 m) higher relative to the land than today, a rock-cut beach was developed at many places round the coasts of the British Isles, and shingle beach deposits were laid down. None of these deposits is known in the Belfast area, but the feature cut at this period by the sea in the slopes north-west of the estuary can be followed from Fortwilliam through the city as far as Peter's Hill. Praeger (1896, pp. 30–54) states that part of the Raised Beach is contemporaneous with the Upper Estuarine Clay but Movius puts the Raised Beach Deposits in the Sub-Boreal period.

Though Fisher and Welch (1933, p. 177) recorded a raised beach section at the Grove, the base being at 24 ft 6 in (7·4 m) above O.D., Cleland (1934, pp. 11–12) refuted the evidence and suggested that the material was 'made ground'.

RECENT MARINE BEACH DEPOSITS

Overlying the Estuarine Clay at the Alexandra Dock a layer of clean yellow beach sand 2 ft (0·6 m) thick, containing derived shells, was noted, overlain in turn by 6 ft 6in (2 m) of mud, the surface of which is at present between tide marks and which is full of littoral burrowing molluscs (see Praeger 1887, p. 30; 1892, pp. 233–4). This is cited by Movius (1953b, p. 16) and Coffey and Praeger (1904, pp. 153, 156) as evidence of a small depression of the land in recent times.

A similar sand has been noted at Ventry Street and possibly Sandy Row, though the latter may be a marginal sandy facies of Estuarine Clay. The sand area in the Newtownards Road area may belong partly to this period, but is probably largely re-sorted glacial sand.   P.I.M.

LACUSTRINE DEPOSITS OF LOUGH NEAGH

Gravel Beaches. Much of the Lough shore is edged by a narrow strip of coarse gravel and boulders which in places reaches a width of 20 yd (18 m) or more. The boulders are often of very large size — up to 4 cu yd (3 m³) — but the bulk of the deposit is of basalt cobbles up to 1 ft (0·3 m) in diameter and clearly the result of the action of the Lough on the boulder clay which often forms steep cliffs at the back of the alluvial flat. Other rocks common to the boulder clay are also found, schists and igneous rocks, from Tyrone, and flint being the most abundant.

On Rams Island the western shore line is of the usual coarse cobbled shingle but the eastern shore, particularly on the long shingle spit at the north end of the island, is of finer gravel and sand. This spit which rises to about 6 ft (1·8 m) above lough level, is of interest as it has been developed by wave action in shallow water without any assistance from tides or currents.

Raised Gravel Beaches. In some area, particularly round Lennymore Bay, the alluvial flat is backed by deposits of sand and gravel which represent shore deposits and storm beaches deposited when the lake stood somewhat higher than at present. These gravels are probably always thin and few good sections through them are seen.

North of the abandoned mouth of the Crumlin River at Waterfoot Bridge the lake flat is separated from the alluvial plain of the river by a low shingle bank ½ mile (800 m) in length. South of the river the raised beach is partly obscured by blown sand but it extends for 1½ miles (2·4 km) to beyond the Glenavy River. Exposures near Bellbrook House reveal from 3 ft to 6 ft (1–2 m) of sand and shingle with iron-pan layers, overlying boulder clay. Another sand ridge limits the alluvial plain at Sandy Bay and there are smaller gravel terraces at several points along the shore.   H.E.W.

Lacustrine Alluvium. The flat terrace, generally limited by a boulder clay cliff, which fringes much of the eastern shore of Lough Neagh is fronted and occasionally backed by lacustrine gravel terraces and is in general covered with a thin layer of lacustrine silt or mud carried in flood conditions across the gravel beach. Lowering of the lake level has dried out these deposits which are frequently heavily overgrown with scrub, but until a few years ago they were water-logged and the alluvium is commonly peaty.

Good exposures of these deposits are unknown, but brown clay with stones which is seen below the lake shingle at a few places is probably alluvial in origin, perhaps reworked boulder clay. Examples of this deposit are seen 1 mile (1·6 km) north of Sandy Bay and on the north of Brankins Island. Stoneless alluvial clay is seen at Johnny's Bay.

The wide expanse of alluvium which surrounds Portmore Lough is also in part lacustrine in origin but it is now covered with fluviatile deposits from the surrounding rivers and is described under that heading. The boundaries mapped between the fluviatile alluvium and the lacustrine deposits are obviously arbitrary.   H.E.W., P.I.M.

Around the inlet at Annaghdroghal a narrow strip of sand and silt with a little gravel borders the Lough. Behind the low gravel bars which bound this deposit occur wide flat areas of black peaty alluvium, from much of which peat cover has been cut off.   H.E.W.

RIVER ALLUVIUM

The principal river, the Lagan, is bordered throughout its length by flood plain alluvium still subjected, in part, to periodic flooding. Derived from the contiguous sand or clay drift through which the valley is excavated, this alluvium is usually a brown, fine, sandy loam or reddish clay with subordinate layers of fine to coarse gravel and organic layers at some horizons. Both in the main Lagan valley, just south of the King's Bridge, and in the Blackstaff or Bog Meadows area the fresh water alluvium merges downstream into the Estuarine Clay. The alluvium may thus be in part contemporaneous with the Upper Estuarine Clay, though deposition is still in progress.

Deposits of flood-plain alluvium occur along all the rivers and streams in the western part of the area. Generally of limited extent they occasionally cover wide areas like the flood-plain of the Stonyford River near Stonyford. In addition there are flat deposits, often of considerable area, in the hollows of the undulating drift-covered country between the Lough and the Lagan valley, usually drained by mere rivulets.

The wide flat area round the northern end of Portmore Lough, from Georges Island to Portmore, is underlain by alluvium which is doubtless partly lacustrine in origin but which was probably derived mainly from the Ballinderry and Roughan rivers and the minor streams which flow westwards into the basin. This alluvium probably overlies peat of the same age as that forming the extensive moss to the south, and the mapped boundaries between the fluviatile alluvium, the peat, and the lacustrine alluvium to the west are necessarily arbitrary.

In County Down the tortuous drainage of the inter-drumlin hollows has resulted in the development of an intricate network of alluvial flats along the courses of the streams. The deposits are often peaty but they are rarely of any great extent and are generally thin. The most extensive areas are along the Raver-net River and around the group of small lakes, often largely filled by alluvial deposits, in which it rises.

DETAILS

From Aghagallon to Portmore Lough the broad flat areas underlain by peat are covered in part by recent stream alluvium. Exposures are few, the best being in the Ballinderry River near Portmore Castle where 6 ft (1·8 m) of alluvial silt, with a lens of peaty material 1 ft (0·3 m) thick, are seen. In the Tunny Cut, an artificial drain 300 yd (275 m) S.E. of Tunny Bridge, 4 ft (1·2 m) of alluvial sand and silt are exposed. Other streams debouching on to the peat flat to the south, notably the Aghalee Burn, have also covered the peat with a layer of alluvium.

Apart from a continuous strip of alluvium along the Aghalee Burn overflow channel, the remaining patches of alluvium in the basalt country are irregular, usually small, disconnected patches between drift mounds. Many of these hollows were probably the sites of former lakes. For example, ¾ mile (1200 m) N.N.W. of Moira, on a fairly large alluvial flat, a crannog is recorded on the six-inch base map [143 647]. To the south of Darganstown [127 608] a mile (1·6 km) long narrow strip is dissected into smaller patches by a sinuous esker.

South-east of Magheralin, both the River Lagan and its tributary, the Pound Stream, have fairly broad alluvial belts. Near their junction, drainage ditches show 4 ft to 5 ft (1·2–1·5 m) of practically stone-free reddish brown silty clay with rare pebbles of chalk and flint. Shallow pits near here may have been dug for brick clay. Eastwards, the Lagan is joined by an unnamed tributary from the south which at one time probably joined the Lagan ¼ mile (400 m) E. of Steps Bridge. An abandoned alluvial flat is now occupied by a minor drainage channel. A well [154 587] on the alluvial flat of the tributary 230 yd (210 m) W.S.W. of The Forest Farm went through 2 ft (0·6 m) of peat and 6 ft (1·8 m) of sand with 'oak' at 5 ft (1·5 m) below surface, before entering boulder clay.

South and south-east of Moira the alluvial plain of the River Lagan averages 400 yd (365 m) wide. Where the Lagan Canal crossed the river by an aqueduct [173 602], four boreholes were put down for the River Lagan Bridge (M1 Motorway). Borehole 36/599 proved 2 ft (0·6 m) of top soil, 23 ft (7 m) of silty clay and 11 ft (3·4 m) of gravel over boulder clay.

Alluvium occurs intermittently northwards along the line of the canal towards the entrance of the Aghalee overflow channel which is partly occupied by the old canal reservoir, the Broad Water, for the first 1500 yd (1375 m).

The alluvial belt of the Lagan continues eastwards to Lisburn. Near Halftown three boreholes were put down for the motorway. The deepest (Borehole 36/340b) sited at [238 631] proved 16 ft 6 in (5 m) of clay and sand, in beds up to 4 ft (1·2 m) thick, overlying boulder clay.

South of Lisburn the River Lagan is joined by a right bank tributary, the Ravernet River. On the alluvial flat of the latter, south of Sprucefield, the alluvium is thin as shown in Borehole 36/334a, 100 yd (91 m) S.S.W. of Sprucefield [262 622], which proved 2 ft (0·6 m) of fill and 2 ft 6 in (0·75 m) of alluvial clay on boulder clay. Farther south the Ravernet alluvial plain widens to about 400 yd (365 m), becomes constricted again at the rock gorge east of Ravernet village and to the east forms a discontinuous ribbon for 6 miles (9·7 km) to the collection of small lakes in the south-east corner of the Sheet. The main Lagan alluvial plain averages only 100 yd (91 m) wide in the vicinity of Lisburn. At the Island Mill, Lisburn [273 643] three boreholes to a depth of 15 ft (4·6 m) appear not to have reached the base of the alluvium, though Young (1875, p. 33) states that the Boulder Clay was pierced at 30 ft (9·1 m) on this site.

Downstream from Lisburn the Lagan flows on a meander-belt usually 100 yd to 200 yd (91–183 m) wide with some wider expanses at confluent streams. At the Motorway bridge [299 679] south-east of Dunmurry, alluvium 12 ft (3·6 m) thick rested on red sand and at Newforge factory [328 694] it is 15 ft (4·6 m) thick. In the Annadale area it passes imperceptibly into estuarine alluvium.

The extensive alluvial plain, drained by the Derryaghy River, which lies west of the railway, may be the site referred to in the Ordnance Survey manuscript of 1837 quoted by Marshall (1933) which stated that Irish Deer horns were found 70 years previously "20 ft (6.1 m) below the surface" in a bog near Lambeg village. The horns were 2 ft (0·6 m) long and 28 in (0·7 m) between the tips. Shells were also found. The area was under cultivation by 1933. Excavations in 1953 for a pipe trench in this area revealed a section in sandy alluvium with some peat. Horns of *Megaceros giganteus* were obtained from this site.

Trench section [281 674] 300 *yd* (275 *m*) *N.E. of Moss Side*

	ft	in	m
Top Soil	1	0	(0·3)
Sand, silty, dark brown with a small amount of peat	0	4	(0·1)
Sand, dark brown with more peat than above	0	4	(0·1)
Sand, yellowish grey with poorly defined bedding	0	11	(0·28)
Peat, dark brown with *Megaceros giganteus*	0	6	(0·15)
Clay, silty, greenish blue, fibrous and full of rootlets, traces of bedding	0	9	(0·23)
Sand, grey with streaks of green	1	0	(0·3)
	4	10	(1·4)

Much of this flat belt is now covered by the Derryaghy Industrial Estate and housing estates, where the alluvium is known to be up to 18 ft (5·4 m) thick.

The Estuarine Clay of the River Blackstaff passes into alluvium in the Bog Meadows area. Much of this flat-lying area has been altered now by the new drainage tunnel to the River Lagan, by the construction of the Motorway and by extensive dumping. Parts, however, are still liable to flooding and retain the original marshy aspect.

Tongues of alluvium extend north along the Clowney River and south-west along Lady Brook.

In the Lower Palaeozoic country, as in the basalt area, alluvium is present in irregular small disconnected patches between drift mounds. Many of these deposits are peaty and represent infilling of small lake basins.   H.E.W., P.I.M.

RIVER TERRACES

Most of the streams and rivers which flow westwards into the Lough Neagh basin have a number of small terraces a few feet above the flood-plain, but the best development is by the Crumlin and Glenavy Rivers where up to four terraces are seen.

The wide flat area at the mouth of the Crumlin River is probably partly lacustrine in origin but it corresponds in level with the third terrace of the river and has about 16 ft (4·9 m) of alluvial deposits resting, in one case, on boulder clay. The material ranges from gravel and sand to fine laminated silts and clays, occasionally showing contorted bedding. A good exposure 600 yd (549 m) W. of Cherry Valley shows that the laminated silts and clays contain small flat lenses of hard pinkish brown clay, which are apparently contemporaneous clay-galls.

Ditch exposures in the area south of the river are mainly of brown silty clay but in one case 650 yd (595 m) N.E. of The Gulf [124 742] a ditch section shows brown silty clay, with a band of clayey peat 1 ft (0·3 m) thick, apparently overlying 8 ft (2·4 m) of red gritty sand and pebbles which rest on dark grey sandy silt.

An area about ¼ mile (400 m) square round Cherry Valley appears to be a fragment of a higher terrace and exposures of laminated silt are probably fluviatile, but in a ditch behind the house, 6 ft (1·8 m) of this material are seen to rest on stiff laminated clay with silt bands which may be of glacial origin.

Between Cidercourt Bridge and Crumlin the river flows with a moderate gradient on rock head between steep banks of drift and there is practically no development of flood-plain alluvium, but up to four terraces are preserved in places with the uppermost 40 ft (12 m) above the present river level, as seen 200 yd (183 m) W. of Glendaragh.

Above Crumlin the river meanders along an alluvial strip rarely more than 200 yd (183 m) wide and consisting mainly of a flood-plain with occasional fragments of one or two higher terraces a few feet above it. The extent of both flood-plain and higher terraces is greatest near Dundrod Bridge, where the flat area south of the Corn Mill is probably a dissected relic of a third terrace.

The Glenavy River has no estuary, and the lower part of its course, below Leap Bridge, is similar to that of the Crumlin River above Cidercourt, with no flood-plain and fragments of four terraces on the steep drift banks. Between Edenvale and Glenavy the river, flowing on or near rock-head, is margined by a series of terraces at different levels which are never more than a few feet above the river and bear witness to the cutting down of the stream through the boulder clay before reaching rock-head. Later terraces often cut into the earlier ones, and a terrace which is just above the flood-plain can be followed downstream to a position where it is actually three levels above the river. Upstream from Glenavy the river meanders along a flood-plain rarely more than 100 yd (91 m) wide with a few fragments of higher terraces and a small alluvial cone. At two places, old courses of the stream marked by alluvial flats diverge widely from the present river. Just west of Ballydonaghy Bridge the old course, which is followed by the road, swings south-westwards, rejoining the present course 600 yd (549 m) downstream, and at Weir House [164 739] the old course, followed by a disused mill race, makes a wide sweep to the south to join the river again at Glenavy, over 1 mile (1·6 km) away.

The southern head-stream of the Glenavy River, the Stonyford River, forms a low terrace a few feet above the flood-plain at Crooked Bridge, but on following this terrace upstream it is seen at Stonyford to pass into the flood-plain of the upper reaches of the river.

The Ballinderry River, like the Crumlin and Glenavy Rivers, flows on rock-head for the lower part of its course, but here the banks are gentle and there are no relics of high terraces. In its upper course, where it flows beside and between the fragments of the Ballinderry esker, there are a few areas of low terrace and two small alluvial cones, but most of the deposits are of flood-plain alluvium.

The River Lagan has very few extensive terraces, probably because of their impersistence in the soft sand and gravel drift. Small areas do occur, however, in the district between Lambeg and Seymour Hill, and on some of the small tributaries in the Ballycarn and Ballylesson areas.

The Ravernet River has two extensive terraces on the north bank north of Ravernet village, extending for over ½ mile (805 m) and up to 300 yd (275 m) wide. Above Legacurry Bridge it is surprisingly free from terraces, only a few small areas occurring on the main stream and its tributaries. This is probably because the drumlins of this area rest on a rock floor and there has been virtually no downcutting by the river during the development of the confused drainage pattern in the district.

The Glen River, in its steep and relatively juvenile course from Glen Bridge to Suffolk, is cutting down through the drift at a much faster rate than most of the other streams in the area and its banks are marked by a series of impersistent terraces at varying levels. South of Suffolk, where the gradient lessens, terrace deposits are extensive and are over 20 ft thick (6 m) in places. H.E.W.

BLOWN SAND

The prevailing westerly wind and the twelve mile fetch across the Lough have resulted in the accumulation of some deposits of sand on the eastern shore. Before the level of the water was stabilized by the construction of the weir at Toome a century ago the level fluctuated considerably, and during dry periods considerable areas of foreshore were exposed. As a result there are widespread deposits of blown sand along the shore, particularly in the bays where the sand accumulated most freely. This sand may be in part derived from the Lough Neagh Clays and from recent river-carried deposits but much is probably from the glacial sands which appear to underlie wide areas in the east and south of the lough and which are worked by dredges as sources of building sand.

The widest area of blown sand is in the area of Lennymore Bay which opens like a funnel to the south-west. The sand covers a strip 0·8 mile (1·2 km) long by up to ¼ mile (400 m) wide running south from Waterfoot Bridge. In addition there is certainly some blown sand on the raised shingle beaches west of Waterfoot and south of The Gulf [124 742]. The thickness of blown sand is rarely great but 300 yd (275 m) N. of The Gulf it was at one time worked as building sand and a ditch exposure shows 3 ft 6 in (1·1 m) of sand on 7 ft (2·2 m) of peat. Peat 'full of hazel nuts' was also found under sand at the farm buildings 700 yd (640 m) further north.

At Sandy Bay the raised shingle beach is overlain by blown sand, at least in places, and at Lady Bay there is a wide area of blown sand between Gulf Bridge and Sandhills School. Here also the sand has been worked. An exposure on the lake shore 200 yd (183 m) S. of Gulf Bridge shows 3 ft (0·9 m) of horizontally bedded sand resting on peat.

At Selshan Harbour there is a belt of sand 100 yd (91 m) or more wide from near Selshan Bridge to Gawley's Gate Post Office and another smaller patch on the south side of the bay towards Derrymore Point. Exposures near Selshan and Gawley's Gate show from 2 ft to 4 ft (0·6–1·2 m) of sand resting on peat.

H.E.W.

PEAT

An area over 3 sq. miles (7·8 km²) in extent south and west of Portmore Lough is underlain by basin peat of unknown thickness. These deposits extend north and west below the alluvium and lake deposits and may well underlie part of Lough Neagh.

Most of the area has been drained and reclaimed for grazing and there is little to be seen of the underlying peat. It is clear that over much of the ground the peat has been cut to drainage level but there is no present working and the only indication of the original level of the peat surface is in the roads which stand up on peat causeways. Local reports speak of 'white marl', probably diatomite, underlying the peat, 4ft to 5ft (1·2–1·5 m) below the present surface, near Gawley's Gate, and diatomite was also reported near Derryola Island and below the alluvium near Tunny Island.

Though no detailed investigation of the peat in this area has been undertaken it is probably of the same age as other basin-bogs round Lough Neagh which were described by Jessen (1949) as of Boreal and Sub-Boreal or Atlantic age.

In the Lagan valley peat was once widely distributed but has been entirely cut away for fuel. Green (1929, p. 134) says that a bog at The Maze was cut before 1800 and that the western part of Lisburn was built on a bog known as Tamneymore Big Moss which once stretched to Derryaghy. Part of this area is described under 'Alluvium'.

Just east of Newport Cottage [233 610], up to 7 ft (2·1 m) of peat were recorded in shallow bores beside the Lagan Canal. The record occurs in a narrow half-mile long north-south belt which ranges from 50 yd to 100 yd (46–91 m) wide. An area 600 yd (549 m) S.E. of Ballyskeagh is now a well-drained alluvial flat with only traces of peat and is crossed by the embanked M1, but it still retains the local name of 'The Moss' as an indication of its former character.

In County Down there are a number of small areas of peat in hollows in the drift-free rock floor north and west of The Temple. These are small in area and are doubtless remnants of formerly more extensive bogs which have been cut to drainage level. A road causeway across a small area ½ mile (800 m) S. of Lough Moss revealed a thickness of 10 ft (3 m) of peat. Larger areas of peat and peaty alluvium occur round Bow Lough and Gills Lough in the south-east corner of the Sheet. These deposits are waterlogged and no information is available about their thickness.

The submerged peat beneath the Estuarine Clay of Belfast has been noted elsewhere, as has the thin late-glacial peat formerly seen in the brickyards on the right bank of the Lagan in Belfast (Chapter 11, pp. 138, 146).

Thin hill-peat on the high ground near the basaltic escarpment at the northern margin of the Sheet now only covers small areas on Divis and Black Mountain and is wasting away. The greatest thickness observed was 2½ ft (0·75 m) although 5 ft (1·5 m) were recorded in 1902.   P.I.M., H.E.W.

LANDSLIP

All round the Antrim plateau the Lias and Rhaetic clays provide a weak foundation for the escarpment of Chalk and Tertiary basalts. Where water can reach the clays, from outcrop or through joints, they become plastic and cause rock falls and landslips. Areas of landslip, caused by this agency, occur all around the basalt plateau, from Binevenagh to Belfast, and are often very extensive and occasionally active.

Only relatively small areas of landslip occur on the Belfast (36) Sheet, high on the slopes between Black Mountain and Divis. Grassed over and covered by scrub, they are apparently stable but any attempt to build on them would be inadvisable. The landslip material is usually a mixture of Chalk and basalt blocks in a ploughed-up matrix of Trias marl, but there are instances of areas of relatively undisturbed basalt and chalk which have moved bodily downwards.

                                                                H.E.W.

INTAKE AND FILL

At the head of Belfast Lough large areas have been, and are still being, reclaimed from the sea. The former course of the Lagan has been diverted and part of many of the sections in trial borings must be regarded as fill. Of 30 ft (9·1 m) of soft grey silty clay at the East Twin the top 10 ft (3 m) are regarded as dredged fill. The outline of the 18th century shore line has been indicated on the one-inch map giving an indication of the pre-reclamation shore outline.

Wide areas of the 'Bog Meadows' have been covered to a depth of several feet by dumping, particularly along the line of the M1 motorway.   P.I.M.

Chapter 13

ECONOMIC GEOLOGY

ANHYDRITE

THE OCCURRENCE of evaporite in the Permian rocks beneath east Belfast was suggested by the records of 'white limestone' in old borehole journals from the area, and confirmed by the Avoniel Borehole (p. 193). This proved several beds of anhydrite including a 16 ft (4·9 m) bed at 380 ft (116 m) below the surface. While it would probably be difficult to work this below the built-up urban area, it is reasonable to assume that it continues down-dip to the north and occurs at greater depth below the harbour area and Belfast Lough. If Permian beds continue towards the Larne Basin, as well they may, there may be a very large quantity of anhydrite beneath south-east Antrim.

BASALT

In the past basalt was widely utilized for rubble masonry and many buildings survive to bear testimony to the durability of this material. It was often roughly squared and dressed.

Today, basalt is used almost exclusively for road material and concrete aggregate and several working quarries produce large quantities annually. The non-amygdaloidal or sparsely amygdaloidal central portions of the lava flows where the rock is hard, compact, and unweathered are utilized where possible. Nevertheless, large quantities of amygdaloidal and lateritized basalt have been removed from time to time as fill. A good deal of this type of material was used in the foundations of the M1 Motorway and a certain amount is being used on private building sites.

Most of the quarries are sited along the basalt escarpment and are sometimes worked in conjunction with the limestone quarries below in which the basalt represents unwanted overburden.

In Spence's Quarry, a mile north of Moira Railway Station, the fairly massive 22-ft (6·7 m) thick first flow above the chalk has been worked since 1955. Formerly, Mr Spence worked Rock Hill Quarry, a mile to the north-east where up to 30 ft (9·1 m) of basalt, probably a flow or so up in the lava succession, was used.

Several quarries in the Maghaberry area such as Richardson's, Aughnagarey, Balmer's Glen and Hull's at Mullaghcarton probably supplied rough filling for Maghaberry Airfield during the 1939–45 War.

Small quarries in the Hollymount dolerite and in the basalt north of Moira were opened up for local use.

A group of quarries round White Mountain and Castlerobin work massive flows low in the Lower Basalts with up to 40 ft (12 m) of sound rock available in some of the working faces. In all cases care is necessary to avoid vesicular and kaolinized bands at the top of flows and sheared material from small crush zones.

There are a great number of abandoned quarries all over the basalt outcrop and in the mugearite at Crew Hill and Ballynacoy. Most of these are small and were opened as local sources of road metal, but large quarries in good quality basalt which offer future potentialities, or suggest possible adjacent sites, exist at Posey Hill, Ballyvanen; Bo Hill; The Rock; Charlton's Folly and Lake View, Aghalee; and Aghacarnan. Mugearite, which is probably too fine-grained and splintery for use as aggregate, could be worked at Ballypitmave and Ballynacoy.

To the north of Collin Glen, workings are confined to the active Black Mountain Quarry and several small abandoned quarries nearby. The latter were opened on the scarp and so had no good access roads and the rock quarried was of the lowest two flows which proved unsatisfactory as a source of road stone. In this district the lava when extruded to form the lowest three flows contained a relatively high proportion of steam and perhaps other gases. Consequently, the basalt of these flows is rubbly rather than massive, highly amygdaloidal and kaolinized and unsuitable for roadstone.

South of Collin Glen and in the neighbourhood of Groganstown there are several extensive quarries in the lowest three flows. Many of these have now been abandoned due to the unsatisfactory nature of the stone and have been replaced by others in the higher flows of more massive and better-quality basalt. These flows form the major part of Collin Mountain and the high ground to the north and to the west at Aughrim. In the upper reaches of Collin Glen a massive slightly porphyritic basalt was noted during the resurvey. This rock forms the second flow here and is now being quarried on a large scale.

BRICK CLAYS

Most of the older buildings in Belfast are built of locally made dark red brick which has weathered well for a century or more. These bricks were probably made in the small brickfields which worked the plastic glacial clays underlying a considerable part of the Lagan valley, or the boulder clay which, where it overlies Triassic rocks, is often relatively stone-free. In the later years of the nineteenth century and the first half of the twentieth, the Keuper Marl became increasingly used as a source of clay for wire-cut bricks. The distribution of pits, the majority of which are abandoned and many of which are built over, is shown in Fig. 19.

Kinahan (1889, p. 370) recorded that bricks and drainage and other tiles were being manufactured then, generally from the washed 'till' and also from Keuper Marl. He further said that it had been the custom in the neighbourhood of Belfast to take ground likely to come in soon as building sites, and make bricks on the spot. Lamplugh (1904, p. 126) said that this was still in progress at Ravenhill in 1902.

Other early records of china production in, for example, the Ballymacarret area of Belfast (Kinahan 1889, pp. 370, 373) probably refer to imported raw material.

The glacial clays yielded bricks of moderate quality and when the Keuper Marl was dug by hand with separation of gypsum veins they also were reasonably good. With increasing mechanization of working in the fourth decade of the present century, however, some works turned out bricks containing very heavy concentrations of soluble salts which not only gave a less durable brick but caused heavy efflorescence and sulphate attack on cement and mortar. The use of gypsiferous marl has now ceased and the manufacturers of marl bricks now exercise close control on the quality of the raw material.

Within the area of the Sheet only the Keuper Marl offers large reserves of brickmaking material. There are limited areas of glacial clays of suitable type but because of their low-lying position they are apparently unsuitable for large-scale mechanical working.

Some details of the composition and properties of bricks from this area are given by Bonnell and Butterworth (1950, pp. 23–9, 62–95). These show that only over-fired bricks made from the glacial clays are durable, all other specimens tested proving to be of moderate or poor durability. The chemical composition varied widely as shown by the following table:

Composition of Belfast Bricks

	Bricks made from Glacial Clay		Bricks made from Keuper Marl	
SiO_2	56·01	63·42	44·84	37·91
TiO_2	1·06	0·70	0·66	0·60
Al_2O_3	19·25	13·75	12·96	12·17
Fe_2O_3	8·28	6·33	4·53	3·33
FeO	0·44	0·31	1·45	2·26
Mn_2O_3	0·08	0·12	0·08	0·10
CaO	3·55	4·18	7·10	10·28
MgO	4·62	5·54	7·33	8·25
K_2O	4·60	4·09	3·84	3·40
Na_2O	1·20	0·95	0·63	0·56
SO_3	0·25	0·36	0·59	2·25

Works in Keuper Marl. Extensive brickfields were developed to work the Keuper Marl in the western and northern suburbs of the city of Belfast during the latter half of the nineteenth century and the first half of the twentieth. Only two of the works are still in production and these use clay brought in from other areas, the rest having all closed down within the last few decades. Very extensive pits were opened in the Springfield, Parkview and Oldpark areas and smaller workings were found at the Skegoneill Brickworks, off Skegoneill Avenue, on the left bank of the Milewater River in Alexandra Park, and at the City Brickworks on the north side of Limestone Road.

All these pits are now abandoned, the last about ten years ago, though Parkview and City Works are still in production. The brickfields are now largely built over and little trace remains of most of them.

At a time when the city pits were being closed a new works was opened just south of Glen Bridge, Collin Glen. This works is still in production but marl is now brought from a pit near Crow Glen, outside the Sheet boundary. In general the uppermost 250 ft (76 m) of the marl are relatively free of gypsum, as in the Forth River area, but at Collin Glen this zone appears to be absent.

Works in Glacial Clays. South-west of the Courthouse, Crumlin Road, an area 150 yd by 200 yd (137 m by 182 m), now long built over, was formerly a brick-

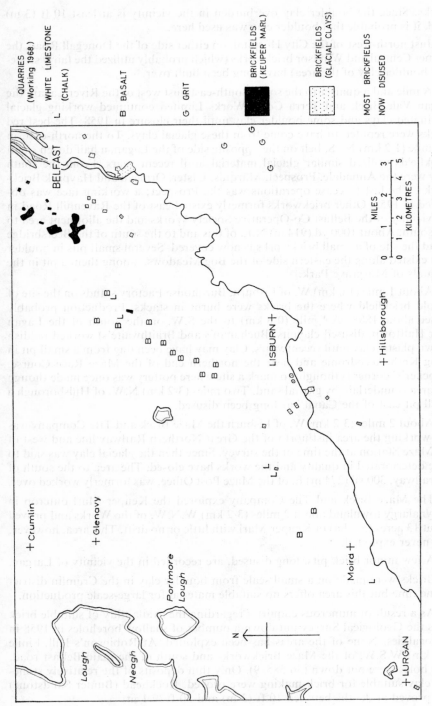

FIG. 19. *Distribution of brickyards, basalt, chalk, and greywacke quarries in the Lagan Valley*

works. Since the boulder clay overburden in the vicinity is at least 10 ft (3 m) thick it is probable that boulder clay was used here.

Just north-west of the City Hospital, on either side of the Donegall Road, the former Central and Windsor brickworks (which probably utilized the fairly stone-free boulder clay of the area) have long been built over.

A mile and a quarter to the south-south-east, just west of the River Lagan, the Lagan Vale Brick and Terra Cotta Works Limited continued working glacial laminated clays and some boulder clay until their closure in 1958. The best red bricks were reputed to have come from these glacial clays. To the north-east, in a ¾-mile (1·2 km) N.–S. belt on the opposite side of the Lagan, a half dozen small brickfields utilized similar glacial material until recent years. From the south they were the Annadale, Prospect, Marquis, Ulster, Ormeau and Haypark Brick-works. The last to cease operations was the Prospect; a working face was observed in 1952. Other brickworks formerly existed east of the Ravenhill Road in the vicinity of the Belfast Co-Operative Society Works and the allotment area to the south. About 1000 yd (914 m) N.E. of this and to the south of the Beersbridge Road the site of a small brickworks is now covered. Several small pits in boulder clay existed along the eastern side of the Bog Meadows, among them a pit in the grounds of Musgrave Park.

About 1 mile (1·6 km) W. of Lisburn, Burnhouse Factory stands on the site of an old brickfield where the bricks were burnt in stacks. Production probably ceased about 1888. A ½ mile (0·8 km) to the S.W. on the south of the Lagan near Halftown, disused claypits (Buchanan's and Braithwaite's) worked reddish brown plastic clay until recent years. Clay may have been dug from a small pit in Long Kesh Aerodrome and near the northern end of the Maze Race Course. 'Crocker's' corner is thought to mark a site where pottery was once made though the area is underlain by glacial sand. Two miles (3·2 km) N.W. of Hillsborough a small pit east of the Canal has long been disused.

About 2 miles (3·2 km) W. of Lisburn the Maze Brick and Tile Company was still working the area just north of the Great Northern Railway line and west of the Maze Station at the time of the survey. Since then the glacial clay was said to have deteriorated in quality and the works have closed. The area to the south of the railway, 300 yd (274 m) E. of the Maze Post Office, was formerly worked over.

The Maze Brick and Tile Company explored the Keuper Marl outcrop in Ballynalargy townland about 2 miles (3·2 km) W.N.W. of the Works and proved about 13 acres (5·2 ha) of Keuper Marl with little or no drift. This area, however, was never exploited.

A few minor brick pits, long disused, are recorded in the vicinity of Lurgan.

Bricks were made on a small scale from boulder clay in the Crumlin district at one time but this area offers no suitable material for large-scale production.

As a result of numerous enquiries regarding the availability of suitable brick clays the Geological Survey put down a number of shallow boreholes in 1958 in six localities. None of the areas has been exploited. At Robinson's Hill, 1 mile (1·6 km) W.S.W. of the Maze Brickworks and south of the main Belfast road, four holes were put down (36/585–9). Only thin deposits of the relatively stone-free clay suitable for brick-making were proved. Rockhead (Bunter Sandstone) was proved at depths between 10 ft (3 m) and 30 ft (9 1 m).

Two boreholes (36/592–3), drilled to the east of the Halftown pits, showed that the reddish brown, fairly stone-free clay continues to the east with little sand overburden and is from 24 ft to 30 ft (7·3–9·1 m) thick. In Boreholes 36/592–3 the following section was proved:

Borehole at Halftown

	ft	m
Soil	1	(0·3)
Sand, yellow	6	(1·8)
Clay, sandy	2	(0·6)
Clay, reddish brown, stone-free. Probably laminated, but much pugged up in core tube	15	(4·5)
Boulder Clay, stiff	6	(1·8)
	30	(9·1)

This clay ranged from 49 to 53 per cent SiO_2, 6 to 7 per cent CaO, and 0·14 to 0·32 per cent SO_4.

Three boreholes were put down south of the road opposite the Charley Memorial P.E. School, Drumbeg and fairly thick relatively stone-free clay was proved as follows:

Borehole 36/590—Drumbeg

	ft	m
Soil	1	(0·3)
Clay, red, sandy	5	(1·5)
Clay, reddish brown, tough and plastic at the top; somewhat silty near the base. Rare pebbles	24	(7·3)
	30	(9·1)

The clay had 48 to 65 per cent SiO_2, 4·6 to 9·6 per cent CaO, and 0·1 to 0·2 per cent SO_4.

The sandy and fairly stone-free boulder clay at Milltown was tested by three boreholes [265 685] which proved it to be too thin in itself—5 ft to 15 ft (1·5–4·6 m)—to justify working. The clay which has 74 per cent silica, 0·5 per cent CaO and 0·5 per cent SO_4, would make an excellent brick and it might be possible to work it with a proportion of the underlying Triassic Sandstone, which is fine grained with thin mudstone bands, to give a high silica brick. It is of interest that bricks made in Leicester from Keuper Sandstone are of excellent quality (Bonnell and Butterworth 1950, pp. 73–4). Two further boreholes to evaluate the Keuper Marl south-west of the existing works at Colin Glen proved 22 ft (6·7 m) of boulder clay at Springbank [274 705] and 23 ft (7 m) [209 697], south of Collin House. Analysis of the marl showed 28 to 46 per cent SiO_2, 13·7 to 23·6 per cent CaO, and 0·3 to 0·5 per cent SO_4.

CERAMIC CLAYS

Clay suitable for the production of earthenware (ball clay) is obtained from the Lough Neagh Clays in County Tyrone (One-inch Sheet 35) and may occur in beds of this formation in the Lennymore and Portmore areas on the east side of the Lough. Drift thickness may make working difficult, and indications are that the clays contain too much iron to be useful.

GYPSUM

Gypsum veins occur in the Triassic (Keuper) beds and in the lower part of the Permian marls. When the Keuper Marl was worked by hand for brickmaking the gypsum was picked out and was used locally, but no beds sufficiently thick for mechanized working are known and there seems no likelihood of future production.

LIGHTWEIGHT AGGREGATE

Some clays and shales when heated almost to fusion point expand to a porous cellular material which, on cooling, is strong enough to be used as aggregate for structural concrete. Tests on Lower Palaeozoic shales from County Down have proved that high quality aggregates can be made from some of them including material from Homra House Glen, Ravernet, within the limits of this Sheet. No commercial production of this material has yet been started (1969).

LIGNITE

Deposits of lignite in the Lough Neagh Clays occur at shallow depth in a few places near the lough shore at Store Quay, and were worked in the past as a source of fuel though the value of the lignite for this purpose is poor. The Department of Industrial and Forensic Science, Ministry of Commerce, has examined samples of this material with a view to the extraction of organic chemicals from it, notably montan wax, with fairly good results.

To determine the extent and thickness of the lignite a series of boreholes was drilled in 1964–5 in the area south-west of Crumlin. Two boreholes, No. 1 [128 752], 550 yd (502 m) S.W. of Cherry Valley, and No. 4 [126 746], 200 yd (183 m) E. of Thistleborough, proved thicknesses of over 40 ft (12 m) of lignite, unbottomed at depths of between 80 ft (24 m) and 90 ft (27 m) from the surface. Other boreholes sited near the known exposure at Thistleborough and at Bellbrook failed to find any lignite at depths down to 100 ft (30 m) though lignitic clay was recovered. Samples from Borehole No. 1 had from 23·4 to 44·6 per cent ash and 50 to 60 per cent moisture. Samples from No. 4 had from 7·6 to 25·5 per cent ash (average 17·6 per cent) and had calorific values from 8077–9979 B.Th.U./ lb (average about 9000).

The thicknesses proved in these boreholes indicate that a total of some 10 million tons of lignite occur in the area between Thistleborough and Cherry Valley beneath overburden of 60 ft to 80 ft (18–24 m) (Fig. 20). With a calorific value akin to that of peat, it seems unlikely that in present conditions this offers much prospect as a fuel, but it is a raw material of potential value.

Lignite occurs in the Lough Neagh Clays at other localities to the south of that investigated, and there are old records of beds up to 25 ft (7·6 m) thick in the Sandy Bay and Portmore areas. No details of quality or locality are available.

LIMESTONE

The Upper Cretaceous Chalk has long been used as a source of high calcium lime. At one time, lime-burning was widespread with kilns widely dispersed over

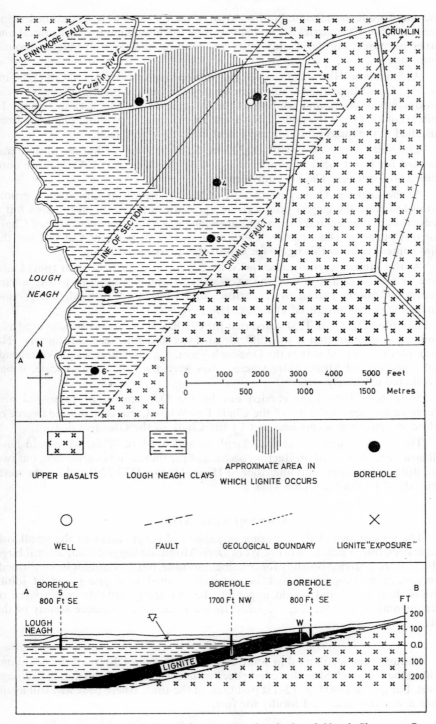

FIG. 20. *Map and section of the lignite deposits proved in the Lough Neagh Clays near Crumlin*

the countryside at individual farms. Most of the quarries were abandoned many years ago but six quarries are still active; two calcining lime for agricultural and other purposes; one producing putty and whiting and three producing ground limestone.

At Small and Hayes's Quarry in Ballymakeonan townland, 1400 yd (1280 m) N. $55\frac{1}{2}°$ E. of Magheralin Church, the quarry was worked on a small scale for the production of putty and whiting at the Whiting Mills, Magheralin—about $1\frac{1}{2}$ miles (2·4 km) from the quarry. Since the recent survey, crushing plant has been installed at the quarry and ground limestone is now produced.

Clarehill Quarry (W), 350 yd (320 m) S. of St John's Church, Moira, was worked for burnt lime. A sample analysed by the Ministry of Agriculture contained 96·7 per cent $CaCO_3$ equivalent.

At the Legmore Quarry north-west of Moira, the Ulster Limestone Corporation Ltd were the largest producers of agricultural limestone in the area but a reduction in the demand for ground limestone has halted the quarrying of Chalk for this purpose. The Legmore Chalk contains 97·98 per cent $CaCO_3$, but the flint content reduces this figure for run-of-the-quarry production.

At the Mullaghcarton Limeworks, lime is burnt in vertical shaft (pot) kilns. One of the two kilns is working at present. Along the outcrop to the north, quarries formerly known as Anderson's and Clark's are disused, but ground limestone is now produced at Laurel Hill [191 646].

Lamplugh (1904, p. 125) stated that the Chalk had occasionally been used as a building stone, but that it was difficult to dress and not very satisfactory. No examples were noted within the One-inch Sheet, but elsewhere, as at Whitehead, Kilwaughter and Moneymore where it has been used on a small scale, it has weathered extremely well.

Some tests have been carried out by the Department of Industrial and Forensic Science on the use of the Chalk for 'White Concrete', but the degree of whiteness required seems best met by the Chalk of the Glenarm district.

The extensive quarries at Knocknadona and south of White Mountain have all now virtually ceased production because of the increasing overburden, but two are still producing ground limestone at Hannahstown [277 720] and on the western side of Collin Glen [268 718].

ORNAMENTAL STONE

The appearance of the exposure of gabbro (dolerite) forming the small volcanic neck fully 1 mile (1·6 m) N.W. of Divis Mountain suggests that several large blocks were quarried possibly for ornamental stone, but this cannot be confirmed. This fresh coarse-grained rock is massively jointed with one group of joints almost horizontal and would appear to be eminently suitable as a source of ornamental stone. One factor against its use is the relative inaccessibility of the site.

PEAT

Though peat is still used in some country districts in Ulster, all the peat bogs within the One-inch Sheet have long been exhausted or have been dug to drainage level. Lamplugh (1904, p. 124) stated that, at the time, where peat was still available it was dug and used locally for fuel.

SANDSTONE

In the past the working of freestone was an important industry in the Belfast area, though most of the better quality Triassic Sandstone came from quarries in the Scrabo district near Newtownards. Nevertheless, in the Lagan Valley, Bunter Sandstone appears to have been at one time wrought on a small scale, notably for bridges for the railway (Kinahan 1889, p. 252) and for the Lagan Canal. Locally, some of the walls round the larger estates were built of this red sandstone.

The quarries are, without exception, now all disused and overgrown. They occur at no specific horizon within the Bunter Sandstone and most of the stone seems to have been used locally.

The sandstone from the Lurganville quarries (p. 36)—the 'Kilvarlin' [sic] quarries of Kinahan (1889, p. 252), 2 mile (3·2 km) S.E. of Moira, was used in the walls of Hillsborough Estate and probably for some of the canal bridges and the aqueduct to the north of the quarries.

At Dunmurry, fairly massive yellowish false-bedded sandstone is exposed in a 25 ft (7·6 m) face near Greenmount and appears to have been quarried on a small scale. Small quarries north-west of Milltown were said to have yielded material for bridges on the Ulster Railway (1839).

Other quarries in soft red or yellow sandstones beside the Falls Road, Belfast, have been exploited for sand rather than for building stone (Lamplugh and others 1904, p. 125).

LOWER PALAEOZOIC GRITS AND SHALES

The grits and greywackes of the Lower Palaeozoic rocks of County Down are used primarily for road metal and concrete aggregate. In the past the grits have been used for dry walling. Lamplugh (1904, p. 124) records the former use of grits as setts in Belfast, but this has long ceased.

To the south of Purdysburn numerous small abandoned quarries are a legacy from the early days of road-making when roadstone was wrought as near as possible to the place where it was required. These quarries were in greywackes, usually fairly coarse-grained. Following the tendency towards centralization of the industry relatively few quarries are required and only one in this district (300 yd (274 m) S. of Charity Bridge and reopened a few years ago) is now active. In general, the greywackes (the 'gritstone' of the quarrymen) of the district make admirable roadstone although care has to be taken in crushing to ensure the production of a sufficient proportion of nearly cubical material.

Massive grits and greywackes are extensively quarried in the area south-west of Carryduff and crushed for road-metal and concrete aggregate. There are large working quarries east of Lough Moss, at Kiln Quarter and north-east of Rockmount House. Over much of the area to the south the presence of bands of grey shale in the grits, undesirable in road-metal aggregate, has rendered unusable much otherwise suitable material, but the development of lightweight aggregate made by bloating shale at high temperatures, suggests that suitable shale may be valuable in the future.

SAND AND GRAVEL

The Malone Sands on the south and east sides of Belfast have been dug in the past at a number of localities both for building purposes and as moulding sand. Nearly all the pits have been built over and few traces of their existence remain.

Over the remainder of the area of One-inch Sheet 36, small pits have been opened from time to time in most of the areas shown as glacial sand and gravel on the map. There is a considerable demand in the building and other industries for sand and gravel of closely specified grades. Homogeneous sands of the required grade are rarely found *in situ* in the poorly graded glacial deposits which may frequently, as in the Malone Sands, have subordinate clay bands. Nevertheless, no machinery has been installed at any of the pits for grading or sorting, and production of sand and gravel has gradually declined despite the fact that fairly extensive reserves of variable quality still exist. Formerly, the Malone Sands were widely used as a moulding sand but the more demanding modern specifications reject these sands as a useful source for that purpose. From only one pit, that at Red Hill to the north-west of Lisburn, was sand still being dug at the time of the survey.

North of Lurgan, along the shores of Lough Neagh, small lacustrine beach ridges or bars of sand and gravel have been partly dug away for local use.

Glacial sand and gravel was dug ½ mile (0·8 km) N.W. of Aghagallon where a face of about 18 ft (5·4 m) can still be seen. South and south-west of Aghagallon, small patches of coarse gravel usually associated with drumlinoid mounds of boulder clay, have been dug on a small scale.

At The Nut Hill, some 2½ miles (4 km) S.E. of Moira, pebbly sand with red sand lenticles was dug both north and south of the road. The deposits here are part of the esker which runs east from this point for 1½ miles (2·4 km). Other pits in the same esker occur near Corcreeny House and at Beechfield Farm where 12 ft (3·7 m) of coarse ill-bedded gravel can be seen. A few pits are seen in the irregular sand and gravel patch between ⅓ mile and 1 mile east of Beechfield Farm.

Northwards, in the vicinity of the Maze and Long Kesh Aerodrome, large quantities of sand and gravel have been raised in the past. Disused sand and gravel pits with faces up to 30 ft (9·1 m) remain within the Maze Race Course while east of Gravel Hill up to 30 ft (9·1 m) of sand and gravel have been dug from the gravel ridge. Broad spreads of red sand with recorded thicknesses of over 25 ft (7·6 m) exist eastwards to the Ravernet River, but have not been exploited.

North of the River Lagan, in the vicinity of Red Hill, the western termination of an esker, large pits have and are being dug in the 30 ft (9·1 m) or so of red sand with gravelly layers. Farther east, in the same esker, pits have been dug at intervals for a distance of two miles (3·2 km).

South of the esker, at Knockmore Hill, a maximum of 20 ft (6·1 m) of sand and gravel have been dug from the hill whilst ½ mile (0·8 km) E.S.E., old sand and gravel pits remain just east of Knockmore Factory.

The Lisburn Esker, between Lisburn and Dunmurry, has been exploited for gravel practically throughout its length. As recently as 1960 the previously undisturbed section beside the road to the south-west of the Linen Research Institute was removed for fill.

Just east of Lambeg Bridge, between Sandymount and Kilroosty Lough, sand was extracted (according to local residents) during the 1914–18 War and sold at 6d. a load for red sand and 1s. a load for grey sand. To the east and north of the road between Ballyskeagh and Drumbeg a few grassed hollows mark the sites of former small pits.

North of Edenderry Mills, near Ballylesson, at Terrace Hill and at Milltown sand pits were formerly worked on a small scale. At Terrace Hill, grassed depressions mark the site of former moulding-sand pits.

In the Belfast area the sand overburden above the laminated clay at Lagan Vale was used until recently, as were similar sands at Annadale (Lamplugh 1904, p. 126). Between the Lagan and the Blackstaff the sites of many old sand pits are now built over. The Sir David Keir building and Ashby Institute at Queen's University were built in one of these excavations. The soft Bunter Sandstone has been dug for sand around Milltown Cemetery on the western side of Belfast.

The small areas of sand and gravel at and near Crumlin and Glenavy have been worked from small pits at a few localities and local farmers still dig small quantities for their own use. Similarly, the gravel spreads around Aghalee and the Ballinderry Esker have also yielded small quantities of sand and gravel. More extensive working in the western extremity of the Lisburn Esker at Lissue near Magheragall has left large pits where 30 ft (9·1 m) or so of well-washed sand and gravel were still exposed at the time of the survey.   P.I.M., J.A.R., H.E.W.

WATER SUPPLY

Details were given in the 1904 Belfast Memoir (p. 145, Appendix II) of 29 deep well sections in the Belfast area. The list was not exhaustive and information on drilled wells has been compiled which adds to the original material and covers wells and deep borings made during the period which has elapsed since 1904. Some of the information is of doubtful accuracy, and in many cases few details are available. This is particularly the case with regard to yields but many of the wells must be pumped greatly below their capacity.

The borings are summarized in Appendix 2. The most important are shown on the One-inch Map. Smaller dug wells are located on the Six-inch manuscript field maps. The well numbers are given according to current Geological Survey practice, the first numeral indicating the One-inch Sheet number and the last the serial number according to the order in which they were recorded.

SURFACE WATER AND PUBLIC SUPPLIES

The supply to Belfast City and district by the Belfast Water Commissioners is drawn from three separate upland sources, two outside the limits of One-inch Sheet 36.

1. *Woodburn supply.* Water from this source is collected in the basaltic upland catchments some 6937 acres (2809·5 ha) in extent, north of Carrickfergus, and is held in a series of seven impounding reservoirs with a capacity of 1183 million gallons. The water is filtered at Dunanney and Oldpark and stored in service reservoirs. Water from springs at the base of the Chalk at Whitewell and Ligoniel was added to this supply. Both springs lie just north of the limits of One-inch Sheet 36. They yield 100 000–150 000 and 100 000 gallons per day respectively, though the latter falls off during the summer months.

2. *Stoneyford supply.* 5348 acres (2166 ha) of basaltic upland constitute the catchment area of this scheme whose two main impounding reservoirs have a combined capacity of 912 million gallons. Supplies from this district are conveyed to Belfast and other areas in the Lagan Valley by the Lagmore aqueduct.

3. *Mourne supply.* Water from the Mourne area is collected on an upland granite catchment of 9000 acres (3645 ha) and stored in the Silent Valley and Ben Crom reservoirs with a capacity of 4700 million gallons. It reaches Belfast in a triple conduit via the service reservoir at Knockbrackan.

A scheme to draw supplies from Lough Neagh at Dunore Point, near Antrim, has come into operation in 1968 with a potential yield of a further 30 million gallons per day.

In 1966 the production figures from Woodburn, Stoneyford and the Mournes were 10·75, 3·2, and 28·19 million g.p.d., giving a total of 42·14 million g.p.d. of which about 35 million gallons were used in the city area.

Other public supplies in the Lagan valley include the Lurgan Water Scheme, and the Antrim Rural District Council's Langford Lodge Scheme, both involving the abstraction of water from Lough Neagh, and the Lisburn Urban District Council's supply drawn from deep boreholes in the Trias. Some private companies in the textile trades have private catchment areas and reservoirs within the district.

GROUNDWATER

Two deep wells at Lisburn are utilized for public supply but most of the deeper wells are used industrially. The majority of the deeper wells are sunk into the Bunter Sandstone and yield water of good quality. In the 19th century the manufacture of mineral water became an important local industry in Belfast, based on this pure water. Many of the earlier industrial wells are now disused, however, particularly in the case of some linen manufacturers who found the water to be too hard for their purposes.

Shallow wells for farm and domestic supplies are ubiquitous in the rural areas though the spread of piped supples is rapidly reducing their importance. Some of them obtain supplies from the drift deposits.

Figures are usually available for the deeper boreholes but no reliable estimates are available for the yield of private domestic dug wells. A private household of four persons will not use more than 140 g.p.d., whilst general farms will use some 4·6 g.p.d. per acre (0·4 ha). Each foot of a 4-ft (1·2 m) diameter well will hold some 314 gallons so that supplies to the owners seem inexhaustible despite a probable low rate of percolation into the well in the case of those sunk into boulder clay.

Lower Palaeozoic. Few deep wells have been drilled in the Ordovician or Silurian, the only examples being Kesh House Farm (total depth 82 ft (25 m) of which 30 ft (9·1 m) were in boulder clay) and a public well at Hilden View bored 100 ft (30 m) into Ordovician rocks. Most of the wells are dug wells, ranging up to 60 ft (18 m) in depth, which appear to derive their supplies from the stony base of the boulder clay and not from the solid rock into which there is little penetration. The permeability of the slates and grits is low and any supplies derived will be due to the large number of fracture planes and joints. The Lower Palaeozoic rocks must be regarded as poor aquifers though, as Hartley records, wells elsewhere in County Down have yielded about 900 or 1 000 g.p.d. where they are sunk below the level of permanent saturation. The chance of obtaining a suitable supply at depth depends on the location of a fractured zone.

Permian and Trias.
Permian and Bunter Sandstone. The sandstones of the Bunter and Permian are by far the most important aquifers in this district. The Permian aquifer has been penetrated only by a few boreholes in east Belfast. Though it is apparently prolific, the area underlain by it is very restricted. The Bunter Sandstone consists of over 1000 ft (304 m) of fairly uniform porous permeable reddish brown sandstone, poorly cemented and with occasional thin marl partings. It is separated

from the Permian Magnesian Limestone and sandstones by some 270 ft (82 m) of marls with anhydrite. The Bunter aquifer is most useful, the groundwater being readily available beneath drift cover, and over 100 deep wells have been sunk by commercial and industrial concerns to tap it. Yields of 10 000 g.p.d. are usual from deep wells though exceptional yields up to 30 000 g.p.d. are recorded.

Keuper Marl. The siltstones, mudstones and marls of this group are about 1000 ft (304 m) thick and are not normally regarded as water-bearing. In fact their impervious nature often results in spring lines where they are overlain by permeable beds, and they reduce the recharge of the underlying Bunter Sandstone where they overlie them on the north side of the Lagan valley. Nevertheless, some wells sunk into the Keuper have obtained supplies of hard water though with small yields. An exception is the 12-in (0·3 m) bore at Glenbank (36/1) which gave over 3000 g.p.h., but in this case the supply may come from a dolerite intrusion 73 ft (22 m) thick at 363 ft (111 m). Skerry (sandstone) beds in the Keuper Marl may also give small supplies.

GROUNDWATER CONDITIONS IN THE BUNTER

Recharge of the aquifer. The term 'recharge' includes all additions to groundwater in the aquifer by direct infiltration, influent seepage from rivers, lakes, or sea, leakage from other aquifers, or artificial recharge. Precipitation is the ultimate source of all infiltration and the quantity of water available is dependent on rainfall. In drift-free areas a permeable aquifer will absorb virtually all the rainfall except that which is lost as evaporation, but in the heavily drift-covered Lagan valley the percentage of precipitation which eventually reaches the aquifer is uncertain. The only quantitative work on this area is that of Hartley (1935) who estimated that 8 per cent of the rainfall penetrated the aquifer, either directly or by influent seepage from rivers, giving a calculated intake of one million gallons per hour from a catchment area of 220 square miles (572 km²). Some of Hartley's assumptions must be suspect. It seems unlikely that the precipitation on the Lower Palaeozoic rocks, which underlie much of the catchment, or on the Antrim Basalts, finds its way underground into the Bunter Sandstone, and the actual Bunter Sandstone outcrop is only about 40 square miles (104 km²) in extent. Again, his figure of 8 per cent for penetration is probably very low for the Bunter area, though not for the whole catchment. In the circumstances, his estimate was as good as could be made, and without figures for run-off, which were not available, any further estimate is equally unsound.

The average rainfall for the Lagan valley area is about 37 in (940 mm) per year and the Meteorological Office estimate potential evaporation to be about 18 in (457 mm), the actual evaporation figure being about 16 in (406 mm) on low ground and 12 in to 13 in (300–330 mm) on high ground.

With these figures, and the estimated flow of the Lagan—as yet not accurately measured—of 100 cusecs in drought and 5000 cusecs in flood it may be calculated that the rainfall, less evaporation, is 9 750 000 g.p.h. and the average river flow is 9 000 000 g.p.h. The difference of 750 000 g.p.h. is presumably infiltration.

An estimate for the infiltration directly in to the Bunter Sandstone outcrop, assuming a penetration factor of 40 per cent—which appears, from empirical observation in England, to be reasonable—gives a figure of 520 000 g.p.h. It may be significant that all three figures, arrived at through calculation on wide variables, are of the same order of magnitude.

Hartley calculated the groundwater flow into Belfast Lough as 300 000 g.p.h. from the gradient of the water table and its cross section. The difference between this figure and the intake represents the amount abstracted from the aquifer by wells and that lost by springs, increments to rivers, and leaking to other aquifers. Hartley calculated the abstraction from the wells to be at least 300 000 g.p.h., though this figure is probably now very much lower. The figure for springs and increment to rivers is indeterminate.

Even at the lowest estimate of intake to the Bunter Sandstone aquifer it is clear that there is a surplus of at least 200 000 g.p.h. available in the Lagan valley. More accurate estimates must await the accurate gauging of run-off from the Lagan and its tributaries and the collection of detailed figures of pumping yields and water levels from all the wells using this source.

Factors affecting abstraction. The variations in yield from wells in the Bunter Sandstone aquifer appear to be due to faulting, the presence of dykes, or lithological variation.

Numerous basaltic dykes trending west-north-west to north-north-west are recorded in surface exposures and many unmapped dykes are recorded in the well journals. Of variable thickness, ranging from a few inches to 60 ft (18 m), the dykes may act as impermeable barriers and effect a compartmentation of the sandstone. Whilst this may have no appreciable effect on standing water levels, it may slow up recharge and thus reduce yields. Hartley noted that of five deep wells cut through dykes the highest recorded pumping figure was 2200 g.p.h. This is far below the theoretical yield and it was notable that a minimum of 6000 g.p.h. was the lowest figure for other wells in the vicinity unaffected by dykes. At McConnel's Distillery (Ravenhill Road), when a newer well was drilled to replace the existing unsatisfactory well, the yield was 23,000 g.p.h. The position was said to be chosen "so as to be altogether away from the dolerite dykes which came into the older well". Hartley claims that 30 per cent of the Belfast boreholes appear to be affected by dykes as their yields are only 20 per cent of the calculated value.

That this factor is not constant is shown by a well at Belfast Ropeworks which encountered a 14-ft (4·3 m) dolerite intrusion but still yielded 22 000 g.p.h.

The well records suggest that the structure of the Trias may be much more complex than indicated on the One-inch Map and several unmapped faults may be present. The degree and complexity of jointing may be associated with the faulting and both will normally tend to increase the supply of water. An exception to this is the faulted area of sandstone, bounded on three sides by marl, to the north-west of Belfast where several of the boreholes have yields lower than average or, in one case, produced no water at all. This area is completely built over, allowing virtually no local infiltration, and the surrounding marl outcrops severely limit recharge from the flanks.

The effects of both dykes and faults are seen in the yields from the public supply boreholes in Lisburn. The first well gave a yield of over 20 000 g.p.h. due to the unusual number of joints clearly seen in the cores (Hartley 1935, p. 19). The second well gave a yield of only 3000 g.p.h. and penetrated a dyke or dykes at three levels. The other wells with satisfactory yields (12 000 g.p.h.) were in normal sandstones though No. 4 did penetrate a single dyke.

Lithological variations may affect both the quantity and quality of the water. At Broomhedge (Megharry's) a 126-ft (38 m) borehole in 1946 (36/470) obtained no water from a hard breccia which contained angular fragments of Silurian in a red sandy matrix probably at a low horizon within the Bunter Sandstone. Small supplies at the Linen Research Institute, Lambeg and at Long Kesh (three bore-holes) were apparently due to the presence of the Permian Marls. The yields in these cases were 250 g.p.h. and 700 g.p.h. respectively, although several wells, notably the Belfast Co-Operative Society's borehole in Ravenhill Avenue, the Castlereagh Laundry and Belfast Ropeworks, derive good supplies either from the marls or other Permian beds below them. Thin marl bands in the Bunter Sandstone at higher levels may also affect yields in some cases.

True artesian conditions have been recorded in only three instances— Burn-house Factory, Bloomfield Estate, and Andrews Flour Mill—but the condition did not persist on pumping. A sub-artesian rise in water level is, however, common when the Bunter Sandstone is pierced below the drift.

Occasionally the yields of wells in the Trias are said to have fallen off. This was frequently ascribed to neighbouring wells using all the water and in some cases deepening 'competitions' appear to have developed. In many cases the decrease was probably due to silting-up of the hole or corrosion of pipes restricting the flow. True cases of interference, or depression of the water table, are rare.

Usually the sandstone is overlain by drift up to 185 ft (56 m) thick. Hartley claims that an exceptional thickness of drift only diminishes the yield by a small amount. This similarly applies to wells commenced in Keuper Marl and sunk into the Bunter, providing the marl thickness is not excessive—i.e. that recharge from the flanks is possible. Both the dip (5° to 10°) and the ground slope give a rapid increase in thickness of marl cover to the north-west and such a thickness will quickly become prohibitive if a supply from the sandstone is required.

Groundwater Chemistry. Though only a few deep wells are still in operation compared with the number which have been sunk, this is seldom a reflection on the quantity available but, rather, due to the hardness of the water and the extension of the public water supply. The water hardness, most of it temporary, produced excessive scaling when used in boilers and when soft water became available the wells dropped out of use. Amongst the advantages claimed for the continued use of borehole water is cheapness, constant low temperature, purity and the independence of the supply from temporary shortages.

In the mineral water industry this purity of water is highly valued and the normal filtration is the only treatment necessary. When forced to use the town supplies, due to well-cleaning operations, difficulty has been experienced due to excessive clogging of filters and the taste of chlorine.

Few wells seem to have the excess of iron which might be expected from the ferruginous Bunter Sandstone. One weaving company employs a Polarite filter to remove any iron contamination which would have a detrimental effect on the cloth, and still pumps the required amount of water at half the cost of the town water.

Waters from the Bunter Sandstone are characterized by the predominance of calcium salts, chiefly calcium carbonate which forms a high percentage of the total solids (temporary hardness). Water from these wells shows little variation. In the Permian Marls and sandstones there is a considerable increase in sulphates and chlorides (permanent hardness) due to the different lithology with its abundance of gypsum.

Salt water infiltration is comparatively insignificant in the Belfast area and there is no indication of a spread of pollution. In the marginal coastal areas, where the sandstone from which the water is drawn crops out in the sea, the potential danger of pollution will always exist but can be minimized by controlled extraction. Saline contamination has been reported at three boreholes in Belfast—Pacific Flour Mill, Dixon's (Whitla Street), and Thompson's Flour Mill—all in reclaimed land seaward of the 1715 coastline.

Elsewhere, the sandstone water has a total hardness of 120 to 1800 mg/l. The water from Belfast Ropeworks well had the exceptional concentration of 29 458 mg/l total dissolved solids. A single record of groundwater temperature states that it was a steady 55°F.

Cretaceous. The White Limestone (Chalk) is a hard non-porous rock and its value as an aquifer is limited to the water which it holds in joints and fissures, some of which are widened by solution. It has usually a very narrow outcrop and its replenishment must come largely from percolating waters from the overlying basalt. The limited outcrop also means that few wells are sunk into the Chalk, but two wells in Mullaghcarton reached it, one through 40 ft (12 m) of basalt cover, and a well at Springvale, 1 mile (1·6 km) N.W. from Mullaghcarton is reliably reported to have entered chalk below drift, 1000 yd (914 m) W. of the known outcrop, and indicated the existence of a Chalk inlier in an area previously regarded as basalt. All wells in the Chalk appear to give good supplies of water.

A deep bore north of White Mountain reached the Chalk at a depth of 300 ft and obtained a yield of 1000 g.p.h. No wells have been sunk to the Hibernian Greensand.

Springs, often of considerable yield, are thrown out from the base of the chalk by the underlying impervious Lias or Keuper Marl at frequent intervals along the outcrop and are used by farms and dwellings from Mullaghcarton to Collin Glen.

Antrim Lava Series. Though individual basalt flows are well jointed, the inter-lava laterite beds are relatively impermeable and the few deep wells in the basalts give rather poor yields. Shallow wells occasionally get small yields from the vesicular bases or tops of flows but most of the shallow wells on the basalt outcrop obtain their supply from the base of the drift and any excavation into solid has been to form a sump.

Lough Neagh Clays. These, and the overlying drifts, are almost watertight and wells in these districts are very poor, being little more than sumps for surface drainage.

Drifts. The Boulder Clay varies considerably with the underlying rock but is usually relatively impermeable. In the past most of the rural dwellings depended on shallow wells dug in this deposit, often carried down to rock-head where, except in the case of the Keuper Marl and Lough Neagh Clays, some water is almost always obtained. The large diameter of these wells permits them to act as storage reservoirs and, though liable to contamination, they have usually been adequate for domestic purposes. With the development of more intensive stock farming these wells are sometimes inadequate and most of the district is now on mains water. In general, the boulder clay on the Lower Palaeozoic rocks, on the Basalt, and on the Bunter Sandstone will yield some water; higher yields are sometimes obtained from sand and gravel lenses in the clay.

Larger supplies are obtained from wells in the sands and gravels, though the low water-tables enforce considerable depths in such areas as Ballynahatty, near the Giant's Ring. Even small areas of sands contain large supplies of water and the villages of Crumlin and Glenavy were built on dry ground with plenty of water from deep wells. Springs from the bottom of the sands and gravels give steady supplies, as for example, at the Causeway End Esker, near Lisburn.

Large and reliable supplies have been obtained from wells, usually not more than 10 ft (3 m) deep, sited on alluvial flats.

Though water from wells in the drift was of great importance before the institution of piped supplies, statements about quantities and yields tend to give an exaggerated idea of volume available. The figure of 35 g.p.d. per head for domestic use or 4·6 g.p.d. per acre (0·4 ha) for general farm use is probably a reliable estimate for most shallow wells. All water from the drift, particularly from the boulder clay, tends to be very hard.   P.I.M., H.E.W.

Chapter 14

SOILS AND AGRICULTURE

THERE ARE WIDE variations in the soils of the area which may be attributed mainly to differences in parent materials and topography. The texture of the surface soil as shown on the map (Fig. 21) ranges from sandy loam to heavy clay loam and there are also areas of peaty and organic loam soils.

The area is largely covered with glacial deposits but these are predominantly local in origin and show a close relationship to the underlying geological strata. McConaghy (1952) indicated that solid geology was one of the major factors influencing soil conditions; Proudfoot and Boal (1960) found this to be so in County Down but only a small degree of correlation with the solid geology was recorded by Fowler and Robbie (1961) in the Dungannon area.

The correlation between the rock types from which the parent material is derived and soil texture is well illustrated by data given by McConaghy (1948).

TABLE 1.

Influence of parent material on soil texture

| Soil Texture (with clay content) | Basalt | Parent Material Trias | | Silurian |
		Keuper Marl	Bunter Sandstone	
Light (under 12 per cent)	11	6	69	18
Medium (12–30 per cent)	44	19	40	69
Heavy (over 30 per cent)	45	75	11	13

In this area, around and to the west of Belfast, the soils overlie drift derived mainly from the Basalt, Trias and Silurian formations. The area is roughly bisected by the Lagan valley in which the soils are developed mainly from Triassic parent material. To the north and west, basal boulder clay predominates while in the south the drumlinoid drift is derived from the Lower Palaeozoic rocks.

Apart from the parent materials, the main factors which influence soil formation are the duration and intensity of weathering processes and the extent to which soluble weathering products are removed from the soil. The extent of interference by man or animals may also be important, especially in areas where relatively advanced systems of farming have been followed for many years.

The net result of the humid conditions in Northern Ireland, with annual rainfall practically always exceeding evaporation, is either waterlogging or a downward movement of water, containing soluble constituents such as lime and

178

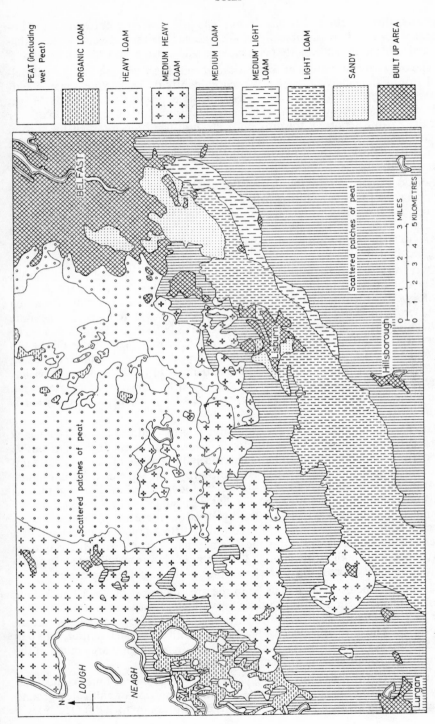

FIG. 21. *Distribution of the main soil types in the Belfast area*

magnesia, through the soil to the drains. The first process will occur where impermeable or slowly permeable layers are present; the second occurs where the soil and subsoil are sufficiently permeable to allow drainage water to escape.

The only extensive areas of freely drained soils in this region are those developed from parent material which is predominantly derived from Bunter Sandstone. Among these soils there is a considerable proportion of sandy brown podzolics but acid brown earths of low base status predominate. It would appear that the natural tendency towards podzolization has been retarded by normal farming processes including the application of lime and manures. The profile from Belvoir Park (Table 2) is an example of such an acid brown earth of low base status.

Soils developed on the Basalt and Lower Palaeozoic strata often show signs of gleying due to imperfect or impeded drainage. Where drainage is satisfactory, the soils are normally classified as acid brown earths, those derived from Silurian material having a low base status while those overlying basaltic drift have a high base status. Descriptions and analyses of typical profiles from each of these types are given in Table 2.

Over most of the area underlain by the Lower Palaeozoic rocks the surface is thickly dotted with drumlins. The drift on these drumlins is usually reddish brown in colour, containing 25 to 30 per cent clay and with many stones which are generally flat and of local origin. The stoniness and topography facilitate drainage of the drumlin soils, and on the drumlins a sequence of soils can be observed: these include medium-textured, greyish brown leached soils on the top and gentle slopes of the drumlins; medium-heavy, deeper, slightly gleyed soils on the lower slopes, and surface water gleys and peats in the inter-drumlin basins. Many of these soils are under tillage, mainly barley, or temporary grassland, while the interdrumlin soils, usually wet, are under permanent grass which is usually rush-infested.

The basaltic boulder clays generally give rise to medium-textured or heavy loams, the former usually being developed on the drumlins. The stiff retentive soils derived from basaltic boulder clay, especially in areas of flat or gentle undulating topography such as south Antrim, have been attributed by Lamplugh (1904) to glacial 'puddling' of the boulder clay. Medium and medium-light loams are also derived from basaltic drift which has been subjected to water sorting, for example, in the Six Mile Water and Braid valleys, but there are no extensive areas of such soils in the Belfast region. Proudfoot and Boal (1960) recorded soil of a lighter texture in association with the Cretaceous outcrop near Moira but the outcrop is not normally wide enough to have much influence on the soils of the area, particularly where it forms the lower part of the basalt escarpment.

The most difficult basaltic soils lie in the area due west of Belfast stretching almost to Glenavy. In addition to their heavy texture, these soils are subject to other adverse factors such as altitude, aspect and drainage problems due to topography. The development of a layer of unhumified organic matter, 2 to 6 in (50–150 mm) deep, on the surface is not unusual. In this higher area much of the land is under rough grazing and some peat occurs.

Further west towards Lough Neagh and south-west towards Lurgan, medium-heavy and medium loams predominate and the proportion of cultivated ground increases in these areas. All the basaltic soils at altitudes up to 500 ft (152 m) will give good grassland when adequately limed and manured, though on the heavier soils attention must be given to drainage and care exercised in grazing management to avoid poaching.

Vastly different soils are derived from the two dominant Trias formations. The Keuper Marl and the boulder clay derived from the marl give heavy stiff reddish soils, which although commonly found on steep slopes, are usually wet and covered with rushes. These soils are practically all under permanent grass and where satisfactory drainage can be provided, they are capable of giving high yields of pasture or hay. Where the parent material is Bunter Sandstone, sandy, light and medium loams predominate, generally light brown or reddish brown in colour, with coarse sandy subsoils of a yellowish brown or brick colour. Over much of the Lagan Valley area, the dissected glacial sands and gravels have a dominating effect and give light well-drained, easily tilled loam soils with patches of heavier soils on the boulder clay. Detailed notes by Kilroe on the soils of the Lisburn district with some mechanical analyses are given in Lamplugh (1904). These soils are extensively used for market gardening and for cereal growing.

The largest peat-soil area occurs near Lough Neagh where some reclamation of cut-out peat bog has been carried out. This land normally produces good crops of potatoes and cereals and is seldom left under pasture for more than three years.   J.S.V.McA., P.I.M.

TABLE 2
Soil Profiles

BROWN EARTH SOILS OF LOW BASE STATUS

Parent Material & Vegetable type	Depth in	Symbol	Horizon Description	pH	Coarse Sand %	Fine Sand %	Silt %	Clay %	Loss on Ign. %	Organic Carbon %	Ca	Mg	'H'	Excl. Cap.	Base Sat. %	SiO_2/R_2O_3 (Clay)
Belvoir Park																
Fluvio-glacial sands Triassic (deciduous forest)	0–12	A	Dark reddish brown loamy sand with very few stones; fairly sharp change to .	5·2			Data not available				2·4	0·8	10	13	24·6	—
	12–18	B/C	Reddish brown sandy loam; very few stones merges to . . .	5·0							0·3	0·1	9	10	4	—
	18+	C	Greyish brown sandy loam with very few stones	5·4							1·2	0·6	7	10	18	—
Hillsborough																
Silurian drift (coniferous forest)	0–10	A	Brown fibrous loam, friable; good crumb structure; merges to .	4·9	18	35	25	15	13	—	0·4	0·08	16	17	2·8	—
	10–18	(B)	Lighter brown friable loam; less fibrous; some stones; merges to . .	5·0	22	37	20	23	9	—	0·1	0·04	13	14	1·0	—
	18–24	C	Light brownish yellow loam, stony; merges to	5·0	27	33	16	20	7	—	0·03	0·05	10	10·5	0·80	—
	24+	C	Reddish coloured layer; friable; stony	4·8	48	18	16	14	7	—	0·02	0·30	8	8·5	3·8	—

BROWN EARTH SOILS OF HIGH BASE STATUS

Parent Material & Vegetable type	Depth in	Symbol	Horizon Description	pH	Coarse Sand %	Fine Sand %	Silt %	Clay %	Loss on Ign. %	Organic Carbon %	Ca	Mg	'H'	Excl. Cap.	Base Sat. %	SiO_2/R_2O_3 (Clay)
Templepatrick																
Basaltic drift (old pasture)	0–8	S	Dark chocolate-brown loam; good crumb structure; merges to . .	5·3		31	30	33	—	4·6	9	10	27	47	43	—
	8–15	S	Lighter coloured loam, fairly stony; fairly sharp change to . .	5·5		35	31	29	—	2·8	11	12	24	47	50	—
	15–30	(B)2	Reddish brown loam; fairly stony .	5·7		44	32	18	—	0·8	22	14	15	53	72	—
	30+	C	Yellowish brown, stony clay loam showing slight mottling . .	6·3		47	33	16	—	0·4	31	16	8	57	—	—

IMPERFECTLY DRAINED SOILS WITH MOTTLING

	Horizon	Depth	Description	pH											
Saintfield															
Silurian drift (Pasture on drumlin)	S	0–9	Medium loam stony, greyish brown with good crumb structure; fairly sharp change to	7·32	50	17	27	11	3·4	17	3	6	26	77	2·1
	(B)	9–12	Orange-brown loam, stony, with traces of mottling	6·7	52	20	23	6	0·6	8	2	8	19	58	2·0
	C$_1$/g	12–24	Stony clay loam, yellowish brown, mottled	6·6	46	21	22	5	0·3	12	5	8	26	70	2·0
	C$_2$/g	24+	Stony clay with grey, yellowish brown and green mottling	6·8	53	23	23	6	0·2	8	6	7	22	68	2·3
Massereene Park															
Basaltic drift (Pasture on flat ground)	S	0–8	Clay loam, hard crumb structure; dark brownish grey with brown mottles round structural units and root channels	6·6	19	9	49	26	6·2	28	7	10	47	79	1·5
	A/g	8–11	Tough sticky clay with blocky or columnar structure; a few stones; grey with intensive mottling towards base	6·5	29	17	38	15	3·0	25	12	12	51	77	1·8
	B/g	11–16	Yellowish brown with grey mottling; compact, merges to	6·8	46	27	26	10	0·6	25	26	12	65	82	2·2
	C/g	16–21	More gravelly, with stones; strongly mottled	6·9	48	32	17	9	0·3	21	29	12	63	81	— —
		21+	Strongly mottled; gravelly	6·7	51	30	16	9	0·2	24	26	12	63	81	— —

[1]'Clay' includes 'loss by solution'.   [2]Recently limed.

Chapter 15

BIBLIOGRAPHY

A comprehensive bibliography of all published work up to 1903 was given in the Memoir on the Geology of Belfast (1904). Only works of this period to which reference is made in the text are included in the following list.

ADAMS, T. D. and HAYNES, J. 1965. Foraminifera in Holocene marsh cycles at Borth, Cardiganshire (Wales). *Palaeontology*, **8**, 27–38.

ANDERSON, T. B. 1965. Evidence for the Southern Uplands Fault in north-east Ireland. *Geol. Mag.*, **102**, 383–92.

ANDREWS, S. and DAVIES, O. 1940. Prehistoric finds at Tyrone House, Malone Road, Belfast. *Ulster J. Arch.*, (3), **3**, 152–4.

ARKELL, W. J. and TOMKEIEFF, S. I. 1953. *English Rock Terms*. Oxford.

BADEN-POWELL, D. F. W. 1937a. On the Holocene marine fauna from the implementiferous deposits of Island Magee, Co. Antrim. *J. Animal Ecol.*, **6**, 86–97.

—— 1937b. On a marine Holocene fauna in north-western Scotland. *J. Animal Ecol.*, **6**, 273–83.

BAILEY, E. B. 1924. The desert shores of the Chalk seas. *Geol. Mag.*, **61**, 102–16.

——, CLOUGH, C. T., WRIGHT, W. B., RICHEY, J. E. and WILSON, G. V. 1924. Tertiary and post-Tertiary geology of Mull, Loch Aline and Oban. *Mem. Geol. Surv.*

BAILY, W. H. 1874. Sketch of the geology of Belfast and the neighbourhood. *Science Gossip*, 169–70.

BARROIS, C. 1876. Recherches sur le terrain Crétacé supérieur de l'Angleterre et de l'Irlande. *Mém. Soc. Géol. du Nord*, **1**.

BARTON, R. 1757. *Lectures in natural philosophy designed to be a foundation for reasoning pertinently upon the petrifactions, gems, crystals and sanative quality of Lough Neagh in Ireland*. Dublin.

BERGER, J. F. 1816. On the geological features of the north of Ireland. *Trans. Geol. Soc.*, **3**, 134–95, 217–22.

BINNEY, E. W. 1855. On the Permian Beds of North-West England. *Mem. Litt. Phil. Soc. Manch.*, 2, **12**, 209–69.

BLACK, M. 1953. The Constitution of the Chalk. *Proc. Geol. Soc. Lond.*, no. 1499 (Session 1952–3), 81–6.

BONNELL, D. G. R. and BUTTERWORTH, B. 1950. Clay building bricks of the United Kingdom. *H.M.S.O.*

BRINDLEY, J. C. 1967. The geology of the Irish Sea area. *Irish Nat. J.*, **15**, 245–9.

BROMLEY, R. G. 1967. Some observations on burrows of thalassinidean Crustacea in chalk hardgrounds. *Quart. J. Geol. Soc.*, **123**, 157–82.

BRYCE, J. 1837. On the Magnesian Limestone and associated beds which occur at Holywood in the County of Down. *J. Geol. Soc., Dublin*, **1**, 175–80.

—— 1845. Notice of a Tertiary deposit in the neighbourhood of Belfast. *Phil. Mag.*, **26**, (3), 433–6.

—— 1853. On the geological structure of the counties of Down and Antrim. *Rep. Brit. Ass.* (for 1852), **22**, 42–3.

—— and HYNDMAN, G. C. 1843. An elevated deposit of marine shells, etc., *in* PORT-LOCK, J. E., *Report on the geology of Londonderry and parts of Tyrone and Fermanagh*, 738–40. Dublin.

BUCKLAND, W. 1817. Description of the Paramoudra, a singular fossil body that is found in the chalk of the north of Ireland, with some general observations upon flints in chalk. *Trans. Geol. Soc. Lond.*, **4**, 413–23.

BULLERWELL, W. 1961. The Gravity Map of Northern Ireland. *Irish Nat. J.*, **13**, 254–7.

BURCHELL, J. P. T. 1934. Some littoral sites of early post-Glacial times located in north Ireland. *Proc. Prehist. Soc. East Anglia*, **7**, (3), 366.

—— 1939. On the northern glacial drifts; some peculiarities and their significance. *Quart. J. Geol. Soc.*, **75**, 299–333.

CARRUTHERS, R. G. 1953. *Glacial drifts and the undermelt theory*. Newcastle-upon-Tyne.

CHANDLER, M. E. J. 1946. Note on some abnormally large spores formerly attributed to *Isoetes*. *Ann. Mag. Nat. Hist.*, **13**, (11), 684–9.

CHARLESWORTH, J. K. 1928. Origin of the relict fauna of Lough Neagh. *Geol. Mag.*, **65**, 212–5.

—— 1939. Some observations on the glaciation of North-East Ireland. *Proc. Roy. Irish Acad.*, **45 B**, 255–95.

—— 1953. *The Geology of Ireland*. Edinburgh.

—— 1957. *The Quaternary Era*. London.

—— 1963a. Some observations on the Irish Pleistocene. *Proc. Roy. Irish Acad.*, **62B**, 295–322.

—— 1963b. *Historical geology of Ireland*. Edinburgh and London.

—— and CLELAND, A. McI. 1928. 'White Trap' at Ligoniel, Belfast. *Irish Nat. J.*, **2**, 75–6.

—— and ERDTMAN, G. 1935. Post-glacial section at Milewater Dock, Belfast. *Irish Nat. J.*, **5**, 234–5.

—— and HARTLEY, J. J. 1935. The Tardree and Hillsborough dyke swarms. *Irish Nat. J.*, **5**, 193–6.

—— and PRESTON, J. 1960. Geology around the University Towns: N.E. Ireland; the Belfast area. *Geologists' Assoc. Guide*, No. 18.

—— and others. 1960. The geology of North-East Ireland. *Proc. Geol. Assoc.*, **71**, 429–59.

CLARK, R. 1902. Notes on the fossils of the Silurian area of North-East Ireland. *Geol· Mag.* (4), **9**, 497–500.

CLARKE, R. E. L. 1935. Recent improvements to Lisburn water supply. *Inst. Civil Eng.* (N. I. Assoc.) March.

CLELAND, A. McI. 1928. An old land surface in a Chalk quarry. *Irish Nat. J.*, **2**, 37–40.

—— 1934. The new Raised Beach (?) section at 'The Grove', Belfast. *Irish Nat. J.*, **5**, 11–2.

CLOSE, M. H. 1867. Notes on the general Glaciation of Ireland. *J. Roy. Geol. Soc. Ireland*, **1**, 207–42.

COFFEY, G. and PRAEGER, R. L. 1904. The Antrim Raised Beach. A contribution to the Neolithic history of the north of Ireland. *Proc. Roy. Irish Acad.*, **25**, 143–200.

CONYBEARE, W. 1816. Geological Features of North East Ireland. *Trans. Geol. Soc. Lond.*, (1), **3**, 121–33.

—— and BUCKLAND, W. 1816. Descriptive notes referring to the outline of sections presented by a part of the coasts of Antrim and Derry. *Trans. Geol. Soc. Lond.*, (1), **3**, 196–216.

COOK, A. H. and MURPHY, T. 1952. Measurements of Gravity in Ireland. Gravity Survey of Ireland North of the line Sligo–Dundalk. *Geophys. Mem.* No. 2 (4). *Dublin Inst. for Adv. Studies,* School of Cosmic Physics, 1–36.

COX, P. A. 1967. New dry dock at Belfast. *Inst. Civil Eng.* (N. I. Assoc.).

DAWKINS, W. B. 1864. Outline of the Rhaetic formation in West and Central Somerset. *Geol. Mag.,* 1, 257–60.

DIXON, F. E. 1949. Irish mean sea level. *Sci. Proc. Roy. Dublin Soc.,* 25, N.S., 1–8.

DORAN, I. G. 1946. Soil mechanics site investigation for the proposed new building at the Royal Victoria Hospital, Belfast. *Inst. Civil Eng.* (N. I. Assoc.).

DOUGHTY, P. S. 1966. Giant Irish Deer remains from Lough Neagh. *Irish Nat. J.,* 15, 187.

DUNHAM, K. C. and ROSE, W. C. C. 1949. Permo-Triassic geology of South Cumberland and Furness. *Proc. Geol. Assoc.,* 60, 11–40.

DWERRYHOUSE, A. R. 1923. The glaciation of North-Eastern Ireland. *Quart. J. Geol. Soc.,* 79, 352–422.

EVANS, E. E. 1937. The site of Belfast. *Geography,* 22, 169–77.
—— 1944. Belfast: the site and the city. *Ulster J. Arch.,* 7, 5–29.
—— 1950. Worked flints from Boulder Clay in Belfast. *Ulster J. Arch.,* 13, 42–3.

EYLES, V. A. 1952. The composition and origin of the Antrim laterites and bauxites. *Mem. Geol. Surv. N. Ireland.*

FARRINGTON, A. 1949. The Glacial drifts of the Leinster Mountains. *J. Glaciol.,* 1, 220–5.

FISHER, N. and WELCH, R. J. 1933. New raised beach section at 'The Grove', Belfast. *Irish Nat. J.,* 4, 177.

FITCH, R. 1840. Notice of the existence of a distinct tube within the hollows of the Paramoudra. *Mag. Nat. Hist.,* (2), 4, 303–4.

FOWLER, A. 1944. A deep bore in the Cleveland Hills. *Geol. Mag.,* 81, 193–206.
—— 1955. Permian of Grange, Co. Tyrone. *Bull. Geol. Surv. Gt. Brit.,* No. 8, 44–53.
—— 1955. *Sum. Prog. Geol. Surv.* (for 1954), 52.
—— 1958. *Sum. Prog. Geol. Surv.* (for 1957), 45.
—— and ROBBIE, J. A. 1961. Geology of the Country around Dungannon. *Mem. Geol. Surv. N. Ireland.*

GARDNER, J. S. 1885. On the Lower Eocene plant-beds of the basaltic formation of Ulster. *Quart. J. Geol. Soc.,* 41, 82–92.

GAULT, W. 1878. Observations on the geology of Black Mountain. *Proc. Belfast Nat. Field Club.* 1, 251–62.

GEIGER, M. E. and HOPPING, C. A. 1969. Triassic stratigraphy of the southern North Sea Basin. *Phil. Trans. Roy. Soc.,* 254B, 1–36.

GEIKIE, A. 1897. *Ancient volcanoes of Great Britain.* London.

GEORGE, T. N. 1960. The stratigraphical evolution of the Midland Valley. *Trans. Geol. Soc. Glasgow,* 24, 32–107.

GRAINGER, J. 1853. On the shells found in the alluvial deposits of Belfast. *Rep. Brit. Assoc.* for 1852, 43–6, 74–5.
—— 1859. On the shells found in the post-Tertiary deposits of Belfast. *Nat. Hist. Rev.,* 6, 135–51.
—— 1875. On the fossils of the post-Tertiary deposits of Ireland. *Rep. Brit. Assoc.* for 1874, 73–6.

GRAY, W. 1868. Glacial markings recently observed round Belfast. *Ann. Rep. Belfast Nat. Field Club,* 5, 34.

GREEN, E. R. R. 1929. *The Lagan Valley* 1800–1850. London.

GREIG, D. C. and others. 1968. Geology of the country around Church Stretton, Craven Arms, Wenlock Edge and Brown Clee. *Mem. Geol. Surv. G.B.*

GRIFFITH, A. E. 1961. A note on some shelly fossils from the arenaceous greywackes of Co. Down. *Irish Nat. J.*, **13**, 258–9.

GRIFFITH, R. 1837. Presidential Address. *J. Geol. Soc. Dublin*, **1**, 146–9.

—— 1838. *Second report of the Commission appointed to consider and recommend a general system of railways for Ireland.* Dublin.

HAMILTON, W. 1786. *Letters concerning the northern coast of the County of Antrim.* London.

HANCOCK, J. M. 1961. The Cretaceous system in Northern Ireland. *Quart. J. Geol. Soc.*, **117**, 11–36.

—— 1963. The Hardness of the Irish Chalk. *Irish Nat. J.*, **14**, 157–64.

HARDMAN, E. T. 1873. On analysis of white chalk from the County of Tyrone, with note of occurrence of zinc therein, and in the overlying basalt. *Geol. Mag.*, **10**, 434–8.

—— 1876. Fossiliferous Pliocene clays overlying basalt near the shore of Lough Neagh. *Geol. Mag.*, (2), **3**, 556–8.

—— 1877a. On the age and mode of information of Lough Neagh, Ireland. *J. Roy. Geol. Soc. Ireland*, **4**, 170–99.

—— 1877b. On the geology of the Tyrone Coalfield and surrounding district. *Mem. Geol. Surv. Ireland.*

—— 1879. The fossiliferous clay beds overlying basalt, Lough Neagh, and the geological age of that lake. *Geol. Mag.*, (2), **6**, 214–6.

HARTLEY, J. J. 1931. The geology of the Belfast District, with some notes on the hydrology of the area. *Official Circular Brit. Waterworks Assoc.*, **13**, (13), No. 96. 719–30.

—— 1934. White Trap of Shore Road, Belfast. *Irish Nat. J.*, **5**, 12–5.

—— 1935. The underground water resources of Northern Ireland. *Inst. Civil Eng.* (N.I. Assoc.).

—— 1937. Lisburn Esker. *Irish Nat. J.*, **6**, 208.

—— 1940. The sub-soils of Belfast and District. *Inst. Civil Eng.* (N.I. Assoc.), 1–12.

—— 1943. Notes on the Lower Marls of the Lagan Valley. *Irish Nat. J.*, **8**, 128–32.

—— 1949. Further notes on the Permo-Triassic rocks of Northern Ireland. *Irish Nat. J.*, **9**, 314–6.

—— 1950. Geological aspects of Dam Sites. *Inst. Civil Eng.* (N.I. Assoc.).

—— and HARPER, J. J. 1937. A recently discovered Ordovician inlier in County Down with a note on a new species of *Pyritonema* .*Irish Nat. J.*, **6**, 253–5.

HELMBOLD, R. 1952. Beitrag zur Petrographie der Tanner Grauwacken. *Beitr. Min.*, **3**, 253–88.

HILL, D. A. 1947. Land utilization in the Belfast Area. *Geography*, **32**, (1), 20–33.

HOLLINGWORTH, S. E. 1942. The correlation of gypsum-anhydrite deposits and associated strata in the North of England. *Proc. Geol. Assoc.*, **53**, 141–51.

HULL, E., WARREN, J. L. and LEONARD, W. B. 1871a. The country around Belfast, Lisburn and Moira. *Mem. Geol. Surv. Ireland.*

—— 1871b. The country around Bangor, Newtownards, Comber and Saintfield in the County of Down. *Mem. Geol. Surv. Ireland.*

HUME, W. F. 1897. The Cretaceous strata of County Antrim. *Quart. J. Geol. Soc.*, **53**, 540–606.

HUMPHRIES, D. W. 1961. The Upper Cretaceous White Sandstone of Loch Aline. *Proc. Yorks. Geol. Soc.*, **33**, 47–76.

JELETZKY, J. A. 1951. Die Stratigraphie und Belemnitenfauna des Obercampan und Maastricht Westfalens, Nordwestdeutschlands und Dänemarks sowie einige allgemeine Gliederungs-Probleme der jüngeren borealen Oberkreide Eurasiens. *Beih. Geol. Jb.*, **1**, 1–142.

JESSEN, K. 1949. Studies in late Quaternary deposits and flora-history of Ireland. *Proc. Roy. Irish Acad.*, **52 B**, 85–290.

JONES, E. 1953–4. The social geography of Belfast. *J. Stat. and Social Inq. Soc. Ireland*, **20**.

—— 1958. The delimitation of some urban landscape features in Belfast. *Scottish Geogr. Mag.*, **74**, 3, 150–62.

—— and EVANS, E. E. 1955. The growth of Belfast. *Town Planning Review*, **26**, 93–111.

JONES, T. R. 1878. Note on the foraminifera and other organisms in the chalk of the Hebrides. Appendix to JUDD, J. W. On the Secondary rocks of Scotland. *Quart. J. Geol. Soc.*, **34**, 739–40.

JOPE, E. M. 1952. Porcellanite axes from factories in N. E. Ireland, Tievebulliagh and Rathlin. *Ulster J. Arch.*, **15**, 31–59.

JUKES, J. B. 1868. The Chalk of Antrim. *Geol. Mag.*, **5**, 345–47.

JUKES-BROWNE, A. J. 1903. The Cretaceous Rocks of Britain 2. *Mem. Geol. Surv.*

KELLING, G. 1961. The stratigraphy and structure of the Ordovician rocks of the Rhinns of Galloway. *Quart. J. Geol. Soc.*, **117**, 37–75.

KELLY, J. 1868. On the geology of the County of Antrim with parts of the adjacent counties. *Proc. Roy. Irish Acad.*, **10**, 235–327.

KINAHAN, G. H.1889. Economic geology of Ireland. *J. Roy. Geol. Soc. Ireland*, **8**.

KING, W. 1853. On the Permian fossils of Cultra. *Rep. Brit. Assoc.* for 1852, 53.

LAMONT, A. 1946. Red flints. *Irish Nat. J.*, **8**, 398–9.

—— 1953. Records of Irish trilobites and brachiopods. *Geol. Mag.*, **90**, 433–37.

LAMPLUGH, G. W. 1901. Names for British ice-sheets of the glacial period. *Geol. Mag.* (4) **9**, 142.

—— 1905. Geology of the country around Cork and Cork Harbour. *Mem. Geol. Surv. Ireland.*

—— and others. 1904. Geology of the country around Belfast. *Mem. Geol. Surv. Ireland.*

LAPWORTH, C. 1878. On the graptolites of Co. Down. *Proc. Belfast Nat. Field Club*, Appendix 4 (1876–7), 125–47.

LINDSTRÖM, M. 1958. Different phases of tectonic deformation in the Rhinns of Galloway. *Nature*, **182**, 48–9.

LOGAN, W., REZAK, R. and GINSBURG, R. N. 1964. Classification and environmental significance of algal stromatolites. *J. Geol.*, **72**, 68–83.

LUMSDEN, G. I., TULLOCH, W., HOWELLS, M. F. and DAVIES, A. 1967. Geology of the neighbourhood of Langholm. *Mem. Geol. Surv. G.B.*

MCADAM, J. 1848. On the cuttings of the Belfast and Ballymena Railway. *J. Geol. Soc. Dublin*, **4**, 36–41.

—— 1850a. Observations on the neighbourhood of Belfast with a description of the cuttings on the Belfast and County Down Railway. *J. Geol. Soc. Dublin*, **4**, 250–65.

—— 1850b. Supplementary observations on the neighbourhood of Belfast. *J. Geol. Soc. Dublin*, **4**, 265–8.

MACLOSKIE, G. 1872. Silicified wood of Lough Neagh. *J. Roy. Geol. Soc. Ireland*, **3**, 163–74.

MCCONAGHY, S. 1948. Some characteristics of the Soils of Northern Ireland. M. Agr. Thesis, Queen's University Belfast (Unpublished).

—— 1952. Soils *in* JONES, E. (ed). *Belfast in its regional setting*. Belfast.

McGugan, A. 1957. Upper Cretaceous foraminifera from Northern Ireland. *J. Palaeont.*, **31**, 329–48.

—— 1964. Lower Cenomanian foraminifera from Belfast. Northern Ireland. *Irish Nat. J.*, **14**, 189–94.

McMillan, N. F. 1947. The Estuarine Clay at Greenisland, Co. Antrim. *Irish Nat. J.*, **9**, 16–9.

—— 1957. Quarternary deposits around Lough Foyle, Northern Ireland. *Proc. Roy. Irish Acad.*, **58B**, 185–205.

Marr, J. E. and Nicholson, H. A. 1888. The Stockdale Shales. *Quart. J. Geol. Soc.*, **44**, 654–732.

Marshall, H. E. 1933. *The parish of Lambeg.*

Meyer, H. O. A. 1965. Revision of the stratigraphy of the Permian evaporites and associated strata in north-western England. *Proc. Yorks. Geol. Soc.*, **35**, 71–89.

Mitchell, G. F. 1951. Studies in Irish Quaternary deposits, No. 7. *Proc. Roy. Irish Acad.*, **53B**, 111–206.

—— 1954. Praeger's contribution to Irish Quaternary geology. *Irish Nat. J.*, **11**, 172–5.

—— and Parkes, H. M. 1949. The Giant Deer in Ireland. *Proc. Roy. Irish Acad.*, **52B**, 291–314.

Moore, C. 1861. On the zones of the Lower Lias and the *Avicula contorta* zone. *Quart. J. Geol. Soc.*, **17**, 483–516.

Movius, H. L. 1953a. Graphic representation of post-glacial changes of level in north-east Ireland. *Am. J. Sci.*, **251**, 697–740.

—— 1953b. Curran Point, Larne, Co. Antrim. The type site of the Irish Mesolithic. *Proc. Roy. Irish Acad.*, **56B**, 1–195.

Parasnis, D. S. 1952. A study of the rock densities in the English Midlands. *Mon. Not. Roy. Act. Soc. Geoph. Supp.*, **6**, (5), 252–71.

Patterson, E. M. 1952. A petrochemical study of the Tertiary lavas of N. E. Ireland. *Geochm. Cosmochim. Acta*, **2**, 283–99.

—— and Swaine, D. J. 1957. Tertiary dolerite plugs of North-East Ireland. A survey of the geology and geochemistry. *Trans. Roy. Soc. Edin.*, **63**, 317–31.

Patterson, W. H. 1892. On a newly discovered site for worked flints in the County of Down. *J. Roy. Soc. Antiq. Ireland*, **2**, (5), 154–5.

Peach, B. N. 1901. On a remarkable volcanic vent of Tertiary Age in the Island of Arran, enclosing Mesozoic fossiliferous rocks. *Quart. J. Geol. Soc.*, **57**, 226–43.

Peake, N. B. and Hancock, J. M. 1961. The Upper Cretaceous of Norfolk. *Trans. Norfolk Norw. Nat. Soc.*, **19**, 293–339.

Pettijohn, F. J. 1957. *Sedimentary Rocks.* 2nd ed. New York.

Pollock, J. and Wilson, H. E. 1961. A new fossiliferous locality in Co. Down. *Irish Nat. J.*, **13**, 244–8.

Portlock, J. E. 1843. *Report on the geology of the County of Londonderry and parts of Tyrone and Fermanagh.* Dublin.

Praeger, R. L. 1888. On the estuarine clays at the new Alexandra Dock, Belfast. *Proc. Belfast Nat. Field Club*, (2), **3**, Appendix **2**, No. 3, 29–51.

—— 1892. Report on the estuarine clays of the north-east of Ireland. *Proc. Roy. Irish Acad.*, (3), **2**, 212–89.

—— 1896. Report upon the raised beaches of the north-east of Ireland, with special reference to their fauna. *Proc. Roy. Irish. Acad*, (3), **4**, 30–54.

—— 1903. The post-glacial deposits of the Belfast district. *Rep. Brit. Assoc.* for 1902, 611.

Preston, J. 1963. The dolerite plug at Slemish, Co. Antrim, Ireland. *L'pool and Manch. Geol. J.*, **3**, 301–14.

PROUDFOOT, V. B. and BOAL, F. W. 1960. Two soil maps of County Down. *Geogr. J.*, **126**, 60–6.

REID, R. E. H. 1958. Remarks on the Upper Cretaceous Hexactinellida of Co. Antrim. *Irish Nat. J.*, **12**, 236–43, 261–8.

—— 1959. Age of the Cretaceous Basal Conglomerate at Murlough Bay, Co. Antrim. *Geol. Mag.*, **96**, 86–7.

—— 1962. The Cretaceous succession in the area between Red Bay and Garron Point, Co. Antrim. *Irish Nat. J.*, **14**, 73–7.

—— 1963. New records of *Gonioteuthis* in Ireland. *Irish Nat. J.*, **14**, 98.

—— 1964. The Lower (pre-*Belemnitella mucronata*) White Limestone of the east and north-east of Co. Antrim. *Irish Nat. J.*, **14**, 262–9, 296–303.

REYNOLDS, D. L. 1928. The petrography of the Triassic sandstone of north-east Ireland. *Geol. Mag.*, **65**, 448–73.

—— 1931. The dykes of the Ards Peninsula, Co. Down. *Geol. Mag.*, **68**, 97–111, 145–65.

ROBBIE, J. A. 1964. The geology of Crow Glen, Co. Antrim. *Irish Nat. J.*, **14**, 224–9.

RUST, B. R. 1965a. The stratigraphy and structure of the Whithorn area of Wigtown-shire, Scotland. *Scot. J. Geol.*, **1**, 101–33.

—— 1965b. The sedimentology and diagenesis of Silurian turbidites in south-east Wigtownshire. *Scot. J. Geol.*, **1**, 231–46.

SAMPSON, G. V. 1814. *Memoir explanatory of the Chart and Survey of Londonderry.*

SAVAGE, R. J. G. 1964. Post-glacial wild boar from Belfast Lough. *Irish Nat. J.*, **14**, 303–5.

SCOULER, J. 1837. Observations on the lignites and silicified woods of Lough Neagh. *J. Geol. Soc. Dublin*, **1**, 231–41.

SEYMOUR, H. J. 1897. (Exhibit at Dublin Microscopical Club) Amygdaloidal basalt from Black Quarry of Squires Hill near Belfast. *Irish. Nat.*, **6**, 277.

SHERLOCK, R. L. 1926. A correlation of the British Permo-Triassic Rocks, Pt. 1, North England, Scotland and Ireland. *Proc. Geol. Assoc.*, **37**, 1–72.

—— 1928. A correlation of the British Permo-Triassic Rocks, Part 2. England South of the Pennines and Wales. *Proc. Geol. Assoc.*, **39**, 49–95.

SIMPSON, A. 1963. The stratigraphy and tectonics of the Manx Slate Series, Isle of Man. *Quart. J. Geol. Soc.*, **119**, 367–400.

SMITH, D. 1868. *Outlines of the rocks of Antrim.* Belfast.

SMITH, D. B. and FRANCIS, E. A. 1967. Geology of the country between Durham and West Hartlepool. *Mem. Geol. Surv. G.B.*

SOLLAS, W. J. 1880. Nodules of the Trimmingham Chalk. *Ann. Mag. Nat. Hist.*, (5), **6**, 384–95, 437–60.

STEPHENS, N. 1957. Some observations on the "Interglacial" platform and the Early post-glacial raised beach on the east coast of Ireland. *Proc. Roy. Irish Acad.*, **58B**, 129–49.

—— 1963. Late-glacial sea-levels in North-East Ireland. *Irish Geogr.*, **4**, 345–59.

STEWART, S. A. 1871. A list of the fossils of the Estuarine Clays of Counties Down and Antrim. *8th Ann. Rep. Belfast Nat. Field Club* (for 1870–1), Appendix **2**, 27–42.

—— 1881. Mollusca of Boulder Clay in N. E. Ireland. *Proc. Belfast Nat. Field Club*, Appendix **5**, 165–176 (for 1879–80).

SWANSTON, W. 1878. On the Silurian rocks of Co. Down. Pt. 1. Correlation. *Proc. Belfast Nat. Field Club*, Appendix **4**, (for 1876–7), 107–23.

—— 1879. On supposed fossiliferous Pliocene clays overlying basalt near the shore of Lough Neagh. *Geol. Mag.*, (2), **6**, 62–5.

—— 1892. A fossiliferous ironstone nodule. *Proc. Belfast Nat. Field Club*, **3**, 401.

SYNGE, F. M. and STEPHENS N. 1966. Late- and Post-glacial shorelines, and ice limits in Argyll and north-east Ulster. *Trans. Inst. Br. Geogr.*, **39**, 101–25.

TATE, R. 1864. On the Liassic strata of the neighbourhood of Belfast. *Quart. J. Geol. Soc.*, **20**, 103–14.

—— 1865. On the correlation of the Cretaceous formations of the north-east of Ireland. *Quart. J. Geol. Soc.*, **21**, 15–44.

—— 1867. On the Lower Lias of the north-east of Ireland. *Quart. J. Geol. Soc.*, **23**, 297–305.

TAYLOR, R. 1824. On the alluvial strata and on the chalk of Norfolk and Suffolk, and on the fossils by which they are accompanied. *Trans. Geol. Soc. Lond.*, (2), **1**, 374–8.

TRUSHEIM, F. VON. 1963. Zur Gliederung des Buntsandsteins. *Erdoel-Z. Bohr- u. Fördertech.*, **79**, 277–92.

TWENHOFEL, W. H. 1937. Terminology of the fine-grained mechanical sediments. *Rep. Comm. on Sedimentation for* 1936-7. Nat. Research Council (U.S.A.), 81–104.

TOMKEIEFF, S. I. and MARSHALL, C. E. 1935. The Mourne dyke swarm. *Quart. J. Geol. Soc.*, **91**, 251–92.

VOORTHUYSEN, J. H. VAN. 1951. The quantitative distribution of Holocene foraminifera in the N. O. Polder. *Proc. 3rd Int. Cong. Sediment. Neth.*, 267–72.

WALKER, G. P. L. 1960a. The amygdale minerals in the Tertiary lavas of Ireland: III. Regional distribution. *Miner. Mag.*, **32**, 503–27.

—— 1960b. An occurrence of mugearite in Antrim. *Geol. Mag.*, **97**, 62–4.

—— 1962a. Garronite, a new zeolite, from Ireland and Iceland. *Miner. Mag.*, **33**, 173–86.

—— 1962b. A note on the occurrences of tree remains within the Antrim Basalts. *Proc. Geol. Assoc.*, **73**, 1–7.

—— 1962c. Low-potash gismondine from Ireland and Iceland. *Miner. Mag.*, **33**, 187–201.

WALLIS, F. S. 1927. The Old Red Sandstone of the Bristol District. *Quart. J. Geol. Soc.*, **83**, 760–89.

WALTON, E. K. 1955. Silurian greywackes in Peeblesshire. *Proc. Roy. Soc. Edin.*, **65B**, 327–57.

—— 1963. Structure and sedimentation in the Southern Uplands *in* JOHNSON, M. R. W. and STEWART, F. H. (ed). *The British Caledonides.* Edinburgh and London, 71–97.

WARREN, P. T. 1963. The petrography, sedimentation and provenance of the Wenlock rocks near Hawick, Roxburghshire. *Trans. Edin. Geol. Soc.*, **69**, 225–55.

—— 1964. The stratigraphy and structure of the Silurian (Wenlock) rocks south-east of Hawick, Roxburghshire, Scotland. *Quart. J. Geol. Soc.*, **120**, 193–222.

WARRINGTON, G. 1967. Correlation of the Keuper by miospores. *Nature*, **214**, 1323–4.

—— (in press). The stratigraphy and palaeontology of the "Keuper" Series of the Central Midlands, England.

WATTS, W. A. 1962. Early Tertiary pollen deposits in Ireland. *Nature*, **193**, 600.

—— 1963. Fossil seeds from the Lough Neagh Clay. *Irish Nat. J.*, **14**, 117–8.

WHEELER, C. F. 1937. Chairman's Address. *Inst. Civil Eng.* (N.I. Assoc.).

WHITEHURST, J. 1786. *On the original state and formation of the earth.* 2nd ed. London.

WILLIAMS, A. 1959. A structural history of the Girvan district. *Trans. Roy. Soc. Edin.*, **63**, 629–67.

WILSON, H. E. and ROBBIE, J. A. 1966. Geology of the country around Ballycastle. *Mem. Geol. Surv. N. Ireland.*

WOLFE, M. J. 1968. An electron-microscope study of the surface texture of sand grains from a basal conglomerate. *Sedimentology*, **8,** 239–47.

—— 1968. Lithification of a carbonate mud: Senonian Chalk in Northern Ireland. *Sediment. Geol.* **2**, (4), 263-90.

WOOD, E. M. R. 1906. The Tarannon Series of Tarannon. *Quart. J. Geol. Soc.*, **62**, 644–701.

WOODWARD, H. B. 1882. The geology of the country round Norwich, *Mem. Geol. Surv.* (1881 on title page).

WRIGHT, J. 1878. Recent foraminifera of Down and Antrim. *Proc. Belfast Nat. Field Club.* Appendix **4**, (for 1876–7), 101–6.

—— 1881. Post-Tertiary foraminifera of North East of Ireland. *Ann Rep.* and *Proc. Belfast Nat. Field Club*, **1**, (2), 428–9; and Appendix, **1**, (for 1879–80), 149–63.

—— 1895. The occurrence of Boulder Clay on Divis. *Proc. Belfast Nat. Field Club*, (2), **4**, 215–6.

WRIGHT, W. B. 1924. The age and origin of the Lough Neagh Clays. *Quart. J. Geol. Soc.*, **80**, 468–88.

YOUNG, R. 1867. The Recent elevation of the land in vicinity of Belfast. *Ann. Rep. Belfast Nat. Field Club*, **4**, 20–2.

—— 1871. Boulder Clay of Belfast district. *Ann. Rep. Belfast Nat. Field Club*, **8**, 32–5.

—— 1890. Some notes on the Upper Boulder Clay near Belfast. *Proc. Belfast Nat. Hist. and Phil. Soc.* (for 1889–90), 57.

Appendix 1

BOREHOLE RECORDS

The more important borehole records in the possession of the Geological Survey of Northern Ireland are listed below in abridged form. Figures in square brackets are Irish Grid references, all in grid square J. Figures in round brackets give the registered number of the borehole in the files of the Geological Survey, which include about 1500 records from this Sheet. Notes in square brackets are additions to the original journals.

Heights above Mean Sea Level are estimated from the nearest spot-height or bench mark by eye or Abney Level and all are approximations. Mean Sea Level is 8·9 ft (2·7 m) above the old Ordnance Datum.

Avoniel Borehole

Site: 350 yd W. of Connswater Bridge, Belfast [361 739]. 6-in map Down 4 S.E. Height above Mean Sea Level 13 ft. Date 1959–60. Drilled by Irish Diamond Drilling Company. Cores examined by P. I. Manning (36/596).

	Thickness ft	in	Depth ft	in		Thickness ft	in	Depth ft	in
SURFACE AND DRIFT					Anhydrite, massive	16	0	380	10
Fill	7	6	7	6	Marl with sand-				
Boulder Clay	26	6	34	0	stone wisps	5	2	386	0
Sand	17	0	51	0					
BUNTER SANDSTONE					**MAGNESIAN LIME-STONE**				
Sandstone, reddish brown with thin marl partings	45	7	96	7	Dolomite, yellow, with anhydrite				
Dolerite dyke	10	5	107	0	patches	68	6	454	6
Sandstone, brown with thin marl laminae	7	0	114	0	**PERMIAN SANDSTONE**				
PERMIAN MARLS					Sandstone, pale with orange				
Marl, brown with sandstone ribs	40	6	154	6	staining	77	6	532	0
Marl, brown with gypsum veins	89	8	244	2	Breccia of quartz and greywacke fragments inter-				
Marl with gypsum veins and anhydrite crystals	120	10	364	10	bedded with sandstones	68	0	600	0

193

Belfast Co-Operative Society Borehole

Site: Ravenhill Avenue, Belfast [354 729]. 6-in map Down 4 S.E. Height above Mean Sea Level 31 ft. Date 1927. Communicated by Le Grand, Sutcliff and Gell (36/32).

	Thickness ft	in	Depth ft	in		Thickness ft	in	Depth ft	in
[SOIL AND GLACIAL DRIFT]					Sandy marl ..	22	0	153	0
					Marl with sandstone	11	0	164	0
Pit [Fill]	8	0	8	0					
Boulder Clay ..	40	0	48	0	[PERMIAN SANDSTONE AND BROCKRAM]				
Gravel	2	0	50	0	Sandstone ..	36	0	200	0
Boulder Clay ..	1	6	51	6	Sandstone, grey ..	6	0	206	0
Gravel	5	6	57	0	Sandstone, red ..	24	0	230	0
Sand	17	6	74	6	Sandstone, grey ..	27	0	257	0
Boulder Clay ..	15	6	90	0	Sandstone, yellow	1	0	258	0
					Sandstone, marly	10	0	268	0
[PERMIAN MARLS]					Sandstone, red ..	75	0	343	0
Marl	38	0	128	0	Conglomerate ..	57	0	400	0
Conglomerate ..	3	0	131	0					

Belfast Municipal Electric Power Station Borehole
(now Electricity Department Offices)

Site: East Bridge Street, Belfast [348 739]. 6-map Antrim 61 S.W. Height above Mean Sea Level 15 ft. Date before 1902. Communicated by the Engineer-in-Charge (36/22).

	Thickness ft	in	Depth ft	in		Thickness ft	in	Depth ft	in
[SOIL AND DRIFT]					[BUNTER SANDSTONE]				
Rough formation [? made ground]	2	0	2	0	Hard gritty sandstone with blue stone under [Triassic Sandstone with ? basalt dyke]	7	0	64	0
Hard and soft concrete [? made ground] ..	4	0	6	0	Red sandstone (Described as 'red gritty sandstone', 'hard red sandstone', 'coarse-grained red and white sandstone', 'fine close sandstone', etc.)	85	0	149	0
[ESTUARINE CLAY]									
Dry sleech clay ..	16	0	22	0					
[GLACIAL DRIFT]					Hard dyke-rock [Basalt dyke] ..	3	0	152	0
Heavy close gravel [? Post-Glacial]	14	0	36	0	Sandstone, fine red	114	0	266	0
Hard sand and gravel	4	0	40	0	Sandstone, fine red, with clay marl ..	19	0	285	0
Heavy clay ..	3	0	43	0	Sandstone, red ..	160	0	445	0
Fine close gravel, shell and stone ..	8	0	51	0					
Red clay with stone	6	0	57	0					

	Thickness		Depth	
	ft	in	ft	in
[PERMIAN MARLS]				
Tough clay and sand	3	6	448	6
Red clay and hard stiff marl ..	39	6	488	0
Sandstone, hard red, with layers of magnesia ..	18	0	506	0
Sandstone, hard red, with clay marl	30	0	536	0
Sandstone, hard red, showing more clay ..	18	0	554	0

Belfast Ropeworks Borehole

Site: Belfast Ropeworks Co. Ltd., Newtownards Road, Belfast [364 738]. 6-in map Down 4 S.E. Height above Mean Sea Level 12 ft. Date 1906. Communicated by J. Henderson & Son Ltd. (36/493).

	Thickness		Depth	
	ft	in	ft	in
[SOIL AND GLACIAL DRIFT]				
Forced material [Fill]	3	0	3	0
Clay, brown ..	21	0	24	0
Clay and stones, brown	13	0	37	0
[BUNTER SANDSTONE]				
Marl, red	23	6	60	6
Sandstone, red, hard	8	7	69	1
Rock, hard [? sandstone] ..	0	4	69	5
Sandstone, limey, extra hard ..	0	4	69	9
Marl, red, soft ..	3	6	73	3
Sandstone, red, hard	4	7	77	10
Sandstone, limey, hard	10	7	88	5
Limestone, hard ..	2	7	91	0
Marl, red	3	3	94	3
Sandstone, limey, hard	28	11	123	2
[PERMIAN MARLS]				
Marl, limey, hard	5	2	128	4
Whinstone, limey, broken, extra hard [dyke] ..	14	5	142	9
Marl, limey, soft..	1	4	144	1
Whinstone, hard [Basalt dyke] ..	2	0	146	1
Marl, red, hard ..	1	0	147	1
Marl, limey, hard	64	5	211	6
Marl, soft.. ..	3	3	214	9
Marl, limey, soft ..	48	3	263	0
Sandstone, dark red, limey ..	8	0	271	0
Sandstone, dark limey	11	10	282	10
Sandstone, red, marly, soft ..	49	1	331	11
Limestone, hard, extra hard [Main Anhydrite] ..	3	1	335	0
Sandstone, limey..	12	10	347	10
[MAGNESIAN LIMESTONE]				
Limestone, white, extra hard ..	5	0	352	10
Limestone, extra hard	22	6	375	4
[PERMIAN SANDSTONE AND BROCKRAM]				
Sandstone, extra hard	19	4	394	8
Sandstone, limey extra hard ..	16	6	411	2
Limestone, hard [? anhydrite] ..	4	3	415	5
Sandstone, limey, extra hard ..	3	8	419	1
Sandstone, red, dark	44	8	463	9
Sandstone, limey, hard	3	0	466	9
Sandstone, red	9	8	476	5
Conglomerate rock, extra hard ..	23	7	500	0

Boxmore Works Borehole

Site: 600 yd 110° from Dollingstown School [111 582]. 6-in map Down 13 S.W. **Height above Mean Sea Level 202 ft. Deepened from 435 ft in 1958. Copied from a record obtained by the late J. J. Hartley (36/39).**

	Thickness ft in	Depth ft in		Thickness ft in	Depth ft in
[SOIL AND GLACIAL DRIFT]			flint [Clay-with-flints] ..	10 0	399 0
Boulder Clay ..	8 0	8 0	[UPPER CHALK]		
			Chalk	15 0	414 0
[ANTRIM LAVA SERIES (LOWER BASALTS)]			[KEUPER MARL]		
Basalt	381 0	389 0	Marl, red with autobrecciation near the base ..	126 0	540 0
Sand composed of fine fragments of					

Broomhedge Borehole

Site: Walnut Vale Farm (Mr. F. Megarry) [195 620] 400 yd 336° from Halfpenny Gate Cross-roads. 6-in map Antrim 67 N.E. Height above Mean Sea Level 127 ft. Date about 1946. Copied from a record obtained by the late J. J. Hartley (36/470).

	Thickness ft in	Depth ft in		Thickness ft in	Depth ft in
[SOIL AND GLACIAL DRIFT]			lar fragments of grey and black slate averaging one or two inches across, embedded in a fine-grained marly matrix	66 0	126 0
Drift	60 0	60 0			
[BUNTER SANDSTONE]					
Breccia, hard, with numerous angu-					

Burnhouse Borehole [Burnhouse Animal Products Ltd.]

Site: 270 yd 50° from Young's Bridge about 2 miles W.S.W. of Lisburn [235 631]. 6-in map Antrim 68 N.W. Height above Mean Sea Level about 105 ft. Date 1936. Information from Smyth, Hayes and Company, drillers (36/471).

	Thickness ft in	Depth ft in		Thickness ft in	Depth ft in
[SOIL AND GLACIAL DRIFT]			[BUNTER SANDSTONE]		
Boulder Clay, red	20 0	20 0	Sandstone, red	120 0	200 0
Clay, blue ..	60 0	80 0			

Castlereagh Laundry Borehole

Site: 1 Redcar Street, Belfast [359 730]. 6-in map Down 4 S.E. Height above Mean Sea Level 36 ft. Date 1911. Communicated by Isler & Company (36/31).

	Thickness ft	in	Depth ft	in		Thickness ft	in	Depth ft	in
[SOIL AND GLACIAL DRIFT]					Sandstone, white	49	6	234	0
Top soil	2	6	2	6	Sandstone, red	42	6	276	6
Clay, brown ..	6	6	9	0	Sandstone white	1	0	277	6
Boulder stone ..	2	0	11	0	Sandstone, red	65	6	343	0
Clay and stones ..	16	0	27	0	Conglomerate ..	12	0	355	0
Marl, sandy ..	12	0	39	0	Sandstone, red	2	0	357	0
Marl	40	0	79	0	Conglomerate ..	1	6	358	6
Sand, hard, brown	11	0	90	0	Sandstone, red	24	6	383	0
Marl	14	0	104	0	Conglomerate ..	4	6	387	6
Sand	2	0	106	0	Rock, hard bastard	2	0	389	6
Gravel	1	6	107	6	Hard fakes [sandy				
Clay and stones ..	2	0	109	6	shale]	3	6	393	0
Sand, hard, brown	22	6	132	0	Marl	7	0	400	0
Clay and stones ..	8	0	140	0	Conglomerate ..	13	6	413	6
					Conglomerate and				
[PERMIAN MARLS]					marl	2	0	415	6
Marl, sandy ..	4	0	144	0	Limestone ? ..	0	9	416	3
Marl	16	0	160	0	Marl, hard, flakey	13	9	430	0
Marl and magnesia	11	0	171	0	Kingal [hard sand-				
Magnesia [? Mag-					stone]	2	0	432	0
nesian limestone]	6	0	177	0	Marl	2	0	434	0
Marl and magnesia	6	0	183	0	Kingal and marl ..	5	0	439	0
					Marl and magnesia	3	0	442	0
[PERMIAN SANDSTONE AND BROCKRAM]					Fakes	2	0	444	0
Sandstone, dark					Marl	4	0	448	0
red	1	6	184	6	Marl and magnesia	7	0	455	0

Clonmore House Borehole

Site: Clonmore House, Lambeg, Lisburn (Lisburn R.D.C. Offices) [274 660]. 6-in map Antrim 64 S.W. Height above Mean Sea Level 191 ft. Date ? 1935. Information taken from diagram on pump house wall (36/426).

	Thickness ft	in	Depth ft	in		Thickness ft	in	Depth ft	in
[SOIL AND GLACIAL DRIFT]					Sand	1	0	124	0
Red clay with					[BUNTER SANDSTONE]				
boulders ..	85	0	85	0	Sandstone, red,				
Gravel	2	0	87	0	with marl ..	5	0	129	0
					Sandstone, grey	5	0	134	0
Clay, red, with					Sandstone, red,				
boulders ..	36	0	123	0	with grey streaks	68	0	202	0

Dunville & Co. Royal Irish Distillery Borehole

Site: Former distillery; now Murray's bonded warehouse, Distillery Street, Belfast [326 734]. 6-in map Antrim 60 S.E. Height above Mean Sea Level 15 ft. Date unknown, probably 1890. Copied from a record in the possession of the late J. J. Hartley (36/494A).

	Thickness ft	in	Depth ft	in		Thickness ft	in	Depth ft	in
[SOIL AND GLACIAL DRIFT]					Sandstone, red and white	27	5	168	11
Made ground ..	1	0	1	0	Sandstone, white	12	10	181	9
Clay, red	11	0	12	0	Sandstone, red	2	0	183	9
					Sandstone, (red, white, brown) ..	65	11	249	8
[BUNTER SANDSTONE]					Sandstone, white	24	3	273	11
Sandstone ..	21	0	33	0	Sandstone, white and red.. ..	74	5	348	4
Sandstone, red	2	0	35	0	Sandstone, red	21	9	370	1
Sandstone, grey	1	3	36	3	Sandstone, red and white	28	10	398	11
Sandstone, red and white ..	102	3	138	6	Sandstone, white	1	1	400	0
Sandstone, white	3	0	141	6					

Glenbank Bleach Works Borehole

Site: Messrs. Ewart's Factory, Glenbank Mills, Ligoniel, Belfast [306 767]. 6-in map Antrim 60 N.E. Height above Mean Sea Level 321 ft. Date 1928. Communicated by Le Grand Sutcliff and Gell Ltd. (36/1).

	Thickness ft	in	Depth ft	in		Thickness ft	in	Depth ft	in
[SOIL AND GLACIAL DRIFT]					Marl, mixed and Gypsum ..	208	0	363	0
Pit (Clay)	6	0	6	0	Whinstone [Dyke]	73	6	436	6
Boulder Clay ..	35	0	41	0	Keuper Marl and Gypsum ..	38	6	475	0
[KEUPER MARL]					Bluestone [Dyke]	3	0	478	0
Marl and Gypsum	68	0	109	0	Keuper Marl and Gypsum ..	22	0	500	0
Marl, red	32	0	141	0					
Marl, green, and Gypsum ..	14	0	155	0					

Great Northern Railway Borehole

Site: At S.E. corner of Windsor Football Ground, Balmoral [323 723]. 6-in map Antrim 60 S.E. Height above Mean Sea Level 31 ft. Date ? 1913. Copied from a record in the possession of the late J. J. Hartley (36/482).

	Thickness ft	in	Depth ft	in		Thickness ft	in	Depth ft	in
					[BUNTER SANDSTONE]				
[SOIL AND GLACIAL DRIFT]					Sandstone, red, with some alternating bands of				
Drift	90	0	90	0	grey sandstone..	510	0	600	0

Irish Distillery Connswater Borehole

Site: Irish Distillery, Connswater (now Gallaher's) Belfast [364 744]. 6-in map Down 4 N.E. Height above Mean Sea Level 9 ft. Date 1897. A fuller record of that published in "The Geology of Belfast and District" Memoir 1904. Copied from a record in the possession of the late J. J. Hartley (36/28B).

	Thickness ft in	Depth ft in		Thickness ft in	Depth ft in
[SOIL AND GLACIAL DRIFT]			[PERMIAN MARLS]		
			Marl, sandy ..	45 6	215 11
Made ground ..	3 0	3 0	Marl, soft.. ..	162 0	377 11
Clay, stony ..	7 0	10 0	Marl, hard ..	77 6	455 5
Sand, red	6 0	16 0	Marl, soft.. ..	4 6	459 11
Clay, boulders ..	28 0	44 0	Marl	24 0	483 11
Boulder ..	2 4	46 4	Grey granite		
Gravel	1 0	47 4	[? Anhydrite] ..	6 0	489 11
Clay, stony ..	2 2	49 6	Limestone [? Anhydrite] ..	2 6	492 5
			Marl	11 6	503 11
[BUNTER SANDSTONE]			[MAGNESIAN LIMESTONE]		
Sandstone, red, soft	120 11	170 5	Limestone ..	25 8	529 7

Langford Lodge Borehole

Site: 480 yd E.N.E. of Gartree Point, Lough Neagh [090 748]. 6-in map Antrim 58 N.E. Height above Mean Sea Level 69 ft. Date 1956–8. Drilled to 3198 ft with cores at intervals in the Interbasaltic Bed, Lower Basalts, Chalk and Rhaetic; coring continuous below 3198 ft. Drilled by John Thom Ltd. Cores examined by A. Fowler and H. E. Wilson (36/429).

	Thickness ft in	Depth ft in		Thickness ft in	Depth ft in
SOIL AND GLACIAL DRIFT			brown limonitic veins and patches.		
Boulder Clay ..	22 0	22 0	Some laterite		
Gravel	8 0	30 0	towards the top	53 0	845 0
ANTRIM LAVA SERIES (UPPER BASALTS)			ANTRIM LAVA SERIES (LOWER BASALTS)		
Basalt, fine- and medium-grained with some zeolite. Lithomarge and red bole indicate the tops of lava flows. There are at least 26 flows	762 0	792 0	Basalt, fine-grained with some medium-grained especially in the lower part. Some of the rocks are brown and partly kaolinized. Lithomarge and red bole indicate the top of lava flows. There are at least 30 flows ..	1743 0	2588 0
INTERBASALTIC BED					
Lithomarge, dark grey and purple with yellowish					

	Thickness	Depth
	ft in	ft in

UPPER CHALK

Chalk, white, hard, with some flint .. 60 0 2648 0

LOWER LIAS

Shale, dark grey .. 30 0 2678 0

RHAETIC

Shale and shaly mudstone, usually dark grey with some redish brown towards base, with a few laminae and thin bands of shaly limestone .. 57 0 2735 0

KEUPER MARL

Marl, red and green, with some dolomitic marlstone and anhydrite 365 0 3110 0

Marl, reddish brown, dark red, red and green, with traces of bedding in places, a few thin bands showing autobrecciation. Anhydrite in thin vertical and horizontal veins .. 45 0 3255 0

Marl, brown, brownish red and green, commonly silty, with bands of dolomitic marlstone, anhydrite towards the top, rare bands of autobreccia .. 505 0 3760 0

BUNTER SANDSTONE

Sandstone, red, micaceous with some bands of marl and a little anhydrite towards the top .. 560 0 4320 0

	Thickness	Depth
	ft in	ft in

Sandstone, reddish brown and brown in places pale grey, fine-grained, micaceous, with some beds of chocolate brown and green marl 106 0 4426 0

Sandstone, reddish brown, with a little grey, fine-grained, micaceous, with thin bands of marl and with, in places, irregular thin lenses and wisps of marl .. 221 0 4647 0

[PERMIAN]

Marl, brown, silty, micaceous, with thin beds, wisps and lenses of sandstone. Much of the sandstone is white or green 111 3 4758 3

Sandstone and Breccia: angular fragments of Lower Palaeozoic grit in a gritty matrix. The bottom 2 in are of slightly disturbed Lower Palaeozoic grit intersected with "veins" of red sandstone .. 8 0 4766 3

[LOWER PALAEOZOIC]

Grit, pinkish brown in the top 26 ft, brownish grey and grey below, fine to mainly coarse-grained, with rare thin bands (1 in thick) of brown mudstone 236 1 5002 4

Linen Research Institute Borehole

Site: Glenmore House (Linen Research Institute) Lambeg [278 662]. 6-in map Antrim 64 S.W. Height above Mean Sea Level 91 ft. Date 1937. Copied from a record in the possession of the late J. J. Hartley (36/472).

	Thickness ft	in	Depth ft	in
[SOIL AND GLACIAL DRIFT]				
Drift	28	0	28	0
[BUNTER SANDSTONE]				
Sandstone, white	5	0	33	0
Sandstone, soft, red	10	0	43	0
Marl	2	0	45	0
Sandstone, red	4	0	49	0
[PERMIAN MARLS AND SANDSTONES]				
Marl	129	0	178	0
Sandstone	1	0	179	0
Marl	10	0	189	0

	Thickness ft	in	Depth ft	in
Conglomerate (small rounded pebbles)	1	0	190	0
Marl	15	0	205	0
Conglomerate (angular or sub-angular)	5	0	210	0
Marl with some sandstone near base	50	0	260	0
Sandstone, red, passing into marl below. Conglomerate at bottom	70	0	330	0

Lisburn No. 1 Borehole

Site: At S.E. end of Duncan's Reservoir, Lisburn [259 656]. 6-in map Antrim 64 S.W. Height above Mean Sea Level 212 ft. Date 1934. Communicated by Isler & Co. Ltd. and R. E. L. Clarke (36/35).

	Thickness ft	in	Depth ft	in
[SOIL AND GLACIAL DRIFT]				
Mould	1	0	1	0
Clay and stones	17	6	18	6
Ballast, black	10	0	28	6
Loam, dry, and stones	3	6	32	0
Clay and boulders	18	0	50	0
Ballast, black	7	0	57	0
Boulder	1	0	58	0
Clay, brown, and stones	10	0	68	0
Ballast and clay	5	0	73	0
Conglomerate [gravel]	2	0	75	0
[BUNTER SANDSTONE]				
Marl, red	11	0	86	0
Sandstone and marl layers	5	0	91	0
Sandstone, red	36	0	127	0
Sandstone, green	2	0	129	0
Sandstone, red	15	0	144	0
Sandstone, green	1	6	145	6
Sandstone, red	22	6	168	0
Sandstone, green	16	0	184	0

	Thickness ft	in	Depth ft	in
Sandstone, red	2	0	186	0
Sandstone, green	3	0	189	0
Sandstone, red, and marl layers	7	0	196	0
Sandstone, red	8	0	204	0
Marl, red	3	0	207	0
Sandstone, red, and marl layers	15	0	222	0
Sandstone, red	12	0	234	0
Sandstone, red, and green, in layers	8	0	242	0
Sandstone, green	12	0	254	0
Sandstone, red, and marl layers	30	0	284	0
Sandstone, green	16	0	300	0
Sandstone, red and green	8	0	308	0
Sandstone, green	42	0	350	0
Marl, red	5	0	355	0
Sandstone, green	21	0	376	0
Sandstone, red and green	17	0	393	0
Marl, red	5	0	398	0
Sandstone, red and green	2	0	400	0

Lisburn No. 2 Borehole

Site: Old Park Farm, Aghlislone, Lisburn [249 669]. 6-in map Antrim 64 S.W. Height above Mean Sea Level 353 ft. Date 1944. Communicated by Isler & Co. Ltd. and R. E. L. Clarke (36/36)

	Thickness ft	in	Depth ft	in
[SOIL AND GLACIAL DRIFT]				
Turf and mould ..	1	0	1	0
Clay and boulders	5	0	6	0
Clay, yellow, and stones	1	0	7	0
Clay, brown, and stones	23	0	30	0
Rock, soft, brown	0	6	30	6
Marl, red	0	6	31	0
Clay, yellow, and soft brown rock	1	0	32	0
Sandstone, red and grey, with layers of red marl ..	16	0	48	0
Marl, hard, red ..	10	0	58	0
Marl, red and grey with layers of sandstone ..	10	0	68	0
Clay, brown and blue, with pockets of fine yellow sand ..	1	0	69	0
Drift, hard, brown	4	0	73	0
Clay and stones, hard, mottled ..	5	0	78	0
Sandstone, green	0	6	78	6
Clay, yellow, and soft brown rock	18	6	97	0
Marl, red	1	0	98	0
Clay, yellow, sandy and large boulders of ballast or dolerite	10	6	108	6
Clay, hard, or drift	2	6	111	0
Clay, yellow, sandy	1	0	112	0
Clay, brown ..	7	0	119	0
[KEUPER MARL ?]				
Marl, red	0	6	119	6
Clay, yellow, sandy	27	6	147	0

	Thickness ft	in	Depth ft	in
Marl, red	1	0	148	0
Clay, yellow ..	1	0	149	0
Marl, red and grey	20	6	169	6
[BUNTER SANDSTONE]				
Sandstone, grey and green ..	33	6	203	0
Basalt or dolerite, hard blue [Dyke]	10	0	213	0
Sandstone, grey ..	10	0	223	0
Sandstone, white with green marl partings ..	1	0	224	0
Sandstone, white or grey	6	0	230	0
Sandstone, white, grey marl partings and nodules of red marl ..	3	0	233	0
Basalt or dolerite [Dyke]	29	0	262	0
Sandstone, white and green ..	98	0	360	0
Sandstone, grey ..	69	0	429	0
Basalt or dolerite [Dyke]	89	0	518	0
Sandstone, grey ..	17	0	535	0
Marl, red, sandy ..	1	0	536	0
Sandstone, grey, with red marl partings ..	7	0	543	0
Sandstone, grey ..	5	0	548	0
Sandstone, red ..	2	0	550	0
Sandstone, grey, with green marl partings ..	40	0	590	0
Sandstone, red, marly	1	0	591	0
Sandstone, hard, grey	12	0	603	0

Lisburn No. 3 Borehole

Site: South side of Duncan's Reservoir bridge and on west side of Reservoir [258 659]. 6-in map Antrim 64 S.W. Height above Mean Sea Level 211 ft. Date 1944. Communicated by Isler & Co. Ltd. and R. E. L. Clarke (36/37).

	Thickness		Depth	
	ft	in	ft	in
[SOIL AND GLACIAL DRIFT]				
Soil and turf ..	1	0	1	0
Clay, sandy, and stones	4	6	5	6
Ballast	5	6	11	0
Boulders	2	3	13	3
Clay, brown, and boulders ..	4	6	17	9
Clay, red, sandy ..	11	0	28	9
Clay, yellow, sandy	4	0	32	9
Clay, brown, stony	5	3	38	0
Clay, red, with thin layers of stone..	7	0	45	0
Clay, yellow, stony, with soft sandstone	7	0	52	0
[BUNTER SANDSTONE]				
Clay, red, with layers of red sandstone ..	5	6	57	6
Clay, yellow, with layers of sandstone	2	9	60	3
Sandstone, soft, red	7	11	68	2
Sandstone, yellow, laminated ..	6	1	74	3
Sandstone, red ..	1	3	75	6
Sandstone, yellow	6	6	82	0
Sandstone, red, with marl partings	2	6	84	6
Sandstone, yellow	0	6	85	0
Sandstone, red, with yellow and black laminations	9	4	94	4
Sandstone, yellow	0	8	95	0
Sandstone, red ..	2	0	97	0
Sandstone, yellow	6	3	103	3
Sandstone, red, laminated ..	1	9	105	0
Sandstone, yellow	1	3	106	3
Sandstone, blue (pale) with marl partings	2	6	108	9

	Thickness		Depth	
	ft	in	ft	in
Sandstone, red ..	3	5	112	2
Sandstone, yellow and blue ..	2	5	114	7
Sandstone, red ..	3	5	118	0
Sandstone, yellow, with vertical joints	8	0	126	0
Sandstone, hard, red, laminated..	26	6	152	6
Marl, red	1	0	153	6
Sandstone, red ..	3	6	157	0
Sandstone, bluish grey	0	8	157	8
Sandstone, red ..	10	0	167	8
Sandstone, pale blue	13	0	180	8
Sandstone, red ..	3	10	184	6
Sandstone, pale grey	4	0	188	6
Sandstone, red ..	3	6	192	0
Sandstone, pale blue	24	0	216	0
Sandstone, red, laminated ..	2	6	218	6
Sandstone, red, mottled with blue sandstone..	1	10	220	4
Sandstone, red ..	11	8	232	0
Sandstone, pale red, very hard ..	2	4	234	4
Sandstone, pale grey, very hard	6	8	241	0
Sandstone, red ..	1	0	242	0
Sandstone, pale blue (probably a fault at 245 ft) ..	15	0	257	0
Sandstone, red, laminated (hard)	12	0	269	0
Sandstone, pale grey	14	6	283	6
Sandstone, grey and red (broken)	5	6	289	0
Sandstone, grey ..	8	0	297	0
Sandstone, red, with grey layers	1	6	298	6
Sandstone, grey ..	1	0	299	6

	Thickness ft	in	Depth ft	in		Thickness ft	in	Depth ft	in
Sandstone, red, with grey stone layers	3	0	302	6	Sandstone, red ..	13	0	388	0
					Sandstone, grey ..	5	0	393	0
Sandstone, grey ..	72	6	375	0	Sandstone, red ..	1	0	394	0
					Sandstone, grey ..	6	0	400	0

Long Kesh Borehole

Site: 1750 yd S.W.W. of Warren Gate Bridge [249 621]. 6-in map Down 14 N.E. Height above Mean Sea Level 110 ft. Date 1955. Drilled by John Thom Ltd. Cuttings examined by P. I. Manning and J. A. Robbie (36/350).

	Thickness ft	in	Depth ft	in		Thickness ft	in	Depth ft	in
SOIL AND GLACIAL DRIFT					Marl, red	2	0	60	0
					Sandstone ..	44	0	104	0
Soil	1	0	1	0					
Sand	29	0	30	0	PERMIAN				
Boulder Clay ..	11	3	41	3	Marl	64	0	168	0
					Conglomerate of Lower Palaeozoic pebbles in red sandy matrix	97	0	265	0
BUNTER SANDSTONE									
Sandstone, with clay bands ..	14	9	56	0	SILURIAN (?)				
Yellow clay: decomposed basalt dyke	2	0	58	0	Greywacke with some shale bands	34	0	299	0

Miller Sons & Torrence Borehole

Site: 8 Maxwell Street, Belfast (now Knockfergus Ltd.) [333 733]. 6-in map Antrim 61 S.W. Height above Mean Sea Level 36 ft. Date 1899. Copied from a record in the possession of the late J. J. Hartley (36/499).

	Thickness ft	in	Depth ft	in		Thickness ft	in	Depth ft	in
[SOIL AND GLACIAL DRIFT]					[BUNTER SANDSTONE]				
					Sandstone, red ..	62	6	145	6
Clay	8	0	8	0	Fakes, red ..	24	0	169	6
					Sandstone, red ..	40	3	209	9
Sand	6	0	14	0	Fakes and sandstone	22	6	232	3
Clay, brown ..	14	0	28	0	Sandstone, red ..	118	1	350	4
Clay, stony ..	28	0	56	0	Sandstone, red ..	27	0	377	4
Sand, hard ..	13	4	69	4	Sandstone, red, faky	18	8	396	0
Sand, hard ..	13	8	83	0	Sandstone, red ..	4	4	400	4

Pacific Flour Mill Borehole

Site: J. Rank Ltd., Northern Road, Belfast Harbour [346 760]. 6-in map Antrim 61 N.W. Height above Mean Sea Level 9 ft. Date 1936–7. Communicated by C. Isler & Company (36/3).

	Thickness ft	in	Depth ft	in		Thickness ft	in	Depth ft	in
[SOIL AND GLACIAL DRIFT]					Sand, red	15	0	171	0
					Clay, red	6	6	177	6
Made up ground ..	4	6	4	6	Rock, very hard, black	4	6	182	0
River silt and timber [Estuarine Clay]	31	6	36	0	Boulders, black ..	3	0	185	0
					Marl, red	3	0	188	0
Peat, brown ..	3	0	39	0	Sand, red	1	0	189	0
Clay, dark brown	3	0	42	0					
Clay and gravel ..	13	0	55	0	[BUNTER SANDSTONE]				
Sand	1	6	56	6	Sandstone, red ..	132	0	321	0
Clay and sand, brown	45	0	101	6	Sandstone, red, white and black	23	6	344	6
Clay, brown, and stones	12	0	113	6	Sandstone, red ..	80	0	424	6
Clay, loamy, brown	3	6	117	0	Sandstone, white..	2	0	426	6
Clay, loamy, brown	32	0	149	0	Sandstone, red, with thin layers of marl ..	75	6	502	0
Clay, hard, red ..	7	0	156	0					

W. A. Ross & Sons Borehole

Site: Victoria Square, Belfast [340 742]. 6-in map Antrim 61 S.W. Height above Mean Sea Level 7 ft. Date ? 1910. Communicated by Isler & Company (36/2B).

	Thickness ft	in	Depth ft	in		Thickness ft	in	Depth ft	in
[SOIL AND GLACIAL DRIFT]					Sand, loamy, with hard layers ..	11	0	176	6
Concrete	0	6	0	6					
Made ground ..	3	6	4	0					
Silt or River Mud [Estuarine Clay] ..	25	0	29	0	[BUNTER SANDSTONE]				
Large stone ..	3	0	32	0	Freestone	76	6	253	0
Clay, red	70	0	102	0	Sandstone, hard ..	132	0	385	0
Sand, loamy ..	19	6	121	6	Shale and sandstone	5	0	390	0
Sand, hard and pebbles	10	0	131	6	Sandstone or freestone	60	0	450	0
Sand, hard (honeycomb)	34	0	165	6	Sandstone ..	45	0	495	0

T. Somerset & Sons Borehole

Site: 7 Hardcastle Street, Belfast [337 734]. 6-in map Antrim 61 S.W. Height above Mean Sea Level 13 ft. Date 1905. Copied from a record in the possession of the late J. J. Hartley (36/500).

	Thickness ft in	Depth ft in		Thickness ft in	Depth ft in
[SOIL AND GLACIAL DRIFT]			Whinstone [boulder]	4 0	96 0
Made ground ..	5 0	5 0	Conglomerate		
Boulder Clay ..	20 0	25 0	[gravel].. ..	7 0	103 0
Gravel	5 3	30 3			
Boulder Clay ..	7 0	37 3	[BUNTER SANDSTONE]		
Sand and gravel ..	4 9	42 0	Sandstone, red ..	155 0	258 0
Sand, fine.. ..	9 0	51 0	Sandstone, very		
Sand and gravel ..	33 6	84 6	hard, white ..	11 0	269 0
Gravel, coarse ..	7 6	92 0	Dolerite dyke ..	15 0	284 0

Thompson's Flour Mill Borehole

Site: Thompson's Flour Mill, Donegall Quay, Belfast [343 747]. 6-in map Antrim 61 N.W. Height above Mean Sea Level 10 ft. Date not known. Information from J. Mellon (36/483).

	Thickness ft in	Depth ft in		Thickness ft in	Depth ft in
[SOIL AND GLACIAL DRIFT]			[BUNTER SANDSTONE]		
Drift	185 0	185 0	Sandstone ? ..	215 0	400 0

Ulster Weaving Company Borehole

Site: Ulster Weaving Company Factory, Linfield Road, Belfast [331 735]. 6-in Antrim 61 S.W. Height above Mean Sea Level 11 ft. Date about 1930. Information from Ulster Weaving Company Limited (36/481.2).

	Thickness ft in	Depth ft in		Thickness ft in	Depth ft in
[SOIL AND GLACIAL DRIFT]			Boulder Clay ..	16 0	59 0
			Marl ? Drift ..	8 0	67 0
Surface	7 0	7 0			
Clay, red	22 0	29 0	[BUNTER SANDSTONE]		
Sand, red	14 0	43 0	Sandstone ..	333 0	400 0

Ulster Curing Company Borehole

Site: 32 May Street, Belfast [341 740]. 6-in map Antrim 61 S.W. Height above Mean Sea Level 8 ft. Date 1898. Copied from a record in the possession of the late J. J. Hartley (36/498).

	Thickness		Depth			Thickness		Depth	
	ft	in	ft	in		ft	in	ft	in
					[BUNTER SANDSTONE]				
[SOIL AND GLACIAL DRIFT]					Sandstone ..	24	4	99	4
					Faikes [micaceous sandstone] ..	8	2	107	6
Made Ground ..	6	0	6	0	Sandstone ..	26	3	133	9
Mud	6	5	12	5	Faikes ..	2	7	136	4
Clay	50	7	63	0	Sandstone ..	8	7	144	11
Sand	1	8	64	8	Faikes	2	2	147	1
Sand and Clay ..	10	4	75	0	Marl	4	6	151	7

Vulcanite Company Limited Borehole

Site: Laganvale, Belfast, about one mile south of Ormeau Bridge [339 710]. 6-in map Antrim 65 N.W. Height above Mean Sea Level about 16 ft. Date 1899. Copied from Lamplugh 1904 (36/23).

	Thickness		Depth			Thickness		Depth	
	ft	in	ft	in		ft	in	ft	in
[SOIL AND GLACIAL DRIFT]					Sandstone, flakes	10	6	298	10
Surface Clay (Boulder Clay) ..	70	0	70	0	Sandstone ..	43	0	341	10
					Broken [? Sandstone]	4	6	346	4
[PERMIAN MARLS]									
Marl, red	200	2	270	2	Sandstone, extra or very hard ..	5	6	351	10
? [MAGNESIAN LIMESTONE]					Whinstone [Basalt dyke]	4	6	356	4
Limestone ..	0	10	271	0	Sandstone ..	64	6	420	10
Marl	1	4	272	4	Quartz, very hard [conglomerate of rounded quartz pebbles] ..	4	2	425	0
[PERMIAN SANDSTONE AND BROCKRAM]									
Sandstone ..	16	0	288	4					

Appendix 2

RECORDS OF WELLS AND BORES FOR WATER

This list covers virtually all the records of water wells in the Triassic rocks of the Belfast district in the possession of the Geological Survey of Northern Ireland. Reliability of records, particularly figures for yields, is variable as many of the wells have never been tested to capacity. All figures are in feet.

6" Quarter Sheet	Survey No.	Name and Address	Approx. Surface Level (ft)	Depth	Drift	Marl	Sand-stone	R.W.L.	Yield g.p.h.	Remarks
ANTRIM 60 NE	36/478	Ardoyne Tram Depot	230	600	—	460+	140 ?	100	—	Disused
	36/6	Brookfield Linen Co., (York St. Spinning Co.), Courtrai Street	120	1. 400 / 2. 150?	—	—	400 / —	—	6000	Closed ca 1926 / Closed before 1914
	36/1	Wm. Ewart & Sons, Glenbank Bleach Works	330	500	41	459	—	32	3105	—
	36/550	Gt. Northern Mineral Water Co., Jumna Street	80	75	25	—	50	—	NONE	Now closed
	36/5	Rosebank Weaving Co., Flax Street	170	610	300	—	310	—	12000	Now closed. Water never used (too hard)
	36/535	Ulster Flax Co. (now R. R. Dunn & Co.), 7 Wilton Sq. South	75	?	—	—	—	—	800	Covered over. No details known
	36/534	Workshop for the Blind, 136 Lawnbrook Avenue	90	30+	—	—	—	—	800	Still used. No details known
ANTRIM 60 S.E.	36/490	Belfast Silk & Rayon Co., Waterford Street	75	301	76	—	225	—	4300	Not in use.

6" Quarter Sheet	Survey No.	Name and Address	Approx. Surface Level (ft)	Depth	Drift	Marl	Sand-stone	R.W.L.	Yield g.p.h.	Remarks
ANTRIM 60 S.E.	36/546	Clonard Mineral Water Co. (now B.B.C. Refreshments Ltd) Falls Road	70	500	?40	—	460	—	500	—
	36/494	Dunvilles Co. Roy. Irish Distillery (now Murrays) Grosvenor Road	24	1. 400	12	—	388	—	—	—
				2. 240	26½	—	213½	¾	—	—
	36/503	Fairbairn Lawson Combe Barbour, 5–15 Howard Street (North)	70	41½	25½	—	16	—	—	Not known
	36/520	Falls Road Tram Depot	130	—	—	—	—	—	—	No details known.
	36/482	Gt. Northern Railway, Adelaide	40	600	90	—	510	—	6000	Not in use
	36/33	J. & T. M. Greeves, Flax Spinners, 3 Conway Street, Falls Road	60	1. 83½	40	—	43½	?40	1500	Closed
				2. 140	—	—	—	40	—	No details
	36/530	Inver Springs, 18 Beit Street	24	—	—	—	—	—	400	Factory now Warehouse
ANTRIM 60 SE	36/545	J. Mackie, Springfield Road	90	500	—	—	—	—	—	Closed over. Site unknown
	36/492	New City Mineral Water Co., 80 Coolmore St (Donegall Road)	45	207	64	—	143	—	800	Still in use
	36/479	Monarch Laundry, Donegall Road	30	1. 400	25	—	375	—	—	—
				2. 75	25	—	50	—	—	Not in use
	36/473	Caffreys—now Mountain Brewery	270	1. 200	65	135	—	—	—	—
				2. 600	—	—	—	—	—	—
				3. 160	—	—	—	—	—	—
	36/531	J. R. Thompson, Roden Street	33	225	40	—	185	—	1000	Still in use
	36/486	Thompson's Mineral Water Co., Irish Direct Trading Co., Donegall Road	50	250	26	—	224	—	800	—

APPENDIX 2

6" Quarter Sheet	Survey No.	Name and Address	Approx. Surface Level (ft)	Depth	Drift	Marl	Sandstone	R.W.L.	Yield g.p.h.	Remarks
ANTRIM 61 NW	36/519	Antrim Road Tramway Depot Salisbury Avenue	145	—	—	—	—	—	—	Not in use. No information
	36/551	Belfast Mineral Water Co. (now R. J. McKinney Ltd), York Road	20	26	—	—	—	—	—	—
	36/537	Belfast Collar Co., 2A Shaftesbury Street	95	40	—	—	—	—	—	Manager says was artesian. Bore closed in 1926
	36/487	Gracey Bros, Egg & Poultry Merchants, 98-104 Great George's Street	20	400	50	—	350	—	3000	Not in use
	36/3	Pacific Flour Mill (J. Rank)	16	502	185	—	317	17	6000	Used for washing wheat
	36/7	Sinclairs, J. & T., Co., Ltd., 7 Tomb St.	18	646¼	139½	—	506¾	—	5-6000	Still in use
	36/483	Thompsons Flour Mill, Gamble Street	19	585	185	—	400	—	—	Water never used, Saline
	36/8	Whitla St (Dixons Saw Mills)	18	400	—	—	—	—	—	Exact site unknown. Water was brackish
ANTRIM 61 SW	36/491	Isaac Andrews & Sons Ltd, Belfast Mills, 71-77 Percy Street	40	400	—	—	—	—	—	Well capped due to overflow; not used now
	36/22	Belfast Municipal Power Station, East Bridge Street	14	554	57	—	497	—	6500	Well not in use
	36/495	W. J. Briggs (now Emerson, John) 23 Pine Street	25	242	58½	—	183½	—	75	Not in use
	36/542	Bamford's Dairy, 26 Landseer Street	50	400	—	—	—	—	—	Nothing known
	36/538	Cochrane & Co., 78 Bankmore Street	22	300	—	—	—	—	2-3000	Premises vacant. Well not known

6" Quarter Sheet	Survey No.	Name and Address	Approx. Surface Level (ft)	Depth	Drift	Marl	Sand-stone	R.W.L.	Yield g.p.h.	Remarks
ANTRIM 61 SW	36/16	Cantrell & Cochrane Ltd (now John Frackleton & Co.) 9–21 Victoria Sq.	18	1. 116' 2. 399'	78 278	— —	38 321	— —	700 700	One in use 1941. Now disused
	36/543	Corry & Co., Cromac Street	20	1 & — 2.	—	—	—	—	—	Wells disused. Contaminated by Gasworks
	36/4	G. Crawfords, Starch Works, 26 Mill Street	30	A. 175 B. 312	25 12	— —	150 300	— —	— —	Not traced. Site probably at Millfield.
	36/501	Durham Street Weaving Company	25	1. 300 2. 400	60? 60?	— —	240? 340	— —	3000	
	36/502	Wm Ewart, 17 Bedford Street	19	25	—	—	—	—	—	Disused; not fit for drinking
	36/528	Finlays & Co, 22-34 Victoria Square	17	50 +	—	—	—	—	3	Disused. Now Churchill House.
	36/547	Fowlers Mineral Water Co., 51-7 Fitzroy Avenue	30	300	—	—	—	—	300	
	36/13	Fulton's Ormeau Avenue	22	420	c.100	—	c.320	—	—	
	36/12	Grattans & Co., 108-10 Great Victoria Street	21	252	c.100	—	c.152	—	3000	Disused. Now Highland House Anti-Laria Co. No. 2
	36/533	Hendron Bros (formerly Wheeler), 8 Eliza Street	16	243	22	—	221	—	—	Spring Well not in use
	36/540	J. Henning & Son, 19-21 Alfred Street	20	—	—	—	—	—	—	Still in use. Output not known
	36/21	Inglis, Eliza Street	14	300	32¾	—	267¼	—	1440	Not in use
	36/496	Ireland Bros Ltd, 17 Adelaide Street (Now Car Park)	17	402½'	79'	—	323½	33	1000	
	36/518	Lyle & Kinahan, 101 Cullingtree Street	21	1. 275 2. 210	75 70	— —	200 140	— 10¾	400 —	Contaminated In use

6" Quarter Sheet	Survey No.	Name and Address	Approx. Surface Level (ft)	Depth	Drift	Marl	Sand- stone	R.W.L.	Yield g.p.h.	Remarks
ANTRIM 61 SW	36/488	McKenna & McGinley, Bath Place .	40	600	—	—	—	—	5-600	In use
	36/539	H. Matier, 1-3 May Street .	11	—	—	—	—	—	850	Disused
	36/18	Millar & Rankin, 50 McAuley Street .	16	350	?35	—	315	—	—	—
	36/499	Miller Sons & Torrence, now Antrim Mfg. Co., Ltd, 8 Maxwell Street .	45	400¾	83	—	314¾	—	?	Not located
	36/15	Murphy & Stevenson, 40 Linenhall St	20	400	—	—	—	—	3000	—
	36/10	J. Neill & Co., College Place .	22	415	110	—	305	—	500	Still in use for washing purposes
	36/549	Northern Publishing Office, 56 Victoria Street .	19	—	—	—	—	—	—	Nothing known
	36/14	Public Baths, Ormeau Avenue .	21	400	90	—	310	—	—	—
	36/11	Belfast Pure Ice & Cold Storage Co., Ltd, 112 Great Victoria Street .	18	429½	68¼	—	361¼	—	3000	In use
	36/529	Robinson & Cleaver, 56 Donegall Place	19	—	176½	—	318½	—	1000	Not in use
	36/2	W. A. Ross & Sons Ltd, 13-19 William Street South .	16	1. 495 2. 421	127	—	294	—	1000	
	36/497	Sandy Row, Tram Depot, Napier Street	45	210½	—	—	—	—	2000	Disused
	36/500	T. Somerset & Sons Ltd, 7 Hardcastle Stteet .	22	284	103	—	181	—	—	Now disused
	36/20	Swanston's Bone Works, King Street Now Abraham Neill Ltd .	19	235	202	—	33	—	—	Nothingknownbyfirm
	36/498	Ulster Curing Co., 32 May Street .	17	151½	75	—	76½	25	—	In use
	36/480	Ulster Spinning Co., Ltd, Linfield Mill	20	1. 775 2. 400 3. 200	—	—	—	—	1500	—
	36/481	Ulster Weaving Co., Ltd, 47 Linfield Road .	20	1. 541 2. 400 3. 350	81 67	—	461 333	—	2000 2000 3500	Disused All in use

6" Quarter Sheet	Survey No.	Name and Address	Approx. Surface Level (ft)	Depth	Drift	Marl	Sandstone	R.W.L.	Yield g.p.h.	Remarks
ANTRIM 64 NE	36/541	Wheeler & Co., 721-25 Lisburn Road	48	300	—	—	—	—	4000	Still in use. Quantity used now unknown
ANTRIM 65 NW	36/23	Vulcanite Co., Ltd, Laganvale	25	425	70	—	355'	—	2700	Not used. Sst. figure includes 200 Lower Marls
DOWN 4 SE	36/525	Anderson's Felt Works now Short Strand Bus Depot	19	35	—	—	—	—	—	
	36/29	Avoniel Distillery, Avoniel Road	22	$344\frac{1}{3}$	$47\frac{2}{3}$	—	$296\frac{2}{3}$	—	12000	Disused, now in Corp RSD
	36/544	Baltic Timber Co., 46-64 Ravenhill Road	19	120	—	—	—	—	—	Not in use. Used occasionally up to 1941
	36/493	Belfast Ropeworks Co., Ltd, 371 Newtownards Road	21	500	37	—	463	—	22000	Still in use c. 10000 gpd
	36/32	Belfast Co-Operative Society, Ravenhill Avenue (Federation Street)	39	400	290	—	310	30	7000	
	36/30	Bread Street (Bloomfield Est.)	21	354	—	—	—	—	—	Not known. Site on 1902 Map under Ropewks. Cord Rm.
	36/26	Campbell, W. J. & Son now J. Hogg & Son, 1A Ravenhill Road, now builder's yard	19	—	35	—	—	—	—	Not known
	36/31	Castlereagh Laundry, 1 Redcar Street	45	459	140	—	319	—	—	In use
	36/523	Davidson & Co., Ltd (East Yard) 36 Mountpottinger Road	21	84	—	—	—	—	1800	
	36/484	Devonshire Laundry, 173/183 Ravenhill Avenue	40	1. 120 2. 43	—	—	—	—	8-9000 8-9000	1938 pumping figure
	36/524	Dobson's Dairy, 195 Beersbridge Road	30	35	—	—	—	—	2000	Disused

6" Quarter Sheet	Survey No.	Name and Address	Approx. Surface Level (ft)	Depth	Drift	Marl	Sandstone	R.W.L.	Yield g.p.h.	Remarks
DOWN 4 NE	36/28	Irish Distillery, Connswater, Belfast, (now Gallaghers)	18	A.529½ B. 748 C. 380	49½ 46 12	— —	480 702 318	— —	— 6500 5000	— —
DOWN 4 SE	36/24	J. J. McConnell's Distillery, Ravenhill Road	14	1.349 2.300	17 —	— —	332 —	— —	3000 1000	— —
	36/25	Millar & Co., 95 Ravenhill Avenue	40	60	60	—	—	—	—	
	36/527	Sirocco Engineering Works, Bridge End	19	56	—	—	—	—	1600	Still in use
	36/522	Strand Spinning Co., 338 Newtownards Road	19	400	—	—	—	—	3360	Well now disused
	36/27	Templemore Avenue Baths	26	334	50	—	284	—	2000+	
	36/526	Ulster Ice & Cold Storage, Laganview Street	19	480	—	—	—	—	2000	Not in use—pipes corroded.

Appendix 3

LIST OF GEOLOGICAL SURVEY PHOTOGRAPHS

Taken mainly by J. Rhodes and J. M. Pulsford with a few by H. E. Wilson

Copies of these photographs are deposited for public reference in the Library of the Institute of Geological Sciences, South Kensington, London, S.W.7 and in the office of the Geological Survey of Northern Ireland, 20 College Gardens, Belfast BT9 6BS. Prints and lantern slides can be supplied at a fixed tariff.

NI 181–2	Bedded alluvium resting on boulder clay: Crumlin River.
NI 183	Boulder Clay overlying gravel and sand: Crumlin River.
NI 184	Bedded sands and gravel overlying sandy silt with scattered pebbles and angular blocks: Crumlin River.
NI 185	Glacial spillway, cut mainly in drift: Aghalee.
NI 186	Irregular dykes cutting Chalk and the overlying basalt: Knocknadona Quarry (Plate 1B).
NI 187	Quarry in Chalk and basalt lava: Rockville Quarry.
NI 188	Chalk/Basalt contact: Knocknadona Quarry.
NI 189	Irregular basalt dykes in Chalk: Knocknadona Quarry.
NI 190	Rock-cut glacial spillways: Mount Gilbert.
NI 191	Rock-cut glacial spillway: Mount Gilbert.
NI 192	Red and green Keuper Marls: Forth River.
NI 193–4	Keuper Marls: Forth River.
NI 195	Keuper Marl with basalt intrusion: Parkview Brickworks.
NI 196	Chalk slips: Black Mountain.
NI 197	Jurassic-Cretaceous junction: Collin Glen.
NI 198	Yellow sandstones of the Cretaceous: Collin Clen.
NI 199	Hibernian Greensand: Collin Glen.
NI 200	Chalk (White Limestone) on glauconitic sands: Glen Bridge.
NI 201	Bunter Sandstone. Marl bands dissected by sandstone wedges in Bunter Sandstone: Milltown, Belfast.
NI 202	Bunter Sandstone: Milltown, Belfast.
NI 203	Dolerite dyke in Lower Basalts: Black Mountain Quarry.
NI 204	Amygdaloidal Dolerite Dyke: Black Mountain Quarry.
NI 205–6	Banded Lower Basalts: Whiterock.
NI 207	Gypsiferous Keuper Marl: Springfield Road, Belfast.
NI 208–12	Malone Sands overlain by Upper Boulder Clay: Physics Building, Queen's University.
NI 213	Bunter Sandstone: Grove Playing Fields, Shore Road.
NI 214	Incipient Columnar Jointing in Lower Basalts: Collinwell.
NI 215	Chalk-Basalt junction: Collinwell.
NI 216	View of the Chalk-Basalt Scarp: Collinwell (Plate 8A).
NI 217	View across the Lagan Valley: Groganstown.
NI 218	Terrace gravels: Glen River.
NI 219	Laminated glacial clays: Lagan Vale Brickworks.
NI 220	Glacial laminated clays passing up into Malone Sands: Lagan Vale Brickworks.

Appendix 4

REVISED LIST OF MOLLUSCA FROM THE BELFAST ESTUARINE CLAY

This list incorporates the list in Praeger (1892), the late A. S. Kennard's determinations of specimens in the Reserve and Study collections of the Institute of Geological Sciences and determinations by C. J. Wood of material collected during the present survey. Praeger's nomenclature required extensive revision: in those cases where the current trivial name differs from that in Praeger's list the Praeger specific name is quoted in addition.

The nomenclature used largely follows the Winckworth list of British Marine Mollusca (*J. Conch.*, **19**, 211–52) (1932) updated with reference to Tebble's Handbook of British marine bivalves (1966) and Bowden and Heppell: Revised list of British Mollusca—1, (*J. Conch.*, **26**, 99–124) (1966); 2, (*J. Conch.*, **26**, 237–72) (1968).

Depths where known from boreholes are given to the nearest six inches.

Key to the symbols employed, together with the registration numbers of specimens in the collection of the Geological Survey of Northern Ireland is as follows:

P = Praeger's List 1892.
K = A. S. Kennard's determination of the Institute of Geological Sciences Collection.
Kb = Dry Dock boreholes (determined by Kennard)—28/8, 28/9, 36/44–50, 36/52–3.
C = Fergus House, Chichester Street [3415 7410]—NIA 722–5.
B = Borehole in West Twin, 1·2 miles N.E. of York Street Station—NIC 6961–6988 (10–12 ft); NIC 6989–7105 (13–35 ft); NIC 7106–7121 (36–7 ft).
Ma = Musgrave Shipyard (Harland & Wolff's) New Welders' Shop stanchion 1280 yd N.67°E. of Co. Down Railway Terminus—NIC 7131–7173 5–15 ft).
Mb = Musgrave Shipyard (Harland & Wolff's) near No. 14 Slip. Pile No. 836. 1700 yd N.65°E. of Co. Down Railway Terminus—NIC 7174–7210 (1–20 ft).
V = Borehole 36/465 No. 1 Victoria Square, Belfast—NIM 808–810 (5–29 ft).
Va = Borehole 36/465 No. 5 Victoria Square, Belfast—NIM 811–852 (5–25 ft).
Vb = Borehole 36/465 No. 6 Victoria Square, Belfast—NIM 853–890 (4–25 ft).
Vc = Borehole 36/465 No. 3 Victoria Square, Belfast—NIC 7720–7730 (5–8 ft). Upper Estuarine Clay; NIC 7731–7784 (9–28 ft) Lower Estuarine Clay.
H = Borehole 36/428A east side of Herdman Channel, Belfast—NIM 285 (16 ft) Modern sediments; NIM 286–290 (18–28 ft) Upper Estuarine Clay; NIM 293–294 (34–6 ft) Lower Estuarine Clay. (NIM 291–2, 30–2 ft Zone doubtful).
T = College of Technology, Millfield, Belfast 700 yd N.50°W. of City Hall— NIC 7122–7130 (5½–8½ ft).
S = Excavation at Stanley's Motor Works, Great Victoria Street, Belfast— [3355 7375] NIC 10299–10372 (10–12 ft).

Gastropoda:

Aclis minor (Brown)–P: *A. supranitida* S. Wood.
Actaeon tornatilis (Linné)–P.
Akera bullata Müller–P; K.
Alvania beanii (Thorpe) *calathus* (Hanley)–B (13–35 ft).
Aporrhais pespelicani Linné *quadrifidus* da Costa–P; K; B (13–35 ft); Ma (5½–15 ft); Mb (1–20 ft).
Bittium reticulatum (da Costa)–P; K; B (13–35 ft); Ma (5½–15 ft); Mb (1–20 ft); Vb (4–25 ft): Vc (5–8 ft, 9–28 ft); H (18–28 ft, 30–2 ft); T (5½–8½ ft).
Buccinum undatum Linné–P.
Caecum glabrum (Montagu)–P.
Calliostoma zizyphinum (Linné) *conuloide* (Lamarck)–K.
Capulus ungaricus (Linné)–P.
Chrysallida indistincta (Montagu)–P.
C. obtusa (Brown)–P: *Odostomia interstincta* Montagu.
Cingula semicostata (Montagu)–P: *Rissoa striata* Adams.
C. vitrea (Montagu)–P.
Clathrus trevelyanus (Johnston)–K.
C. turtonis (Turton)–P.
Colus gracilis (da Costa)–P.
Cylichna cylindracea (Pennant)–P.
Diadora apertura (Montagu)–P: *Fissurella graeca* Linné.
Diaphana minuta Brown–P: *Utriculus hyalinus* Turton.
Emarginula reticulata J. Sowerby *mülleri* Forbes & Hanley–K.
Eulima trifasciata (J. Adams)–P: *E. bilineata* Alder.
Gibbula cineraria (Linné)–P; K; B (13–35 ft).
G. magus (Linné) *tuberculata* (da Costa)–P; K.
G. umbilicalis (da Costa)–P; Mb (1–20 ft); Vc (9–28 ft).
Haedropleura septangularis (Montagu)–P; B (13–35 ft); Vc (5–8 ft).
Hermania scabra (Müller)–P.
Hinia pygmaea (Lamarck)–P; B (13–35 ft); Mb (1–20 ft); Va (5–25 ft); Bb (4–25 ft); Vc (5–8 ft, 9– 28ft).
H. reticulata (Linné)–P; K.
Hydrobia ulvae (Pennant)–P; Kb (36/45 at 45 ft); B (13–35 ft); Mb (1–20 ft); Va (5–25 ft); Vb (4–25 ft): Vc (9–28 ft); H (16 ft, 34–6 ft).
Lacuna pallidula (da Costa)–P.
L. parva (da Costa)–P: *L. puteolus* Turton.
L. vincta (Montagu)–P: *L. divaricata* Fabricius; K; Va (5–25 ft); Vc (9–28 ft).
Leucopepla bidentata (Montagu)–P.
Littorina aestuarii Jeffreys–P: *L. obtusata* var *aestuarii.*
L. littoralis (Linné)–P: *L. obtusata* Linné; Ma (5½–15 ft); Mb (1–20 ft); Va (5–25 ft); Vb (4–25 ft); Vc 5–8 ft, 9–28 ft); H. (34–6 ft).
L. littorea (Linné)–P; K; Kb (36/49 at 24½ ft); Ma (5½–15 ft); Mb (1–20 ft); Va (5–25 ft); Vc (9–28 ft); H (30–2 ft, 34–6 ft); T (5½ ft to 8½ ft); S (10–12 ft).
L. neritoides (Linné)–P.
L. saxatilis (Olivi)–K; B (13–35 ft); Mb (1–20 ft); Va (5–25 ft); Vb (4–25 ft); Vc (5–8 ft, 9–28 ft); H (16 ft); T (5½–8½ ft).
L. saxatilis (Olivi) *rudis* (Maton)–P.
L. saxatilis (Olivi) *tenebrosa* (Montagu)–P.
Lora rufa (Montagu)–P.
L. turricula (Montagu)–P.
Lunatia catena (da Costa)–P.
L. poliana (Chiaje) *alderi* (Forbes)–K; Ma (5½–15 ft).

Mangelia attenuata (Montagu)–P.
M. brachystoma (Philippi)–P.
M. coarctata (Forbes)–P: *Pleurotoma costata* Donovan.
Margarites helicinus (Fabricius)–P.
Neptunea antiqua (Linné)–P.
Nucella lapillus (Linné)–P; K.
Ocenebra erinacea (Linné)–P; K; B (13–35 ft).
Odostomia acuta Jeffreys–P; Va (5–25 ft).
O. albella (Lovén)–P.
O. conoidea (Brocchi)–P.
O. eulimoides Hanley–P: *O. pallida* Montagu.
Patella vulgata Linné–P.
Patelloida virginea (Müller)–P.
Patina pellucida (Linné)–P: *Helcion pellucidum* Linné.
Philbertia gracilis (Montagu)–P.
Philine aperta (Linné) *quadripartita* Ascanius–P; K; Mb (1–20 ft).
Retusa alba (Kanmacher)–P: *Utriculus obtusus* Montagu; Va (5–25 ft); Vb (4–25 ft).
R. mammillata (Philippi)–P.
R. nitidula (Lovén)–P.
Rissoa albella Lovén–P.
R. inconspicua Alder–P.
R. lilacina Récluz–P: *R. violacea* Desmarest; B (13–35 ft).
R. membranacea (Adams)–P; B (13–35 ft); Ma (5½–15 ft); Mb (1–20 ft); Va (5–25 ft); Vb (4–25 ft); Vc (5–8 ft, 9–28 ft); H (16 ft, 34–6 ft); S (10–12 ft).
R. parva (da Costa)–P.
R. sarsii Lovén–P: *Rissoa albella* var *sarsii*.
Rissoella opalina (Jeffreys)–P.
Tricla lignaria (Linné) *brownii* Leach–P; K.
Trivia monacha (da Costa)–P; K.
Turbonilla elegantissima (Montagu)–P: *Odostomia lactea* Linné.
Turritella communis Risso–P; Ma (5½–15 ft); H (18–28 ft).

Bivalvia:

Abra alba (W. Wood)–P; K; B (13–35 ft); Ma (5½–15 ft); Mb (1–20 ft); Va (5–25 ft); Vc (5–8 ft); H (18–28 ft, 30–2 ft).
A. tenuis (Montagu)–P.
Acanthocardia echinata (Linné)–P; K; Mb (1–20 ft).
Aequipecten opercularis (Linné)–P; K; Ma (5½–15 ft); Mb 1–20 ft); H (18–28 ft).
Anomia ephippium Linné–P; S (10–12 ft).
Arctica islandica (Linné)–P.
Barnea candida (Linné)–P; K.
Cerastoderma edule (Linné)–P; K; Kb (36/46 at 21 ft, 36/50 at 23 ft, 36/45 at 45 ft); C; Ma (5½–15 ft); Mb (1–20 ft); Va (5–25 ft); Vb (4–25 ft); Vc (9–28 ft); H (16 ft, 34–6 ft); T (5½–8½ ft).
C. (Parvicardium) exiguum (Gmelin)–P; K; B (13–35 ft); Ma (5½–15 ft); V (5–29 ft); Va (5–25 ft); Vc (9–28 ft); S (10–12 ft).
C. (P.) scabrum (Philippi)–P.
Chamelea gallina (Linné)–Ma (5½–15 ft).
Chlamys distorta (da Costa)–P: *Pecten pusio* Linné.
C. varia (Linné)–P; K; Kb (36/48 at 21 ft); B (13–37 ft); Mb (1–20 ft).
Clausinella fascinata (da Costa)–P.
Corbula gibba (Olivi)–P; K; B (13–35 ft); Ma (5½–15 ft); Mb (1–20 ft); Va (5–25 ft); H (18–28 ft, 34–6 ft).
Cultellus pellucidus (Pennant)–K.

Dosinia exoleta (Linné)–P.

D. lupinus (Linné)–P: *Venus lincta* Pulteney.

Ensis ensis (Linné)–P; K.

Erycina nitida (Turton)–P.

Gari depressa (Pennant)–P.

G. fervensis (Gmelin)–P; K.

Gastrochaena dubia (Pennant)–P.

Hiatella arctica (Linné)–P: *Saxicava rugosa* Linné.

Laevicardium crassum (Gmelin)–P: *Cardium norvegicum* Spengler; K.

Lima hians (Gmelin)–P.

Lucinoma borealis (Linné)–P; Mb (1–20 ft); Vc (9–28 ft).

Lutraria lutraria (Linné)–P.

L. magna (da Costa)–P.

Macoma balthica (Linné)–P; K; Ma (58–15 ft); Mb (1–20 ft); V (5–29 ft); Va (5–25 ft); Vb (4–25 ft); H (16 ft).

Modiolarca tumida (Hanley)–P: *Modiolaria marmorata* Forbes.

Modiolus adriaticus Lamarck–P.

M. modiolus (Linné)–P; B (13–35 ft).

Mya arenaria Linné–P; K.

M. truncata Linné–P; K; Mb (1–20 ft).

Mysella bidentata (Montagu)–B (13–35 ft); Ma ($5\frac{1}{2}$–15 ft); Mb (1–20 ft); Va (5–25 ft); Vb (4–25 ft).

Mysia undata (Pennant)–P; K; B (13–35 ft); Ma ($5\frac{1}{2}$–15 ft); Mb (1–20 ft); Vc (5–8 ft); H (18–28 ft).

Mytilus edulis Linné–P; K; Kb (36/46 at 21 ft, 38/48 at 21 ft); C; Ma ($5\frac{1}{2}$–15 ft); Mb (1–20 ft); V (5–29 ft); Va (5–25 ft); Vc (9–28 ft); H (16 ft, 18–28 ft); S (10–12 ft).

Nototeredo norvegicus (Spengler)–P.

Nucula nucleus (Linné)–P; K.

N. sulcata Bronn–P; K; B (13–35 ft); H (18–28 ft).

Ostrea edulis Linné–P; K; Kb (36/46 at 21 ft, 36/48 at 21 ft, 36/50 at 23 ft); C; B (13–27 ft); Ma ($5\frac{1}{2}$–15 ft); Mb 1–20 ft); Va (5–25 ft); Vc (5–8 ft); T ($5\frac{1}{2}$–$8\frac{1}{2}$ ft); S (10–12 ft).

Panopea plicata Montagu (?)–P.

Paphia aurea (Gmelin)–P; C.

P. rhomboides (Pennant)–P: *Venus virgineus*; K; B (13435 ft); Ma ($5\frac{1}{2}$–15 ft); Va (5–25 ft); Vb (4–25 ft); Vc (9–28 ft); T ($5\frac{1}{2}$–$8\frac{1}{2}$ ft).

Pecten maximus Linné–P; K; Ma ($5\frac{1}{2}$–15 ft); Vc (5–8 ft).

Pharus legumen (Linné)–P.

Pholas dactylus Linné–P.

Pododesmus (*Monia*) *patelliformis* (Linné)–P.

Saxicavella jeffreysi Winckworth–B (13–35 ft).

Scrobicularia plana (da Costa)–C; Ma ($5\frac{1}{2}$–15 ft); Mb (1–20 ft); V (5–29 ft); Va (5–25 ft); Vb (4–25 ft); Vc (9–28 ft); H (34–6 ft).

Solecurtus chamasolen (da Costa)–P: *S. antiquatus* Pulteney.

Solen marginatus Montagu–K.

Sphenia binghami Turton–P.

Spisula elliptica (Brown)–K; Ma ($5\frac{1}{2}$–15 ft).

S. solida (Linné)–P; Kb (28/8 at 21 ft).

S. subtruncata (da Costa)–P; K; T ($5\frac{1}{2}$–$8\frac{1}{2}$ ft).

Tapes decussatus (Linné)–P; K; Kb (36/48 at 21 ft, 36/49 at $24\frac{1}{2}$ ft, 36/45 at 45 ft); Ma ($5\frac{1}{2}$–15 ft); Mb (1–20 ft); Vc (9–28 ft).

Tellimya ferruginosa (Montagu)–P; H (18–28 ft).

Tellina fabula Gmelin–P; K.

T. squalida Montagu–P.

T. tenuis da Costa–P; K.

Thracia convexa (W. Wood)–P.

T. phaseolina (Lamarck)–P: *T. papyracea* Philippi; K.

T. pubescens (Montagu)–P.

Thyasira flexuosa (Montagu)–P; B (13–35 ft); Mb (1–20 ft); Vb (4–25 ft); Vc (5–8 ft, 9–28 ft).

Turtonia minuta (Fabricius)–P.

Venerupis pullastra (Montagu)–P; K; Ma (5½–15 ft); Mb (1–20 ft).

Zirfaea crispata (Linné).

Appendix 5

FORAMINIFERA FROM THE BELFAST ESTUARINE CLAY

Thirteen samples of clay [registered numbers SAM 3865–3877] collected from No. 3 borehole, Churchill House, Victoria Square, Belfast [341 742], and one sample [SAM 3887] from a temporary exposure at Stanley's Motor Works, Great Victoria Street [336 737], were examined.

In the hand specimens all samples appeared as a grey clay, with abundant plant material in those below 9 ft (2·7 m) Macrofossils were seen and in addition some larger specimens of the foraminifer *Ammonia beccarii* (Linné) var. *batava* (Hofker) were readily discernible under the hand lens.

The treated samples, save for the two highest in the borehole, contained abundant plant material. Very fine-grained quartz sand was present in proportions varying from almost none to the entire residue.

Preliminary examination of the fauna showed that the various varieties of *A. beccarii* [excluding *A. beccarii beccarii* itself] were extremely abundant, accounting for over 90 per cent of the specimens identified in one sample and over 80 per cent in three others. Forty other species and varieties are recorded and in order to show the relative abundance of these species clearly, the percentage on the accompanying diagram (Fig. 22) has been based on the convention that the total number of specimens other than *A. beccarii* vars. is equal to 100 per cent.

The entire fauna is post-glacial, the change in emphasis of the species being indicative of changes in environment. From the faunal diagram the borehole appears to pass through four phases of deposition and detailed examination shows that the change from each to the next is gradual.

The highest sample, at 5ft to 6 ft (1·5–1·8 m), contains by far the greatest number of species, with *A. beccarii* vars. less than 50 per cent of the total fauna and a predominance of *Ammoscalaria? runiana* (Heron-Allen and Earland), *Elphidium excavatum* (Terquem), *E. selseyense* (Heron-Allen and Earland), *Protelphidium depressulum* (Walker and Jacob) *sensu* Brady [= *P. anglicum* Murray] and *Verneuilina media* Höglund amongst the other species.

From a depth of 7–22 ft (2·1–6·7 m) the percentage of *A. beccarii* vars. rises from 68 at 7–8 ft (2·1–2·4 m) to a peak at 15–16 ft (4·6–4·9 m) of 92, falling to 63 in the sample taken at 21–22 ft (6·4–6·7 m). There is a marked diminution in the number of species in the remaining fauna, reaching a low in the sample at 15–16 ft (4·6–4·9 m) with only three species represented. Throughout this range the dominant species in the remaining fauna are again *E. excavatum* and *P. depressulum*. The absence of *A.? runiana* between the depths of 11–16 ft (3·3–4·9m) is particularly noticeable.

The depth range 23–28 ft (7–8·5 m) is marked by a reduction in the percentage of *A. beccarii* vars. to 14–19. A slight increase in the number of species at 23–24

222

	5-6 ft.	7-8 ft.	9-10 ft.	11-12 ft.	13-14 ft.	15-16 ft.	17-18 ft.	19-20 ft.	21-22 ft.	23-24 ft.	25-26 ft.	27-28 ft.	29-29½ ft.	Temporary Exposure
Total number of specimens identified	585	368	340	308	567	499	647	576	397	360	384	271	1	505
Ammonia beccarii (Linné) vars. percentage of total	43	68	74	81	88	92	77	87	63	14	13	19	-	50
Number of specimens other than A. beccarii vars.	248	108	105	57	65	36	85	77	141	308	330	213	1	255
Ammoscalaria? runiana (Heron-Allen and Earland)	◐	○	◐			○	○	●	◐	○	•			○
Bolivina ordinaria Phleger and Parker	•													
Bolivina pseudoplicata Heron-Allen and Earland	○									•		•		
Bulimina aculeata (d'Orbigny)	•													
Bulimina elongata d'Orbigny	•	•	○											
Bulimina elongata spinose var.	•													
Bulimina gibba d'Orbigny	•		•						•					
Buliminella elegantissima (d'Orbigny)	○		○					•	•					•
Cibicides lobatulus (Walker and Jacob)														
Cyclogyra involvens (Reuss)	•													
Discorbis williamsoni Cushman and Parr	•													
Elphidium bartletti Cushman	○	○												
Elphidium crispum (Linné)	○	•												○
Elphidium crispum spinose var.	○	○	•											
Elphidium discoidale (d'Orbigny)	○					○								
Elphidium excavatum (Terquem)	⊙	○	○	◐	●	●	●	●	●	⊙	◐	◐	◐	•
Elphidium granosum (d'Orbigny)			•							○	○			•
Elphidium incertum (Williamson)	○	⊙	⊙				○				•			⊙
Elphidium macellum (Fichtel and Moll)	○	○	○	○	○	•				•				◐
Elphidium selseyense (Heron-Allen and Earland)	⊙	⊙	◐	○	⊙		○	○	•		•			⊙
Elphidium sp.	•					○		○						○
Eponides sp.	•													
Fissurina lucida (Williamson)	○								•					
Fissurina sp.	•								•					
Fursenkoina sp.	•													
Gyroidina sp.								•						
Jadammina macrescens (Brady)	○	•							○	•	•			•
Lagena clavata (d'Orbigny)	•													
Lagena laevis (Montagu)	○	•							•					
Miliammina fusca (Brady)	○		◐					○		○	•	•		◐
Miliolinella subrotunda (Montagu)			•											
Patellina corrugata Williamson			•											
Protelphidium asterizans (Fichtel and Moll)	○		•											
Protelphidium depressulum (Walker and Jacob)	●	◉	●	◐	◉	●	●	●	◉	●	◉	◉		◉
Pseudopolymorphina sp.						○								•
Quinqueloculina seminulum (Linné)	•										•			
Rotalids indet.	•													
Textularia earlandi Phleger									•					
Trochammina inflata (Montagu)									○	•	○	◉		
Verneuilina media Höglund	◐	◐	○					○						○

KEY:-

•	less than 1 %	○	1 - 5 %	⊙	6 - 10 %
◐	11 - 20 %	●	20 - 50 %	◉	greater than 50 %

FIG. 22. *Distribution of forminifera in the Estuarine Clays of the Belfast area*

ft (7–7·3 m) is again followed by a reduction. The dominant forms are *A. ? runiana*, *E. excavatum*, *P. depressulum* and *Trochammina inflata* (Montagu).

The lowest sample at 29–29½ ft (8·8–9·0 m) was barren apart from a single specimen of *T. inflata*, the residue being almost completely plant fragments.

The sample from the temporary exposure contained a fauna of 50 per cent *A. beccarii* vars. and 16 species were represented in the remaining fauna, dominated by *P. depressulum* and *Miliammina fusca* (Brady), but followed closely by the numbers of *Elphidium incertum* (Williamson) and *E. macellum* (Fichtel and Moll).

Since the assemblage is a post mortem one some anomalies in what are regarded as the normal environments of associated species are apparent from the distribution chart.

Adams and Haynes (1965, p. 30), in discussing the ecological significance of a Holocene fauna from the Dovey estuary, list five species the predominance of any of which indicates a brackish-water environment. These are *T. inflata*, *Jadammina macrescens* (Brady), *M. fusca*, *P. depressulum* and *A. beccarii*.

Of the species here recognized from Belfast, Voorthuysen (1951, p. 268) regards only *T. inflata* as living most exclusively in a brackish-water environment and *A. beccarii*, *E. excavatum* and *P. depresulum* as species which prefer a lagoonal-brackish-water habitat. Referring to *A. beccarii*, Voorthuysen (*ibid.*, p. 268) states " . . . the fauna consists here of almost exclusively one species, *Streblus beccarii*. This latter species, however, is not a typically brackish-water form, but only a species which is able to maintain itself in a habitat of rather low salinity." He goes on to suggest that a deposit containing an extremely high percentage of *A. beccarii* indicates well oxygenated, brackish shallow water. An increase in the salinity then gives rise to an increase in the number of species.

Applying this interpretation of faunas to the present evidence, the borehole would appear to have penetrated the following sequence of events. The lowest sample represents terrestrial conditions which were succeeded by a marsh cycle (see Adams and Haynes, op. cit., p. 33) with *J. macrescens* together with the open tidal flat species *E. excavatum*. This in turn was succeeded by low salinity open water conditions with *A. beccarii* vars. predominating and the highest evidence available suggests increasing saline conditions with a rise in the number of species.

There is an anomalous occurrence of the marsh form *M. fusca* in these highest samples and also in the sample from the temporary exposure. This latter sample would appear to correlate with the upper part of the borehole.   M.J.H.

INDEX

Printed in Northern Ireland for Her Majesty's Stationery Office
by Nelson & Knox (N.I.) Ltd., Belfast. Dd. 471368. K7. 9/70. Gp. 3100.

LANCASTER UNIVERSITY LIBRARY

**Due for return by end of service on date below
(or earlier if recalled)**